Wheels Stop

Outward Odyssey
A People's History of Spaceflight

Series editor
Colin Burgess

Rick Houston

Foreword by Jerry Ross

Wheels Stop

The Tragedies and Triumphs of the Space Shuttle Program, 1986–2011

UNIVERSITY OF NEBRASKA PRESS • LINCOLN & LONDON

Manufactured in the United States of America

♾

Library of Congress
Cataloging-in-Publication Data
Houston, Rick, 1967–
Wheels stop: the tragedies and triumphs of the Space Shuttle Program, 1986–2011 / Rick Houston; foreword by Jerry Ross.
pages cm. — (Outward odyssey: a people's history of spaceflight)
Includes bibliographical references and index.
ISBN 978-0-8032-3534-2 (cloth: alkaline paper) 1. Space Shuttle Program (U.S.)—History. 2. Space shuttles—United States—History.
I. Title.
TL789.8.U6S6646 2013
629.44'10973—dc23 2013022160

Set in Adobe Garamond Pro by Laura Wellington.

To the memory of the crews of STS-51L and STS-107

To the memory of my mother and father, Betty J. and Sidney L. Houston

In honor of the thousands of men and women who worked to make the Space Shuttle an American treasure

Contents

Illustrations

Foreword

All these years later, I can still close my eyes and be right back on the end of that robotic arm, suspended high above *Atlantis*'s payload bay and 160 nautical miles above the surface of the earth. The view is one of immeasurable beauty.

It was 8 April 1991, and my STS-37 crewmate Jerome "Jay" Apt and I were on our second and last extravehicular activity (EVA), or spacewalk, of the flight. The rest of the crew was working with Jay, and I had a few moments to myself. We were on a night pass, and as I leaned back, I turned off my helmet-mounted lights and peered into the deep black vastness of space.

All of a sudden, a strange sensation washed over me. In that instant I was one with the universe, doing exactly what God had designed me to do. The sensation was unexpected but it was one of peace, comforting and reassuring. Although I was in one of the most hostile environments to which mankind has ever been exposed, I knew I was exactly where I was supposed to be, doing precisely what I was supposed to be doing. Spacewalking is incredibly complicated, but it just came naturally to me. An engineering background, growing up as a farm boy in Crown Point, Indiana, and my innate mechanical aptitude came together, pieces of an intricate jigsaw puzzle that somehow managed to fit together perfectly for me as an EVA astronaut.

Named to the astronaut office in May 1980, I had no way of knowing that I would one day become the first person to be launched into space seven times. From 1984 when I began training for my first mission, STS-61B, I was in training for a flight about half of the next eighteen years. The process of getting ready to fly in space was rigorous, but I lived for it.

As an astronaut, I had the job I had always wanted. I did not see myself moving up the management ladder. Over the course of my career at NASA, I was offered several management positions. I turned them down, telling whoever would listen that as long as I could fly, that is what I wanted to

1. Jerry Ross, having the time of his life. Courtesy NASA.

do. Executives and headhunters from the private sector came to me with some pretty attractive job proposals as well, but I wanted to fly in space. I took to spaceflight like a duck to water.

As time passed, I learned to recognize the telltale signs when some of my astronaut friends were nearing the end of their space-flying careers. While training for their upcoming flights, I would hear them complain about having to go to this class or that class or to a simulator training session. They would even complain about "having to do" one of the coolest things astronauts get to do on earth, a space-suited EVA training run in the Neutral Buoyancy Lab (NBL) water tank. Sometimes they didn't even want to do another of our most cherished things, to go flying in a T-38. If an astronaut couldn't get excited about getting ready to fly in space, I knew it was going to be his or her last mission.

Fortunately for me, I never got to that point. Ever. I was like a kid in a candy shop on my seventh flight, STS-110 in April 2002, just like I had been on my first. I loved the training and I loved to fly. I looked forward to the training opportunities, even though I had done most of them many times over. From the first flight to the last, I enjoyed every minute of training, because training meant I was getting ready to fly!

Between my flights I worked on developing shuttle EVA tools and procedures, evaluating International Space Station (ISS) EVA hardware and tasks, and developing ISS assembly procedures. The EVA work was almost as challenging as training for a flight. While serving as the astronaut office EVA and robotics branch chief, I worked sixty or more hours a week. I brought my computer and paperwork home and worked until it was time to go to bed and got right back up to do it all again the next day. I spent many Saturdays at work, but that is what it took to get the job done. I was glad to be the one doing it.

I have been asked how I would like my career to be remembered. That is an easy question for me to answer. I would like to be remembered as a person who had a dream as a young kid and was allowed to pursue and achieve it. I had more fun than it is probably legal to have. I loved *every* aspect of what I got to do. I loved the people I got to do it with. I loved the challenges of the flights. I loved seeing the earth from space.

The beauty of the earth from space is not describable. If I had one wish, it would be that everyone could have the chance to see Earth from on orbit. If they did, I believe people would view themselves, the earth, and those around them in a totally different way. They would be much kinder to one another, and they would try to protect this incredible planet that we all live on.

My seventh trip to orbit gave me sole possession of the world record for most spaceflights for less than two months. Then my friend, colleague, and 1980 NASA Group 9 astronaut classmate Franklin R. Chang Diaz tied the mark as a mission specialist on STS-111. Our very first flights—STS-61B and STS-61C—had also been separated by only a few weeks way back in late 1985 and early 1986. For the rest of our careers, I would fly and Franklin would fly soon thereafter. I would fly, he would fly. I would fly, he would fly. Only once, when NASA had some long delays in getting my STS-88 flight off the ground, did Franklin's sixth flight launch before mine. We kept tabs on each other and would occasionally tease the other about getting way too old to fly any more. But Franklin and I were both eager to fly an eighth mission and then some.

With STS-88, the first ISS construction flight, I broke the U.S. records for most spacewalks and for the most time spent outside on spacewalks. I increased both records on my final flight, but with the mountain of EVAs required to build the ISS, it was only a matter of time before my EVA records would be surpassed. Michael E. Lopez-Alegria broke both records during his long duration stay on the ISS. It is my hope that *all* spaceflight records will continue to be broken. New records will mean that human beings are pushing further into space. That is what we ought to be doing, but I do not see it happening anytime soon the way things currently are.

That is my story. But what about the Space Shuttle? The shuttle was a great flying machine, and we will probably never again have the capabilities the shuttle gave us. It was by far the most advanced vehicle ever flown into space. The shuttle was man's first attempt to make flying into space more routine and cheaper. Neither goal was achieved, but the shuttle experience headed us in the right direction. However, the shuttle did greatly increase the number of countries directly involved in the exploration of space, sending their citizens and experiments into orbit. From a human perspective, perhaps that is one of the more important things the shuttle accomplished.

I firmly believe that it was time to end the Space Shuttle program. Statistics say that if we continued to fly the shuttle, there was a fairly high probability of losing another vehicle and crew. I believe those stats. It seemed every couple of flights we would find something new that surprised us and caused new concerns.

I agreed with what the Bush administration planned to do, to stop flying the shuttle and to focus NASA's energies, people, and money on going to the moon and beyond. As long as we continued to fly the shuttle, NASA did not have the resources to go beyond low-Earth orbit. It was time to move on, but unfortunately, we did not move quickly enough. NASA got too many redirections, too much help, and here we are, mired in indecision.

In this book, author Rick Houston does a great job of pulling together a number of diverse sources, including flight controllers, shuttle workers, and many of my astronaut colleagues to fashion a compelling narrative of what it was like to be involved in the Space Shuttle program from the aftermath of the *Challenger* tragedy through wheel stop of STS-135. Believe me when I say that it was an amazing journey, and now with Rick's help, you can come along for the ride.

Jerry Ross

STS-61B, STS-27, STS-37, STS-55, STS-74, STS-88, STS-110

Author, Spacewalker *(Purdue University Press, 2013)*

Acknowledgments

This book changed my life, and thankfully, I am a lesser man for it.

Research led me to Johnson Space Center in June 2010. There, I was shown around by astronaut Douglas G. Hurley. The highlight was to be a run in JSC's motion-base simulator, but when I attempted to strap in, the left lap belt of my seat on the flight deck would not latch because of my oversized belly. I tried harder, sucking in my gut. That small voice that had always taunted me began to laugh. Although we continued with the one belt undone, it was far and away the most embarrassing moment of my life.

I got back home to North Carolina and headed to the walking track behind our local YMCA. After participating in four 5k events, I enrolled in a class at the Y to learn how to actually train for one. I will always be thankful for the encouragement of instructors Crystal Joyner, Julie Winters, Wendy Hayden, and Jennifer Helton and the exhausted camaraderie of Team Attila, my fellow running newbies Mandy Marxen, Leslie Gough, Brandy Whitaker, Mary Hitchcock, and Debbie Ours. They all pushed me much, much farther than I ever could have gone on my own.

Twenty-one months after the embarrassment of my run in the motion-base simulator, I was able to have my photo taken in the commander's seat of *Atlantis* at Kennedy Space Center. I get emotional even typing these words, but this time, the safety harnesses fit. Words cannot fully express my gratitude to former astronaut Jon McBride, Brian Emond of NASA public affairs, and shuttle technician Jay Beason for providing me with an experience I shall never forget—and to my buddy Gary Milgrom for acting as my photographer for the shoot.

Mixed in along the way to my weight loss was the continued research and writing of this book. NASA public affairs officer Gayle Frere helped line up several interviews with current astronauts, while Andy Turnage of the Association of Space Explorers was more than gracious in forwarding many interview requests to retired space travelers.

2. Nearly two years after not being able to fully secure the seat belts of the motion-base simulator in Houston, author Rick Houston strapped into the commander's seat of *Atlantis*. Ninety-five pounds lighter this time, the harnesses fit! Author's collection/Gary Milgrom/Atlanta Pixel Photo.

Astronauts and others involved in the human spaceflight program have been almost universally accommodating. Many reviewed their portions of the manuscript and answered follow-up e-mails. I would even go so far as to now consider at least a handful of them friends.

Late in 1999, I stumbled across a new website called collectSPACE.com. Incredibly, there were hundreds, if not thousands, of space geeks out there who were just like me. The site's creator, Robert Z. Pearlman, was and is our benevolent and all-knowing king. Had it not been for collectSPACE, this book would not be happening—at least not with me as its author.

Years later, I came across a post from none other than Colin Burgess on the collectSPACE message boards. He was in need of authors for an upcoming University of Nebraska Press project called *Footprints in the Dust*, a book that was to detail the lunar missions of the late 1960s and early '70s. Not long afterward, I had my assignment. I was to handle the lead chapter, on the historic flight of *Apollo 11*.

My favorite Australian must have liked what he read, because he offered me a shot at writing the book you now hold in your hands. Thank you, Colin, the best friend I have ever known whom I have never actually met face-to-face. Thanks also go to Rob Taylor, Sara Springsteen, and Courtney

Ochsner of the University of Nebraska Press for their patience with me. Colleen Romick Clark, who had the unenviable task of serving as copyeditor, whipped the manuscript into shape.

This book would never have happened without the support of many other people. First and foremost, I want to express thanks to my Lord and Savior, Jesus Christ. He has carried me through the darkest times of my life. Also, three families took me in as one of their own as I walked my life's valleys. My best friend, Joe Estep, his mom, Sandi, and sister Jennifer are the kind of people you want on your side when you are in a tight spot, and they have been for me. Willard, Judy, Joe, and LeeAnn Knight—LeeAnn is now LeeAnn Burton—gave me room and board when I needed it the most. James, Lib, Jamie, and Amy Reynolds—Amy is now Amy Bottomley—were with me as I chased a dream. Words cannot fully express how much I love and care for each one of these special people in my life.

The very same and even more goes for my immediate family. I have a son from my first marriage, Richard, who is a young adult now. He graduated second in a high school class of more than six hundred students, leading me to the inevitable conclusion that such intelligence evidently skipped a generation in my case. There is absolutely nothing in this world that he cannot accomplish once he sets his mind to it.

My twin sons, Adam and Jesse, are constant sources of amazement and inspiration. Being Jesse and Adam's parent has always been a team effort with the love of my life, my wife, Jeanie. A simple text, phone call, or touch of her hand on a bad day makes everything so much better. I love Jeanie, Richard, Jesse, and Adam with every ounce of my being. They are my reasons for being.

I need to also mention the following: Chris Bergin of NASASpaceflight.com; Dick Conway; Duane Cross (sorry, Hoss, but this is my book—the Yankees stink!!!); Jo deJournette; Michael, Missy, Kaleb, and Haley Dickerson; Joy and Kent Doub; Matt Erickson; Gray Garrison (thanks for making my dreams of driving the pace car come true!); David and Melissa Gentry; Tad and Jodi Geschickter; Dick and Linda Gordon; the entire Graham clan—Jim, Una, Tammy, Heather, Philip, Carolyn, David, Kristie, Jason, and Angie; Artie, Cindy, Chase, and Bryce Greer; Cheryl Gundy of the Space Telescope Science Institute; Pierre, Carol, and Katy Hamel; Bill Harwood of CBS News; Mark and Renee Hayes (Mark, have I ever mentioned how much I would dearly love to own the hood off Richard Petty's car that now hangs in your garage?!?); Suzanne Heffner; Ed Hengeveld; Charlene Hubbard and her son Dusty; Gene and Emma Hubbard; Jim Hunter; Norman Jameson; Donald and Frances Johnson; Jonathan, Angie, and Kendra King; Jimmy, Penny, Patrick, and Jessica Lancaster; Whitney Levens; Christian Lomax; the entire Maplewood Baptist Church family; Joe Menzer; Deborah Evans Price; Dennis Punch; Denver Rakes and his sister Lauren; Tom and Jean Reavis, the very best parents-in-law for which a man could ever ask; Steve Richards (I figure if you're in the acknowledgments, it can only help in my quest to pry Aunt Bea's radio from your possession); Charles and Nancy Scott; Morgan and Cindy Shepherd; Walter Shore, for putting in my hands the very earliest hard copy of this manuscript; Briana Smith; Chuck, Michelle, Ian, and Mattie Tavano; Ike, Linda, and Nathan "Scooter" Trivette; Rodney, Ginger, Nathan, and Anna Wagoner (thank you so much for taking the boys off my hands in my moments of manuscript desperation!); Steve Waid; Todd Wilkerson; Leslie Wilkinson; Deb Williams; Anna Wood and her son Tyler, the only person I know who knows more about *Star Wars* than I do; and Cory and Shannon Yost.

Introduction

For as long as I can remember, I have been fascinated by the concept of human beings willingly strapping themselves to the pointy end of a rocket to be blasted off into the vastness of space. This was the highest form of adventure, and on top of that, astronauts looked so incredibly cool in their spacesuits. Just imagine being a little kid and watching the Apollo moonwalkers doing their thing on the lunar surface. They had a dune buggy and everything!

I do not know that I fully comprehended exactly what it was that they were doing—that they were on the moon, of all places, exploring, conducting experiments and looking like they were having the time of their lives. Best of all, it seemed that every other word they said was "Houston." In my young mind, these spacemen were talking to *me*, because that was *my* last name. I was hooked, big time. There could be no greater job in the world than to be an honest-to-goodness, superhero, Buck Rogers astronaut. Sadly, NASA never had much use for overweight astronauts with not an ounce of technical or scientific ability. I always felt that I could have been used as ballast, if nothing else.

Doug Hurley became an avid NASCAR fan when his cousin, Nan, married Greg Zipadelli, one of the sport's most prominent crew chiefs. I wrote a story on Doug's interest for NASCAR.com several years ago, and we kept in touch via e-mail.

During my tour of JSC, Doug and I spent some time on one of the three shuttle mock-ups in Building 9. He had to run back to his office while Debbie Vi-Vi Nguyen, my NASA public affairs escort, took a call on her cell down on the mid-deck. Me, I was up on the flight deck, all by myself for several glorious minutes. I could flip all the switches I wanted and twist any knob to which I could possibly get. I was not actually in micro-gravity, working the robotic arm as my crewmates scurried about outside servicing

3. After their run in the motion-base simulator at JSC on 22 June 2010, astronaut Doug Hurley and author Rick Houston pose for a photo. He's smiling here, but it was a day that changed Houston's life. Author's collection.

the Hubble Space Telescope, but in my mind's eye I was. I sat in the commander's seat, pushing the rudder pedals this way and that with my feet and taking the control stick as I imagined myself aiming for the runway to end another heroic shuttle mission. This was every space geek's dream, but the best was yet to come.

One of my favorite quotes in this book came from Charles J. Camarda. Charlie was just shy of his forty-fourth birthday when he was selected as an astronaut, my age now as I write this. When he flew for the first and only time on STS-114, he had worked at NASA as an engineer or astronaut for more than thirty years. He described his reaction to the launch almost exactly as I had always imaged mine would be. "You're gonna have to ask Wendy Lawrence," he began. "She was sitting right next to me, and what Wendy Lawrence says is that I was just laughing and grabbing her hand and shaking her the whole time we were going up." Can you not just picture him—or better yet, yourself—on such a thrilling and dynamic ride? There, in the simulator that afternoon, I was about to get an inkling of what Charlie and Doug had experienced on their flights. We were flat on our backs, ready to rock and roll.

"Once you hit T-zero, you're going someplace as soon as they light the boosters," Doug said over a comm loop that on DVD playback sounded just like we were actually on the launch pad, seconds away from lighting the candle.

I could not wait and asked where the clock was located.

"Twenty-eight seconds," Doug responded, pointing to the instrument panel.

I let out a huge sigh, trying to contain my nervous anticipation. Knowing what I felt like then, I cannot begin to imagine what the actual moments before liftoff must have felt like.

"At ten seconds, the navigation system will initialize," Doug continued. "You'll see the gyro spin. Stand by for engines.

"There's the engine start."

All around me, the cabin began to shake.

"Three at 100 percent."

This was really happening, and then the imaginary Solid Rocket Boosters (SRBs) lit.

"There's T-zero. Here we go!"

All heck broke loose. Something grabbed my seat and started to rock it this way and that. If the start of main engines had been something, ignition of the SRBs added an entirely new dimension to the ride. "Wow . . . ," I said, again with the less-than-stellar commentary. And then, I actually giggled.

Doug continued making his calls: "One-oh-two, one-oh-two, auto, auto . . . and Houston, *Endeavour*. Roll program. There you go. We're rolling."

"One-oh-two" referred to an operational sequence mode. It went from 101 at the start of main engines, to 102 at the start of the SRBs, then on to 103 when the twin boosters separated, and finally to 104 after main engine cutoff. "Auto, auto" meant that we were on autopilot. Doug in an instant had checked for both in two separate locations—on the display in front of us and on the eyebrow panel just above the windows. On ascent, "one-oh-two, one-oh-two" and "auto, auto" are important milestones. It meant that performance on this particular run was nominal for the time being—there were no surprises. I had not screwed anything up too badly. I had not *touched* anything. After clearing the tower and rolling, I felt my most important innards bounce off my esophagus.

"We're off to the races, already at 6,000 feet, going 320 knots, Mach 7."

Again, I could figure nothing more eloquent to say than "Wow." Other than that, I was completely speechless, taking it all in, everything that was happening around me.

Doug, ever the professional, flipped a page on the checklist on the instrument panel in front of him. "The next thing you're going to be looking for is for the engines to throttle back going through maximum aerodynamic pressure. There they go. So you're at 70 percent of rated thrust."

He quickly described that moment from STS-127. "When the engines go back to 104, it felt like we lit the afterburner," Doug said. "It was incredible. I mean, it just—bam."

Doug then said something that hit me like a ton of bricks.

"This is where they would say, 'Go at throttle up.' Roger, go with throttle up."

I could not help but think of *Challenger*, and a cold chill overtook me for a moment. This was not playtime; this was not the best amusement-park ride ever. This was serious business, and when Doug and his *real* astronaut colleagues did this sim for real, they were training for situations in which their lives would be on the line, their behinds hanging far over the ragged edge. We continued through the separation of the SRBs, the light flashing in the cabin and Doug noting that the actual event was even more spectacular than that.

The ride smoothed out, but as the acceleration to orbital velocity continued, I felt pressed back in my couch. I saw a gauge that told me that we were at three gs. I realized it probably was not the actual force being applied to my body, but I also knew that I could barely lift my head. Finally, eight and a half minutes after lifting off from the pad came main engine cutoff—MECO. If this had been an actual flight, my crewmates would have been congratulating me on "officially" becoming a flown astronaut.

Over the course of the next ninety minutes, I was an astronaut, training for the upcoming flight of STS-173 (73 was my jersey number as a member of the DuPont Senior High School Bulldog football team, in case you are wondering). Doug took me through another ascent and a total of five landings at both ends of the Shuttle Landing Facility at KSC, at night and at Edwards Air Force Base in California. It took one, maybe two, of those landing attempts to realize fully that as a shuttle pilot, I was a far better journalist.

"Pull the nose up!" Doug ordered, a tone of amusement in his voice. "Stay right. Stay on center line. Come back to the right, there. Pull it up. Pull it up. Keep pulling! There you go. Keep pulling . . . I've gotta arm the gear."

If this had been the real deal, rescue vehicles would have already been dispatched to pick up the pieces of the shuttle that I was about to leave scattered in the general vicinity of the runway.

"Pull it up. Pull it up. Pull it up! Hold it right there. Now . . . just land the thing. A little more nose up. Nose up! Nose up . . . Okay, hold it there. Don't go too high . . . don't go too high!"

The sound of tires screeching on touchdown shot through the flight deck. We bounced.

"Shewwwwwwwwww . . . ," Doug exhaled, laughing as he did so. "We landed."

And then we landed again. The bounce had taken us what seemed like a half a mile or more down the runway, and to be quite honest, I'm not so sure we did not plop down a third time. This was much harder than it looked on television. On another attempt, I all but brought us down on a dusty dirt road in the backwoods of Georgia. I gave up and took up two other crucial tasks—deploy the landing gear at three hundred feet and the drag chute as we slowed. Those required the touches of a couple of buttons. Easy. That, I could do very well. I now consider myself a landing gear and drag chute deploying professional. I am afraid, however, that I took advantage of Doug's hospitality. After each of the landings, he asked, "Want to do another one?" Come on, what in the world was I supposed to say? No? You have got to be kidding.

The best part of the run, however, was the conversations between Doug and me as we waited for the test conductor to reset everything for yet another go. There we were, in the motion-base simulator on the grounds of Johnson Space Center, talking NASCAR. Unbelievably, it turned out that the test conductor, Daniel K. Nelson, himself had actually driven races on dirt tracks in the Midwest.

I have participated in the Dale Jarrett Racing Adventure driving school at Talladega Superspeedway in Alabama, reaching a speed of nearly 178 miles an hour on the backstretch. I have served as the pace-car driver prior to a Busch Series event at Florida's Homestead-Miami Speedway, and I have been the real-deal pace car driver during several races at the legendary Bowman Gray Stadium short track. I have tracked down foul balls at two Major League Baseball games and four Minor League contests. I have shaken hands with Neil Armstrong and spent an amazing NASCAR weekend in Phoenix, Arizona, as host to *Gemini 11* and *Apollo 12* toast of the town Dick Gordon. Over the course of a career that has allowed me to do many cool things, that afternoon alongside Doug Hurley in the simulator at JSC ranks very near the top of my professional life's most memorable moments.

I had just had a small glimpse into the life of an astronaut, and as a result, I came away with an even more profound respect and appreciation for those who worked in the shuttle program. It also focused my resolve to put together the kind of book that would honor their professionalism, the one that you now hold in your hands. Not everyone will ever get the chance to do a simulator run alongside an astronaut, but it's my hope that *Wheels Stop* will be the next best thing.

Somewhere along the way while putting this book together, a wonderful thing happened. The people who dedicated their lives to the Space Shuttle program evolved from flawless superheroes into actual human beings, with frustrations and emotions just like the rest of us. Maybe it was the conversation I had with at least one NASA insider who wound up shedding a tear during our interview. It might have been when I asked one astronaut about STS-107. I did not say another word for a solid forty-five minutes, while his thoughts tumbled out one right after another. Maybe it was becoming acquainted with Milt Heflin and exchanging what is now hundreds of e-mails, many times if for no other reason than to just shoot the breeze.

In this book, I have tried my dead-level best to honor the efforts of the people who worked on the Space Shuttle program. If you are looking for a highly detailed account of its engineering and scientific achievements, this is a book that will most likely be disappointing to you. Instead, the goal here has been to tell what the people who worked in and around NASA during the shuttle era were able to experience. This is their story. Enjoy!

Wheels Stop

1. Back in the Game

The Space Shuttle *Discovery* sat expectantly on Launch Complex 39, pad B, at Kennedy Space Center, ready for a seventh flight into the heavens. The winged spacecraft, with her five-man crew tucked inside, was like a Kentucky Derby favorite corralled into the starting gate, snorting and pawing at the ground, anxious to get started. The orbiter was not the only one anticipating the chance to get this day under way.

Frederick H. Hauck—the commander of this mission, designated STS-26—had been the first of the crew to enter the maw of the beast and strap in. To his right at the controls of *Discovery* was pilot Richard O. Covey. Also on the upper flight deck were mission specialists David C. Hilmers and John M. Lounge. Hilmers was seated directly behind the center control console with Lounge to his immediate right, close enough for the two of them to touch and with Lounge looking squarely at the back of Covey's couch. George D. Nelson, the only STS-26 crew member without a background in military aviation, was down on the mid-deck by himself, seated right next to the hatch. Rick, Dick, Dave, Mike, and Pinky were ready to shake, rattle, and roll their way into orbit. Each was a spaceflight veteran, having flown at least one previous shuttle mission.

On this particular morning, going to all that trouble had seemed an exercise in futility. The crew of STS-26 had made the eight-mile trip from the Operations and Checkout Building to the launch pad not really expecting to fly. An upper-level wind shear had been detected, and if it had persisted, the takeoff would have been scrubbed. Even while stepping out of the crew van at the pad and looking up at *Discovery*, this monstrosity that seemed very nearly alive, they were skeptical. Once they were settled in, the weather delay remained. Then fuses in the cooling system of the new partial-pressure launch and reentry suits worn by Covey and Lounge had to be replaced, and this did not help matters in the least. Surely, this was going to be a no-go.

Maybe not. The upper-level winds calmed to within acceptable limits, and one by one, controllers were polled during the T-minus nine-minute hold.

Go.

Go.

Go.

Go.

The journey was on after all. As the countdown came out of its cautionary hold, the cockpit became a beehive of activity. This was it. Just then, Covey spotted something through the windows on his side of the cabin. "Uh-oh," Covey quipped, not thinking. Hauck shot back, "Covey, what's wrong?!?" Covey, whose first flight launched in the rain, had seen what he thought to be a shower spreading over the waters of the nearby Atlantic Ocean. Hauck's reaction was swift. "Dick," Hauck admonished his pilot, "do not use the words 'uh-oh' again as long as we're flying this machine together." Covey remembers the rebuke very well: "Rick let me know that was not anything he ever wanted to hear from me again."

The phrase "Uh-oh," attributed to pilot Michael J. Smith in mission transcripts, was the last discernible utterance of the doomed crew of the Space Shuttle *Challenger*, which was destroyed seventy-three seconds after launch on 28 January 1986. After a difficult thirty-two months of speculation and introspection, STS-26 was to be the first shuttle flight since that terrible morning.

NASA had never before dealt with anything like the loss of *Challenger* and its crew of seven—five men and two women—assigned to mission STS-51L.

American astronauts had died in the line of duty, of course. Virgil I. "Gus" Grissom, Edward H. White II, and Roger B. Chaffee were killed during training for *Apollo 1*, but they were on the ground when the fire that claimed their lives broke out, well out of sight of the press and public. A generation would pass before audio recordings of those terrifying moments were released, and even then, the event was left mostly up to the imagination. The claim remained intact—dented maybe but still intact: NASA had never lost an astronaut during an actual spaceflight. Several other would-be spacefarers died in various aircraft and automobile accidents, but when it came down to it, those were the kinds of tragedies that could have befallen anyone.

NASA came to be seen as the agency that could handle any complication through the superheroic problem-solving teamwork of its astronauts, flight controllers, engineers, and other assorted employees and contractors. The Mercury program famously experienced what could have been disastrous problems on the flights of John H. Glenn Jr., M. Scott Carpenter, and L. Gordon Cooper Jr., and a fourth astronaut, Grissom, nearly drowned when his spacecraft sank soon after safely splashing down in the Atlantic Ocean. Neil A. Armstrong and David R. Scott tumbled end over end on *Gemini 8* before miraculously regaining control. Twin lightning strikes hammered the launch of *Apollo 12*, followed months later by the epic tale of *Apollo 13*. Returning from the Apollo-Soyuz Test Project, the American crew of Thomas P. Stafford, Donald K. "Deke" Slayton, and Vance D. Brand endured near fatal levels of nitrogen tetroxide as their spacecraft neared splashdown. Brand briefly lost consciousness, while Slayton became nauseous. All three men were hospitalized for two weeks for observation. In the *Challenger* debacle, the agency would in a very real sense become a victim of its successful close calls. Coming so close to death only to avoid it at the very last second seemed to be the rule at NASA, rather than a miraculous exception.

The Soviet Union had not been so fortunate. Vladimir Komarov died 24 April 1967 when his *Soyuz 1* spacecraft smashed into a field in Orenburg Oblast after suffering a number of failures, the last and ultimately most catastrophic of which was a tangled parachute. A little more than four years later, on 30 June 1971, the crew of *Soyuz 11* perished when a faulty valve opened and depressurized the capsule after undocking from the *Salyut 1* space station. Killed within seconds were cosmonauts Vladislav Volkov, Georgi Dobrovolsky, and Viktor Patsayev. These losses, though, were sustained behind the veil of the Iron Curtain that separated the Soviet Union from most of the rest of the world. Many of the details concerning both mishaps came through Western intelligence sources, and those that did not came at a frozen snail's pace, if at all. Komarov's death was not announced by the Soviet news agency TASS until more than seven hours after the accident, in the form of a terse one-sentence statement. Other releases provided a few more details, but not many. Another monumental mishap took place on 3

July 1969, when an unmanned Soviet N1 rocket exploded with the force of a small nuclear bomb, utterly obliterating the Baikonur launch complex. It took American satellite imagery for the West to know anything of the event.

Challenger could not possibly have endured a more public fate. Tens of thousands of spectators had gathered on the central Florida coast that cold January morning to watch as commander Francis R. Scobee, Mike Smith, and mission specialists Ellison S. Onizuka, Judith A. Resnik, and Ronald E. McNair, along with payload specialist Gregory B. Jarvis and teacher observer S. Christa McAuliffe, began their journey into space. McAuliffe, a New Hampshire high school instructor, won out over thousands of Teacher in Space candidates and planned to conduct two live lessons during the flight. If images of the accident itself are the most nightmarish, footage of McAuliffe's parents, Ed and Grace Corrigan, in a KSC viewing stand and students at McAuliffe's Concord High School are very nearly as troubling.

It would have been difficult for the public to handle had the images been shown just once, but they were shown over and over and over again by countless media outlets. The footage hammered home the incredible loss. It had indeed been "obviously a major malfunction," as reported by public affairs officer Steve Nesbitt in the first few moments after the shuttle's breakup. A quarter of a century later, amateur videos of the accident popped up here and there on the Internet, and one, licensed by the *Huffington Post*, features the almost giddy excitement of Steven and Hope Virostek, a Rhode Island couple who had retired to Titusville, Florida, just a few miles from KSC. Their joy for McAuliffe—*C'mon, Chris! Go Chris, go! Beautiful!*—turned to stark and unabashed horror in the seconds after vehicle breakup. Just when viewers thought they had seen every angle and experienced every emotion, this video closes with Hope Virostek's brokenhearted rendering of the Roman Catholic Church's Eternal Rest prayer—"May their souls, and the souls of all the faithful departed through the mercy of God, rest in peace."

Unimaginably, evidence suggests that at least some of the crew survived the initial breakup of the orbiter. Off-nominal switches on his side of the instrument panel indicated Smith tried to somehow regain control of the inextricably out-of-control spacecraft, while three of the four recovered personal egress air packs containing an emergency supply of breathing air showed signs of being activated. A crew-survivability analysis conducted by *Skylab 2* astronaut. Joseph P. Kerwin, serving as a biomedical specialist at

Johnson Space Center in Houston, was released six months after the tragedy. Commissioned by Richard H. "Dick" Truly, the commander of STS-8, who was serving at the time as NASA's associate administrator for space flight, the study concluded that:

> The cause of death of the *Challenger* astronauts could not be positively determined.
> The forces to which the crew was exposed during orbiter breakup were probably not sufficient to cause death or serious injury.
> Finally, the crew possibly, but not certainly, lost consciousness in the seconds following orbiter breakup due to in-flight loss of crew module pressure.

The report continued, outlining the final moments of the STS-51L crew: "After vehicle breakup, the crew compartment continued its upward trajectory, peaking at an altitude of 65,000 feet approximately 25 seconds after breakup. It then descended striking the ocean surface about two minutes and forty-five seconds after breakup at a velocity of about 207 miles per hour. The forces imposed by this impact approximated 200 gs, far in excess of the structural limits of the crew compartment or crew survivability levels."

If that was the horrifying end result, by what means had the accident taken place? A report released 6 June 1986 by the Presidential Commission on the Space Shuttle *Challenger* Accident—better known as the Rogers Commission, after Chairman William P. Rogers—found that the failure of a rubberized O-ring seal in the right SRB allowed pressurized gases and white-hot flame to "blow by" the device, making contact with and burning through metal struts holding the huge External Tank in place. The failure had, in effect, lit the fuse on a 1.6 million–pound bomb.

Looking even deeper into the incident, it became maddeningly clear that the accident might very well have been avoided at any of a number of different points. The night before the fateful launch, temperatures on the central Florida coast dipped into the low twenties. Ice covered many parts of the pad, and when drains were opened, they, too, froze and caused overflows. Winds spread the water still further, exacerbating the icing problem. At launch, the air temperature stood at thirty-six degrees, more than fifteen degrees colder than any previous send-off. Still, management pressed forward with plans to send the flight on its way. Six times already, the mission had been delayed for one reason or another.

Officials of Morton Thiokol—makers of the SRB—recommended on the night of 27 January 1986 that the launch of STS-51L be scrubbed once more due to their concerns over the weather's impact on the structural integrity of its hardware. This was not a flip-of-the-coin close call, either. As far back as July 1985, Morton Thiokol mechanical engineer Roger M. Boisjoly had expressed in a memo serious concerns over the design of the SRBs. More warning memos followed, with little to no effect. Boisjoly and Allan McDonald, Morton Thiokol's liaison for the SRB project at KSC, vehemently argued against launching due to the inclement weather conditions. "I expressed deep concern about launching at low temperature," Boisjoly told the Rogers Commission. "At that point in time, I was asked to quantify my concerns and I said I couldn't. I couldn't quantify it. I had no data to quantify it, but I did say I knew that it was away from goodness in the current database." Reliable data existed for launches that took place at or above fifty-three degrees—anything else was a crapshoot.

Nineteen years and one day separated the *Apollo 1* and *Challenger* tragedies, and "go fever" had helped doom both crews. During the Apollo program, the rush was to meet President John F. Kennedy's mandate of "landing a man on the moon and returning him safely to the earth" by the end of the 1960s. The successes of NASA's Mercury and Gemini programs turned out to be both blessing and curse, in that the momentum they created bore down on the agency with an almost unstoppable force. There was no time to stop and evaluate how things were going, because the moon was waiting. During five years of shuttle flights leading up to the destruction of *Challenger*, a similar rush to the launch pad developed. This time, the problem was not getting to the lunar surface and back. Instead, there was a schedule to be kept. The shuttle flew nine times in 1985, with at least eleven more missions scheduled for 1986. There were commercial payloads to be launched and repaired, and money to be made doing it. Indeed, the Rogers Commission found that "the relentless pressure to increase the flight rate" was a major cause of the mishap.

Overnight, a harsh new climate took hold. Said Hauck, whose father, Philip, was himself a navy man who survived the Japanese attack on Pearl Harbor just eight months after the birth of his son:

The Challenger *explosion crushed all of us who worked in the astronaut office. Beyond the unfathomable loss of our friends and the horrific pain inflicted on their families, our own professional world was disintegrating. When would we fly again?* Would *we fly again? It didn't occur to me that this was an avoidable accident. Every bit of my experience at NASA since I arrived in mid-1978 bespoke professionalism and caution. I'd never seen a more dedicated group in my aviation career. By this time, I had flown two missions on the shuttle, and I always knew I was putting my life at risk when I launched. But deep down inside, I knew NASA had the formula for success. I was wrong.*

Compounding the problem was the intense scrutiny NASA was under following the accident—some of it justified and some of it entirely ignorant and mean-spirited. One publication, the most vile of supermarket tabloids, went so far as to fabricate a transcript of the *Challenger* crew's last moments. For nearly thirty years, the agency had been treated for the most part with kid gloves. That was no longer the case. The gloves were now off. "I wasn't prepared for the public pillorying NASA received at the hands of the press, Congress, and the public," Hauck said. "It was as if they were jilted lovers. We were all the bad guys. In some ways it reminded me of the atmosphere surrounding my return from flying combat in Vietnam—being the object of pent-up anger and frustration. Those were tough days."

The future STS-26 commander was not alone in his concern for the future of the program. Serving as capsule communicator (CapCom) at the time of the accident, it was Covey who had the haunting "Go with throttle up" exchange with Scobee moments before the orbiter broke up. After catching a glimpse of the bulbous cloud of debris beginning to form on a monitor to his left, Covey's eyes widened and his jaw dropped, the very definition of a shocked expression. Just five months before, Covey had spent a week in space as the pilot of STS-51I. This did not require the analysis of a spaceflight veteran or expert controller. The pictures said it all: something had just gone very, very wrong. The mood around Johnson Space Center was unlike anything the agency had ever experienced, remembered Covey:

Everybody was reacting basically to two things. One was the fact that they had lost a Space Shuttle and lost a crew, and two, the Rogers Commission was extremely critical, and in many cases rightfully so, about the way the decision-making processes and culture had evolved. So those two things together are hard for any institution to accept, because this was still largely a workforce that had come through the Apollo era into the shuttle era, and had been immensely successful in dealing with the issues that had come through both those programs to that point. So to be told that the culture was broken was hard to deal with, and that's because culture doesn't change overnight. A lot of people didn't believe this was an accurate description of the situation and environment that existed within the agency, particularly at the Johnson Space Center.

Covey flew before *Challenger* and after, and the difference was telling. "The general difference was a loss of innocence," Covey said. "We all knew the risks associated with flying the shuttle, but when we lost *Challenger*, we hadn't really realized that risk in any way. There was more of a sense of, 'Hey, the risk is real. We need to learn how to deal with that risk in a little bit different way when we go and fly.'"

Lounge, who flew with Covey on STS-51I, went one step further in admitting, "I think we were pretty naïve before *Challenger*. We knew that it was a complex system and there were a lot of ways it could fail. But it was kind of academic, right? It was the typical fighter-pilot attitude, 'It can't happen to me. It happens to those other guys, but it will never happen to me.' And then we saw that tragedy unfold on national television. Suddenly, it was a very real thing that could actually happen."

Each member of what would become the STS-26 flight crew had a deeply personal connection to most if not all of the *Challenger* astronauts. Hauck, Covey, and Nelson were selected to the astronaut corps in 1978, and four of their classmates—Scobee, McNair, Onizuka, and Resnik—were on board *Challenger* when it broke up. Lounge and Hilmers joined the astronaut corps two years later, along with Smith. Covey and Onizuka were in the same test-pilot class, and Covey had also flown the chase plane for an early shuttle landing with McNair as his backseater. Nelson flew with Scobee on STS-41C in April 1984.

Ironically, Nelson's first spaceflight had been on board *Challenger* during the STS-41C mission and his second, STS-61C, landed just twelve days before the accident. The orbiter that carried Nelson on his second flight was *Columbia*, which would meet her own sad end seventeen years later. Unlike the four men who would one day be his STS-26 crewmates, Nelson had never been an aviator in the navy or marines, or a pilot in the army or air force. He was an astronomer. That did not matter after *Challenger*. There was little difference in how pilot astronauts and mission specialists handled the loss. "You take those kinds of things at two levels," Nelson said. "One is the professional level—you hate to see the program fail like that. Then, we're all human beings, so at a personal level, we all mourned over the loss of our friends. We had to deal with those loss issues in the same way that everybody does when they lose someone in an accident. It was a terrible loss. I still miss those guys."

There might have been a more outspoken astronaut in NASA's history than Story Musgrave, but if there was, it was almost certainly behind closed doors well out of earshot of the public. Musgrave flew twice before the loss of the STS-51L crew, both times on board *Challenger*, and when he described the agency as "schizophrenic" before the accident, he did so without flinching. From top to bottom, from the highest branches of the United States government through the agency itself, Musgrave saw the accident as the result of an "organizational failure" prior to the accident:

They had to kill someone. They weren't going to wake up until they killed somebody, and that's all there was to that. They thought they were creating a bus. It was very difficult coming out of the huge success of project management in the sixties. Kennedy said, "Go," and we did it. We've won the race to the moon, let's all go home. So NASA was placed in the very difficult position of having to justify spaceflight. They became very defensive, and they wanted to become very practical. The money's coming from Congress. Wherever the money comes from, NASA's going to try to keep those people happy.

Although the shuttle was the most complex machine ever constructed, Musgrave remembered NASA telling Congress that the odds of an accident were 1 in 100,000. In other words, he said, the shuttle would fly every day for 273 years without anything of any substance going wrong. "That alone is proof that they were out of touch with what they were doing," Musgrave insisted. "They got to believe their own message, and so they thought the shuttle could not have an accident."

And, of course, there was the decision to launch itself, the 800-pound gorilla in the room. Details on what got NASA to that point would come out in the subsequent investigation into the *Challenger* debacle, but launching in inclement conditions was the most obvious causal effect, the tip of the iceberg, if you will. "We knew about the cold weather," Musgrave continued. "We had the plot of O-rings in cold weather. Micromanagement caused the problem, because people from headquarters got involved in the decision process. They never should have. That should have been left up to launch control to make those decisions."

It was only after the accident laid bare so many diverse issues that the culture in and around NASA began to change. Musgrave flew twice before the disaster and four times after—and in so doing, became the only person to venture into space on board all five shuttle orbiters. "One hundred percent, the entire organization changed," he said. "They had to wake up. They woke up with *Challenger* and then they came to acknowledge what they'd actually produced. It was just a total change at all levels."

There were thousands of NASA and contractor employees across the country who had nothing to do with the decision to launch *Challenger*. Still, they felt the loss very, very deeply.

One such person was Tom Overton, who had worked at NASA for twenty years when *Challenger* went down. Growing up in Des Moines, Iowa, in the 1950s, he watched in fascination as German rocket designer Wernher von Braun discussed the potential of spaceflight on Disney television programs such as *Man in Space*, *Man and the Moon*, and *Mars and Beyond*. Even back then, an abiding passion for spaceflight was taking hold.

Overton could not possibly have known then that he would one day lunch with his hero, von Braun, during the heady days of Apollo. After spending four years in California as a draftsman with Douglas Aircraft Company, Overton was asked if he would be willing to move to Florida to work with NASA in developing the Saturn S-IVB stage. The assignment was originally planned to last eighteen months, but more than four decades later, he was still living in Titusville, Florida, with his wife, the former Sharon Muehlenthaler, who also worked on the Apollo program for a couple of years as an employee of IBM. By the last Apollo flight, Overton was a salaried engineer and helped manage the switch-over of the Vehicle Assembly Building (VAB), launch pads, and Mobile Launcher Platform to accommodate the Space Shuttle as well as construction of the Orbiter Processing Facility (OPF). His concentration was in the VAB, where he worked on the lifting sling that attached to the shuttle and rotated it to the vertical position, where it was then mated to the SRB/External Tank stack.

If that wasn't enough, for the first four flights of the Space Shuttle, Overton acted as a forward observer. The job meant that during launch preparations and the event itself, he was in a bunker maybe 2,500 feet or so from the pad. He would eventually oversee scheduling the shuttle workflow through the OPFs and on to the launch pad. Almost from the day he started at NASA, Overton also served as an escort for a wide range of guests who visited KSC for tours, launches, and landings. That job brought him into contact with celebrities and dignitaries such as aviation legend Charles Lindbergh, actor Jimmy Stewart, pro football Hall of Famer Johnny Unitas, Jimmy Carter, Hillary Clinton, George H. W. Bush, and Laura Bush.

Overton spent part of the morning of the disaster with Christa McAuliffe's parents and members of her son Scott's fourth-grade class. He was standing next to Barbara R. Morgan, McAuliffe's backup, at the turn basin a stone's throw from the VAB and press site when the accident took place. Overton, who missed seeing only a handful of shuttle launches in his forty-five-year career at NASA, knew immediately that there was no hope for the crew:

My first comment was, "Oh, my God . . . we have lost them." Barbara looked at me and said, "What do you mean?" I said, "They are gone. We have lost them." Then, she realized what I was saying. I remember looking back at Christa McAuliffe's mother and father, and they were like, "What happened here? Are we going to see the parachutes pretty soon?" It was like they were expecting NASA to take care of every contingency and that somehow they would come through this thing safe and sound. I'll be honest with you. I absolutely did not even sleep that night. I was tossing and turning and feeling like, "Did we contribute to this? How did this happen? Am I a murderer?" Weird things go through your mind. It was very, very traumatic. I had gone through something similar with the Apollo 1 *fire, but I was little bit more removed from that. When you have family right there staring up at it and it blows up in front of you, it was actually more traumatic for me than the* Columbia *landing because you see it right there in front of your eyes.*

Once the shock and depression wore off, Overton felt a certain consensus taking shape within the agency. First was the need to identify, exactly, went wrong. Once that happened, whatever went wrong needed to be fixed. Within a week to ten days, the focus landed primarily on recovered pieces of the SRB that had a hole burned in the side. Remarkably, a manifest was built for a Return to Flight mission within just six months of the accident. Then, the process lengthened to a year . . . then fourteen months . . . then two years and counting. That wasn't necessarily a bad thing, according to Overton. "We started saying, 'I hope these guys that are designing this thing not only fix this, but they fix everything that could possibly go wrong. We do not ever, *ever* want to experience anything like this again,'" he said. "In fact, that is kind of what happened. We had one major problem, and we fixed ten things since we had the time."

This is how much the climate in the NASA community changed in the aftermath of *Challenger*: Prior to the mishap, record processing time for a vehicle was twenty-five days, from landing to rollout to the launch pad in preparation for the next flight. Afterward, it went to ninety days. In the last stages of the Space Shuttle program, it took approximately 120 days to roll an orbiter in and out the doors of the OPF—not due to a reduction of manpower, necessarily, but due to the work involved in getting the spacecraft qualified for spaceflight.

Before the loss of his friends on board *Challenger*, Hauck had been deep in training for STS-61F, his second flight as commander, with Lounge and Hilmers his mission specialists and Roy D. Bridges his pilot. The risky mission's primary goal was to launch the *Ulysses* interplanetary probe from the shuttle's payload bay by means of the thin-skinned, pressure-stabilized upper-stage Centaur rocket. To make the ride uphill, plans called for the shuttle's three main engines to be throttled up to an unprecedented 109 percent of rated thrust.

Consider the implications if the flight had been somehow aborted, forcing it to make a return-to-launch-site or transatlantic emergency landing. The shuttle would have had a cargo bay full of the Centaur's liquid oxygen and liquid hydrogen, which in turn meant that it would have to be dumped in the midst of the crisis. After landing, any leftovers would continue to be pumped out either side of the spacecraft. To make matters even more interesting, a second Shuttle-Centaur mission (STS-61G), led by David M. Walker, would have been launched just four days later to head the *Galileo* probe to

4. STS-26 commander Rick Hauck. Courtesy NASA, via NASASpaceflight.com L2 Forum.

Jupiter. It was, at the very best, a touchy proposition, so much so that John W. Young, the *Apollo 16* moonwalker who served at the time as the chief of the astronaut office, dubbed both STS-61F and 61G as "Death Star" missions.

So dire was the situation that after one frustrating meeting in early January 1986, Hauck said that he told his crew, "NASA is doing business different from the way it has in the past. Safety is being compromised, and if any of you want to take yourself off this flight, I will support you." Just a few weeks before the launch of *Challenger*, and with the benefit of hindsight, Hauck's move can very clearly be viewed as a chilling reminder of the culture in place in and around NASA. "Of any mission, I think that was probably the most frightening one," Hilmers admitted. "If any accident was going to happen, that was one that was waiting to be set up. It was after *Challenger* that we realized, 'Man, what were we thinking to try to do that mission?'" Yet to take Hauck up on his offer and back out was simply unthinkable. "It was a rhetorical question because he knew that no one was going to back out," Hilmers continued. "You say you're going to do something, so you do it. I think all of us would probably rather have died than back out."

The Shuttle-Centaur project was scrapped almost immediately after the accident, but Hauck, Lounge, and Hilmers weren't left in the lurch for long. Hauck was told in the summer of 1986 by Truly and George W. S. Abbey, the pragmatic director of the flight crew operations directorate, that he would have command of the first post-*Challenger* mission. Lounge and Hilmers were all but givens for the crew, and Hauck remembers Abbey asking for his input on a pilot. Bridges had moved into command of a test-flight wing at Edwards Air Force Base, and because Covey had worked with the proposed flight as its ascent CapCom, he seemed a good choice to fill the slot. Although Nelson had taken a leave from NASA to move to the University of Washington, he rounded out the crew. And through it all, none of them could say anything for a good six months or so. In the interim Hauck was temporarily transferred to Washington DC to serve as NASA's associate administrator for external relations.

NASA administrator James C. Fletcher wanted someone with flight experience on the shuttle, with credibility beyond that of a mere political appointee, to help put out fires in a number of different areas. Hauck seemed a good fit for the role, but it was clear he would rather fly the shuttle than fly a desk. "After Dr. Fletcher met with the astronaut office as a group, his aide asked me if I would meet with him at his motel room," Hauck recalled. "That's where I first met him face to face, and that's where he asked me if I would come up and do this. I told him that I would hate to give up my opportunity to fly and that I hoped he would permit me to do that. He said he would." Rather than the two years originally proposed by Fletcher, Hauck got an assignment that lasted about five months.

The fact that Covey was the first to fly two missions as a shuttle pilot, rather than one in the right seat before moving up to command of a flight, did not seem to matter to the graduate of the United States Air Force Academy. Covey was the last member of his class to fly. After that, he could deal with just about anything. He was not being singled out as a flier who needed the practice; it was simply the way the chips fell.

After *Challenger*, the remaining shuttle fleet was grounded, so Nelson headed to the astronomy department at the University of Washington in Seattle. "I'm a kind of restless guy and don't like sitting around not doing much," he commented. "So I decided that I would go off and do some science, just to stay busy."

Before he left, Nelson got the ball rolling on a formal family support plan that outlined how NASA looked after the families of astronauts during a mission—as well as what to do in the event of a *Challenger*-like disaster. "It's still being used," Nelson said. "It's expensive. It costs NASA money, but it seems like if you're going to spend half a billion dollars launching the Space Shuttle, you ought to do something for the crew's family." Shortly after taking part in a practice countdown in December 1986, Nelson got a call naming him to the next flight. He remembers it as being "completely out of the blue."

The crew of STS-26 was officially announced on 9 January 1987, and the news was met with consternation, if not an outright fury, in some corners of the astronaut office. Certainly, some of the sentiment could probably be traced back to jealousies that are inherent in any such highly competitive atmosphere. The astronaut office had always teemed with the fighter-jock mentality of always—*always*—wanting to be first. That kind of almost pathological fire dated all the way back to the earliest days of America's manned spaceflight program, and it continued to exist throughout the final days of the shuttle program. Alan B. Shepard Jr. made the country's first suborbital hop into space, leaving six disappointed astronauts in his wake. After that came NASA's first orbital flight . . . then the space agency's first *Gemini* flight . . . first spacewalk . . . and, a few years later, the ultimate plum—the first moonwalk. Even after it was determined that *Apollo 11* would make the first moon landing attempt, the issue of who would be first out of the lunar module *Eagle* was not settled until after Buzz Aldrin made a well-documented run at the assignment.

There had always been a certain sense of frustration with the crew selection process itself. Was it based on nothing more than a flip of the coin? Politics? Schmoozing the right people? Surely, it did not come down to ability alone, did it? Maybe, maybe not. Astronaut Mike Mullane insisted years later the factors at work were "the fierce competiveness that is in the DNA of most military pilots (and backseaters!) and the insane secrecy surrounding astronaut crew selection." He went on to elaborate: "Had our management been open about that process, it would have gone a long way toward mitigating our discontent at not being selected for a particular mission. In spite of our competitiveness, we were all mature enough to understand we can't all be first. But the void left by management's secrecy resulted in speculations that office politics, favoritism, etc., were key factors in the crew selection. That speculation made it hard to swallow being a runner-up."

Mullane admitted in his 2006 book *Riding Rockets: The Outrageous Tales of a Space Shuttle Astronaut* that some in the astronaut office were "livid" over Nelson's selection, in particular:

While Pinky was well liked, he had taken a sabbatical to the University of Washington after Challenger. *The rest of us had stuck around to do the dog work and be brutalized by [John] Young in the process. In our minds Pinky hadn't paid the dues to have received such a prize as the first post-*Challenger *mission. It was also a sore point that his last mission had been the flight prior to* Challenger, *so he had the additional plum of having back-to-back missions. Norm Thagard was certain Abbey had picked Nelson just to show the rest of us how unfair and capricious he could be.*

There were basically two schools of thought when it came to Abbey—one camp loved him and the other absolutely did not. James D. Wetherbee—a man who commanded five shuttle missions, the most in the history of the program—was wholly insistent that Abbey was a driving force in saving NASA following the *Challenger* mishap. "He was taking such critical and great actions after the *Challenger* accident," Wetherbee began. "I learned and I watched him, and that helped me seventeen years later after *Columbia*. I saw the way he recovered that organization not only immediately after the accident, taking the right actions, doing the right things, but then over the many years after that."

Asked about the criticism leveled at Abbey, Wetherbee responded with a still more impassioned defense of his mentor:

The guy was phenomenal. He was unbelievably valuable to human spaceflight. His principles were so high, with his wisdom, if you didn't understand that, I suspect it's a human tendency to react and think that he was wrong. Everything George did was to help human spaceflight be successful, and those who didn't understand that or misunderstood George, I don't spend a lot of time worrying about them because they just flat didn't get it in my opinion. George cared deeply about people. I think that was something a lot of people didn't understand, was how much he did care about people and families. I saw this behind the scenes. I was working with him for many, many years as his deputy and saw the kinds of things he did quietly, with no thought of wanting the publicity, the thanks or the rewards. He just did it quietly to take care of people because it was the right thing to do.

For his part, Nelson said that whatever ill sentiment that might have existed toward him was not overt. "I'm sure there were rumblings about it and people trying to understand why I was George's pet or whatever," he admitted. "It wasn't a tense or stressful situation. My rationalization is that I was already trained up and that I was an EVA expert, which was the one thing they were missing on that crew. I'm sure everyone was disappointed not to be on that crew, but I never felt like anyone held me accountable or was mad at me for it." With two spacewalks already to Nelson's credit, both he and Lounge trained for any contingency EVAs that might be needed during their upcoming flight. None were.

As it was, STS-26 was dubbed in some corners as the "Quiche Mission" because of what some perceived as a relatively sparse flight plan. It nevertheless received intense interest from the media because, by gosh, these were superheroes who calmly and fearlessly faced down danger. The only things missing were their costumes and capes. Artists painted pictures. The *New York Times* did an entire insert on the flight, while CBS News anchor Dan Rather did interviews for the program *48 Hours*. President Ronald Reagan addressed NASA employees on 22 September 1988, one week before the flight. "I still look back at that," Covey said. "I look at the words he said, and it was extraordinary. Presidents coming to JSC just didn't ever happen, and he'd come for that. . . . He's coming back basically to get the troops fired up and to show the support of the nation and his support for the things they've been doing."

There were many practical perks to being a prime crew for so long. "You get everything," Covey continued. "When we wanted to go fly the Shuttle Training Aircraft on these days and these times, yes, we got it. We want the simulator, we got it. We want to go do this, yes, we got it. So we got everything we wanted. The other astronauts suffer from that, to a large degree. They don't get priority. Then, the next crew is only a month or so behind you, and so they're only prime crew for a month. We were prime crew for eighteen months, two years or something. It was ridiculous."

Others in the astronaut corps noticed the ruckus being raised over the Return to Flight mission and teased the STS-26 crew about it. During a reception in honor of Apollo-era astronauts, Mullane and STS-27 crewmate William M. Shepherd mocked the crew by inhaling helium and then introducing themselves to the guests of honor, saying, "Hi there, I'm Rick Hauck, commander of STS-26. Would you like my autograph?" Hauck and Covey were introduced at another formal affair in rock star–like fashion, complete with spotlights and a smoke machine. Following a meeting between the two crews the next day, STS-27 commander Robert L. "Hoot" Gibson was asked if he had anything to discuss. At the prompt, Mullane and Shepherd set off fire extinguishers, while Jerry Ross cued the song "I'm Proud to Be an American" as Gibson and pilot Guy S. Gardner rose dramatically from their seats wearing clip-on bowties.

Years later, Hilmers recalled such attention:

I knew that other members of the astronaut office probably were a little bit—I don't know if they didn't like it or whatever, but we had a lot of what I thought was good-natured fun poked at us. That was fine with me. I didn't have any problem with it. I tried not to make a big deal out of what I was doing. What I wanted to do was try to be the best crew member I possibly could be. The rest of it was just kind of stuff I had to do, as far as publicity, press conferences and things like that. There probably was more attention placed on us than there should have been. The next mission, STS-27, was a secret Department of Defense mission. From that standpoint and also being the second one after Challenger, *I think that they felt like they were playing second fiddle, whereas they had a very important mission to do as well. That was kind of unfortunate. I don't know what I could have done differently to tone down the publicity or to damp down any resentment that might have been there from other astronauts.*

Office politics aside, it would take nearly twenty-one more months following the crew announcement for NASA to fly again. Truly oversaw the Return to Flight program, and he gave the upcoming flight's commander an open invitation to sit in on meetings throughout the comeback process.

5. Gag crew photos were a NASA tradition dating back decades. Here, the astronauts of STS-26 pose in their finest Wild Wild West attire during a trip to New Orleans. Courtesy David Hilmers.

Hauck served mainly as an observer, but he nevertheless was encouraged to express his thoughts on how things were going. He spoke right up during a meeting in which an upcoming SRB test was being discussed. "There were some people that did not want to do the test, because they were afraid that if it blew up, that it could be the end of the program," Hauck said. "I remember saying, 'I'd rather have it blow up on the test stand than with me sitting on it.' In fact, the test did go forward."

STS-26 would forever be linked to *Challenger*—its mission patch honored the crew's fallen colleagues with seven stars in the shape of the Big Dipper. It also featured the red vector symbol from the NASA "meatball" logo, indicating a desire to build on the agency's traditional strengths, as well as a sunrise representing its new beginnings. The symbolism notwithstanding, the crew looked forward to getting on with the business of manned spaceflight. "Our biggest fear was not about launching," Lounge said. "It was about not launching. We really wanted to finally get off the ground." Hauck tried to soothe his family's fears by telling them that STS-26 would be the safest mission NASA had ever launched. Hauck felt the agency knew most everything that was to be known about the vehicle, and from what he observed, no stone was being left unturned while digging into the smallest detail of every system of the spacecraft. Still, he could not bring himself to guarantee that he was coming home. As hard as hundreds of thousands of people could try, the shuttle was never, ever going to be completely safe.

Meanwhile, *Discovery* bounced back and forth like a ping-pong ball following the loss of its sister craft. Between April and October of 1986, the orbiter was moved between the VAB and OPF no fewer than seven times. She was powered down in February 1987 and did not reenergize again for nearly six months. Officially designated OV-103, the craft moved back into OPF bay 1 on 30 October 1987, after which workers made more than two hundred modifications and outfitted its payload bay for the Tracking and Data Relay Satellite that would be released during the flight of STS-26. Finally, *Discovery* was rolled to the VAB on 21 June 1988 for mating to its SRB/External Tank stack in preparation for the flight. Thirteen days later, on Independence Day in America, the craft at long last made its way to the launch pad. Not since August 1985—when she flew Covey, Lounge, and the rest of the STS-51I crew—had *Discovery* been in space.

Once launch day for STS-26 finally did arrive on 29 September 1988, for all the quiet professionalism they had shown in the nearly three years since *Challenger*, it would have been nearly impossible for the flight's astronauts not to pause for even a moment of reflection. Hilmers waited on the launch platform gantry as the rest of his crewmates boarded *Discovery*, which allowed him time to consider the dichotomy of the situation. "It was a beautiful day, and I remember watching the birds go by and just looking out at the tranquil scene," Hilmers said. "It's always kind of amazing to me that in just an hour or two after getting in, this would be one big blast zone. This peaceful scene turning into all the fire and smoke was always kind of a big contrast in my mind." Once he settled into place on the flight deck, Hilmers had even more time to think. He reviewed his responsibilities during the launch phase, how it would all take place—and yes, about what *could* happen:

I don't think it's a sense of fear. I think it's more of a sense of respect for what could go on and the fact that for quite a bit of that launch time, there's not much that you can do to change anything, particularly during the first couple of minutes. Maybe I went back a little bit and thought in my mind about the changes that had been made, trying to give myself some reassurance that we really thought of everything. Of course, you think of your own mortality, too, and those that are waiting for you. Also, there's a realization that there's millions of people watching and praying for us. You could sense them.

Covey dealt with the dangers of aviation in general and spaceflight in particular long before he got to the launch pad that morning. That was the way it was supposed to work, because if he had not reconciled himself with those kinds of doubts, his shaking knees would not have carried him to the pad. Covey put it this way:

If you're going to do something that's dangerous, then you have already rationalized the danger to something that you can accept, and you feel confident in your ability to respond to those things to keep something bad from becoming worse. If you didn't feel that, then you would never go to the launch pad. You'd be too scared, because it's frightening enough. So we had long gone through this mantra, and believed it, that this was the safest shuttle flight that's ever going to fly. So, my recollection is, as with every launch, I'm more worried about doing something that makes me or the crew look bad than I am about the absolute dangers we're facing. You know, it's one of those things. I can die, okay, but I don't want to screw up and then die.

On Hilmers's first flight, STS-51J in October 1985, the crew had worn blue jumpsuits that today seem antiquated at best. When he began to feel uncomfortable, Hilmers simply undid his safety restraints and sat up on the back of the seat. Once it was time to fly, he scooted down and strapped himself back in. That was not possible in these new bright-orange partial pressure suits.

Some shuttle crews liked to keep the mood light while waiting out the countdown, while others were all business. STS-26 tended to be more serious, if for no other reason than the sheer volume of all that was riding on the flight. Covey, though, did not seem to be letting anything weigh him down—no, not Covey, who actually took a quick nap right before systems checks started really kicking in. "I don't think I've been on a flight where I didn't go to sleep on the launch pad for some period of time," Covey said. "It may be five minutes, ten minutes, but you have to relax. I don't miss anything critical. They talk to me, I wake up. But I know that I can sleep on the launch pad, and I attribute that to the fact that it is such a high-adrenaline type of thing that if you really can relax, then your body just kind of goes really quick."

Down on the mid-deck, Nelson was not asleep—instead, he was by himself with nothing to do. He had one instrument to monitor, a basic altimeter. His only assigned job was one he absolutely, positively did not want to perform: if the launch had been somehow aborted and the crew had to bail out, it would have been up to Nelson to blow the hatch, deploy an escape pole that had been added since *Challenger*, and then help everyone out the door. Given the alternative, he was just fine sitting there, patiently waiting.

After a delay of one hour and thirty-eight minutes due to the suit repairs and the upper-level wind conditions, STS-26 launched at precisely 11:37 a.m. Eastern Daylight Time, two years, eight months, twenty-three hours, and fifty-nine minutes after *Challenger* left the very same pad. The flight was absolutely problem-free for all of seven to ten seconds, at which point an alert message popped up on a computer screen. Hauck described the situation as "not an immediate problem, but anything abnormal that soon after liftoff elevated my heart rate." The alert read "Fuel Cell pH," meaning that the acid/base ratio in one of the three power plants that generated *Discovery*'s electricity was slightly out of balance. It soon corrected itself and mission control told the crew not to worry about it.

On a flight deck that was already thousands of feet up and climbing, there was no series of blaring alarms sounding in the cabin. But even in the midst of the frenzied launch phase, it did not take much for Hauck or anyone else on board to spot the alert message. "Believe me," Hauck said rather wryly. "Anything that shows up anywhere that you are not expecting—you notice!"

Added Hilmers, "The first kind of thought was, 'Oh, my goodness, what's going on? Are we going to have another *Challenger*-type problem?' I remember how difficult it was to even get the checklist out and then secondly, being able to read it with all the vibration that was going on." Yet another problem took place a couple minutes into the flight when the flash evaporator subsystem failed due to some frozen coolant. Although that malfunction led to warm cabin temperatures in the mid-eighties for the first couple of days of the flight, neither glitch was life-threatening. It was nothing like *Challenger*.

Hauck once flew off the deck of the USS *Enterprise* with fellow naval aviator John O. Creighton, who now served as the ascent CapCom for STS-26. When Creighton announced that *Discovery* was about to throttle up from 65 to 104 percent of main-engine thrust, Hauck replied with a quick "Roger, go" seventy-six seconds into the flight. Most every Space Shuttle commander before and most afterward replied, "Roger, go with throttle up." Those had been Scobee's last discernible words, but Hauck would later insist his shortened reply had nothing to do with the *Challenger* tragedy. On the flip side of that coin, Covey would remember his fallen comrades each and every time the call was made on subsequent flights. "I go back to *Challenger* every time I hear 'go with throttle up,'" he remarked. "Certainly, we were very conscious of passing that point during the ascent of STS-26. We didn't say anything, but we certainly all were thinking about it."

Exactly 124.8 seconds after liftoff, and with *Discovery* traveling 4,127 feet per second at an altitude of 151,816 feet, the twin SRBs were freed from the sides of the External Tank. "SRB Sep" provided a momentary relief as the mission punched through the *Challenger* threshold, but there was still plenty of time for things to go very seriously wrong. "Up until the *Challenger* accident, the acknowledged highest-risk components in the Space Shuttle were the Space Shuttle main engines," Hauck said. "They're so complex, so much energy being generated, such tight tolerances. Yes, I was glad when we got off the SRBs, but we still had over six minutes of Space Shuttle main engine operation that had always been acknowledged as one of the bigger risks." Seated right behind Covey during the ascent, Lounge was keeping track of critical launch events on a kneeboard. He would keep the checklist page as a memento, a tangible reminder of his one launch that seemed to take forever:

The launch off the launch pad and into orbit was, most nominally, an eight-and-a-half-minute event, where the engines are accelerating faster and faster and faster. My first flight, that went by in just a blur—the solids ignited and then, suddenly, we're in space. After the Challenger *accident and all the training we did and understanding what really could go wrong, the next time I launched on 26, that eight and a half minutes seemed like it took about an hour. Every second, every incremental Mach number, just slowly crept by. My third flight, which was a couple years later on STS-35, that eight and a half minutes seemed like eight and a half minutes.*

One of the last items on Lounge's launch checklist was main engine cutoff (MECO), which took place 513.42 seconds into the flight. *Discovery* was thundering along more than sixty-eight miles up at nearly five miles per second—or a little more than 17,639 mph. Seventeen seconds later, the External Tank separated. STS-26 had arrived, safe and sound, in space. When Hilmers asked Hauck for permission to celebrate, it was granted—for all of twenty seconds or so. "We just let off these big whoops, cries of relief," Hilmers said. "We just kind of really let it out at that moment. I had this feeling of, 'I made it. We did it.' Even though the mission wasn't over, even though we had a whole bunch to do, we succeeded in one of our big objectives."

While each man had differing responsibilities, each would grant that the basic mission of STS-26 was relatively simple—it was a test flight in which the most important goal was to launch, fly, and return safely. Officially, the flight's primary payload was the five-thousand-pound NASA Tracking and Data Relay Satellite-3 (TDRS-3), designed to enhance the agency's communications between the ground and orbiting spacecraft. A little more than six hours into the flight, before even the first sleep cycle, TDRS-3 was deployed by Lounge with Hilmers's assistance. If the mood inside *Discovery* lightened after reaching orbit, it loosened even more after the satellite was released into space. After all that had happened since *Challenger*, after everything that America's human spaceflight program had been through, for all the changes that had been made, no one could have faulted the STS-26 crew for letting its hair down, so to speak, and relaxing a little bit.

The good humor was personified in the wake-up calls made by comedian Robin Williams, who mimicked a signature line from his recently released movie *Good Morning, Vietnam* by bellowing instead, "Goooooooood morning, *Discovery*!!!" Hauck wasn't expecting the celebrity to be involved, and he would later wind up with a tape of Williams's outtakes while recording the messages. They were even funnier, if more than likely somewhat R-rated. There were also wake-up calls parodying the theme song from the television show *Green Acres* as well as the Beach Boys; finally, just before beginning preparations for its de-orbit burn and reentry, the crew decked itself out in Hawaiian shirts nearly as loud as the launch had been.

After donning the shirts, shorts, and sunglasses, Hilmers took to "swimming" through the weightlessness of the mid-deck, while Covey "surfed" on Nelson's back. In the midst of the levity, though, there was a determined

6. The shirts brought to orbit by the crew of STS-26 were loud and proud, and they were in honor of workers in the Orbiter Processing Facility. Commander Rick Hauck is in the middle, and clockwise from bottom right are pilot Dick Covey, Mike Lounge, Dave Hilmers, and George "Pinky" Nelson. Courtesy NASA, via NASASpaceflight.com L2 Forum.

intent. Return to Flight had been a project entailing far more than just the five men on orbit at long last. Dick Truly helped get NASA on its feet again, and so had William P. Rogers, Roger Boisjoly, Tom Overton, Joe Kerwin, Mike Mullane, Story Musgrave, and thousands of others. The outlandish attire worn by the STS-26 crew honored workers in the OPF who had celebrated their casual-dress days with what they called their "Loud and Proud" shirts. "Spaceflight is serious business," Covey said. "Return to Flight was even more serious. But, in the end, there were a whole lot of people who were devoting their life and their efforts to safely return us to flight. There's different ways you can recognize that. A lot of those things were important to do to get past the idea, 'You can't go to space now and enjoy it.' We could."

Even a rough-and-tough career naval aviator like Hauck knew that it was vital to let off a little steam. He ended up keeping his loud-and-proud shirt as a reminder of the flight, just like Lounge's checklist. "I think psychologically, human beings need to get some relief from tension," Hauck began. "You've got to have fun, but you have got to earn your fun and earning your fun means getting the primary objectives completed. You don't want to be perceived as screwing around and then wind up screwing up your primary objectives. That would be worse than death."

The flight would not have been complete without some sort of tribute to the *Challenger* crew, and Hilmers went to Hauck with the idea. Rather than ad-libbing a few remarks on the spur of the moment, Hilmers put a few thoughts to paper before the flight. Each crew member worked with him on the script, and on the fourth and final day of the flight, they honored their friends and co-workers. With a camera pointed out one of *Discovery*'s window at the earth below, Hilmers began: "We'd like to take just a few moments today to share with you some of the sights that we've been so privileged to view over the past several days. As we watch along with you, many emotions well up in our hearts—joy, for America's return to space; gratitude, for our nation's support through difficult times; thanksgiving, for the safety of our crew; reverence, for those whose sacrifice made our journey possible."

Lounge then took over: "Gazing outside, we can understand why mankind has looked towards the heavens with awe and wonder since the dawn of human existence. We can comprehend why our countrymen have been driven to explore the vast expanse of space. And we are convinced that this is the road to the future—the road that Americans must travel if we are to maintain the dream of our constitution, to 'secure the blessings of liberty to ourselves and our posterity.'"

Covey was next: "As we, the crew of *Discovery*, witness this earthly splendor from America's spacecraft, less than two hundred miles separates us from the remainder of mankind. In a fraction of a second, our words reach your ears. But lest we ever forget that these few miles represent a great gulf, that to ascend to this seemingly tranquil sea will always be fraught with danger, let us remember the *Challenger* crew whose voyage was so tragically short. With them, we shared a common purpose. With them, we shared a common goal."

Nelson spoke next: "At this moment, our place in the heavens makes us feel closer to them than ever before. Those on the *Challenger* who had flown before and seen these sights, they would know the meaning of our thoughts. Those who had gone to view them for the first time, they would know why we've set forth. They were our fellow sojourners. They were our friends."

Finally, Hauck concluded the brief memorial: "Today, up here where the blue sky turns to black, we can say at long last to Dick, Mike, Judy, to Ron and El, and to Christa and Greg, 'Dear friends, we have resumed the journey that we promised to continue for you. Dear friends, your loss has meant that we could confidently begin anew. Dear friends, your spirit and your dream are still alive in our hearts.'"

Covey, Hilmers, and Lounge all returned to space on subsequent shuttle missions, but Hauck and Nelson would never again "explore the vast expanse of space," giving them even more reason to pause and soak everything in.

At age forty-seven, Hauck had already been on three "wonderful" flights and he simply was not looking forward to getting back in line for another two-year training cycle. He would savor every moment possible while on orbit. "I loved looking out the window," Hauck admitted. "I loved floating. I loved that camaraderie, and I knew that I would miss it. But all things have to come to an end someday." During a post-flight press conference, Hauck was asked about the possibility of retirement, leaving him really no other choice than to go ahead and make the announcement. The next day, host Jane Pauley asked him on NBC's *Today* show about his decision not to fly again. "We were on all the talk shows," Hauck remembered. "I remember Jane Pauley said, 'I hear that you've decided you're going to be looking for a job, Captain Hauck.' I thought, 'Isn't that wonderful to be able to tell however many viewers there are on national TV that I'm on the job market?' You can't get better advertising than that."

When asked if knowing he was more than likely not going to fly again had an impact on his experience of STS-26, Nelson wavered:

That's a good question. My first response would be I don't think so, but thinking back on it, maybe it did. I tried to pay particular attention to things during the flight. One of the things I learned from Story Musgrave was to make a checklist, a list of things you wanted to notice, to try and look out for, just to maximize the personal experience of being in space. So I had a really nice list of that kind of thing. Whether I did it because it was going to be my last flight or whether I just really liked the idea from Story, it was probably a mixture of those.

Hauck brought *Discovery* to a stop on runway 17 at Edwards Air Force Base in California on 3 October 1988, four days, one hour, and eleven seconds after she had left Earth. After hauling an American flag with them out of the spacecraft, the crew was surprised to be met at the bottom of the steps by George H. W. Bush, then just a month away from being elected president of the United States.

The astronauts of STS-26 were back home again, having safely returned from a journey that put NASA right back where it belonged. In space.

2. Cloak and Dagger, Science on Orbit

Jerry Ross might have done an EVA during the top-secret flight of STS-27, and then again, it is altogether possible that it never actually took place. Maybe Bill Shepherd was involved, and maybe he was not. No one outside those on a need-to-know basis never really knew with absolute, 100 percent certainty. More than a quarter of a century later, Ross himself would not say one way or the other. He refused to confirm, deny, or give a wink or nod. The secret, if there was one to begin with, was safe with him.

Forget about personal glory—with nine on-the-record EVAS to Ross's credit, one more had all sorts of historical implications. Mike Lopez-Alegria was second on the all-time list of most spacewalks with ten, while Ross stood third, just one behind. That is the kind of thing trivia buffs care to discuss, but we may never actually know the story behind the story. STS-27 was a clandestine Department of Defense (DoD) mission that, according to one NASA website, deployed the *Lacrosse-1* spy payload during its four-day flight in December 1988. The site described it as a side-looking radar, all-weather surveillance satellite launched in support of the U.S. National Reconnaissance Office and the Central Intelligence Agency. It was manufactured by Lockheed Martin Aeronautics, Denver, and completed operations in March 1997. If the information was accurate, the cat was at least partially out of the bag. And, still, Ross would not budge, choosing instead to simply discuss the legacy of the six DoD shuttle missions that were flown in the two and a half years following the *Challenger* tragedy:

The shuttle program was a national asset. As such, it was used to the betterment of mankind and the betterment of our country in many ways. Some of those had to do with scientific research and looking out into the universe, and studying through the space laboratories physics, chemistry, physiology, and lots of other things. Then, there was the other aspect of helping our country stay secure and strong, and placing national assets into orbit that are able to give us insight into what was happening around the world and to keep us from getting surprised with another Pearl Harbor or something equivalent.

So hush-hush were the DoD flights, when STS-27 commander Robert L. "Hoot" Gibson gave a video presentation on the flight and its payload before the Joint Chiefs of Staff at the Pentagon, the tape had to be carried by a courier with a locked briefcase chained to his wrist. When the five-member crew and two lead flight directors were presented with the National Intelligence Achievement Medal, they at first could neither keep nor tell anyone of the honor. Gibson explained:

That's the way it stayed for about four and a half years. Then in 1993, they gave us our medals, gave us the citations and said that anything in the citation is now unclassified. The citation says that we launched aboard Atlantis *and deployed a major new intelligence satellite for the United States with the shuttle arm. We separated away from it, and it had a problem. We re-rendezvoused with it and assisted with fixing it, separated again and left it. And it went on to be a major success. Still today, those are the only things we're allowed to say about STS-27. It was the first time we were allowed to acknowledge that we got off a satellite up there, even though anyone who walked out at sunrise could see it going over.*

The morsels of what we do know and what we *think* we know are tantalizing—the deployment of a super-secret intelligence satellite; some sort of problem; re-rendezvous; and repair. The scenario sounds like one heck of a story—and one that most will never fully know and appreciate.

What was obvious was the damage *Atlantis* sustained when it was struck by debris off the cap of the right SRB during its ascent on 2 December 1988. Some, on the Orbital Maneuvering System pods, was plainly visible through the orbiter's aft windows. The underside heat shield was checked as well with the shuttle's robotic arm, and although its camera was not of the highest resolution and the lighting was not the greatest due to its orbital path, the crew was able to see what appeared to be substantial damage. The main tire was also leaking, so *Atlantis*'s belly was pointed toward the sun for most of the flight to keep it warm and as pressurized as possible.

"We were concerned about it, but we had absolutely nothing we could do about it other than worry, which we tried not to do, because we had other things to keep ourselves occupied," Ross said. Mike Mullane was also on the flight, and he remembered Gibson floating up to the windows to do some sightseeing and remarking, "No reason to die all tensed up." In *Riding Rockets*, Mullane described what seemed to be a certain indifference on the part of mission control to the problems STS-27 was facing. Regardless, Mullane was willing to let those on the ground off the hook because, really, what could they have done to save the crew?

It's difficult for me to imagine what options we would have had if the ground had determined the hit was fatal. I expect there would have been a "failure is not an option" movement to save us, but, in the end, I can't see that anything would have been possible. I checked the processing timeline for STS-29, the next mission after ours. It wasn't rolled out of the OPF until 23 January 1989. I don't know, but I'll bet there wasn't even a stack of SRBs in the VAB *ready to go. I can't imagine either* Discovery *or* Columbia *could have been made ready to get to us before our oxygen was gone.*

When *Atlantis* landed at Edwards Air Force Base on 6 December 1988, astronauts, engineers, and NASA administrators alike were stunned to discover some 707 damaged tiles and one that was missing altogether on the lower right-hand side of the fuselage. This caused some melting to the structure itself, in a cavity over a small antenna access door. It was at that weakened point where disaster could have struck just as it did on the leading edge of *Columbia*'s left wing years later, but this time, it did not. Standing on the Edwards tarmac looking at the heavily pockmarked underside of the shuttle that had just brought him home, Ross knew full well that he and the rest of the STS-27 crew had just dodged a very big bullet. That knowledge was reaffirmed for him following the program's second fatal accident. "After *Columbia*, they re-analyzed some of the damage we had," Ross said. "I have been told that we would've been a *Columbia*, too, had there not been a double thickness of metal underneath where one complete tile came off the orbiter. There happened to be an antenna underneath, and therefore that double thickness of metal is what kept us from burning through." With nearly three hundred tiles sustaining craters of an inch or more in diameter, it was by far the most-damaged vehicle ever successfully returned to Earth.

7. STS-27 pilot Guy Gardner (foreground) and mission specialists Bill Shepherd, Mike Mullane, and Jerry Ross (far right, partially hidden) inspect *Atlantis*'s badly damaged underside following landing. Courtesy NASA, via Ed Hengeveld.

As sobering as the damage must have been, remember too that this was just the second flight after seven lives were lost on *Challenger*. Just four months after STS-27 made it back, STS-29 flew on schedule. Ross could not believe it. "With almost no hesitation whatsoever, they pressed on for the next flight," Ross stated flatly. "So, in a lot of ways, I'm not sure NASA had learned their lessons very well." The agency's reasoning behind continuing to fly, Ross continued, was that the coating material on the SRB nose caps had been changed prior to the flight and that is what caused the debris to be shed. The coating was changed back, so, therefore, there was no reason to stop flying. Ross could not add much more than an ironic chuckle.

Gibson went one step further. If another disaster had taken place during *Atlantis*'s reentry, he felt it would have meant game over for the shuttle program. "We had spent all that money and all that time rebuilding and revamping and we launched one successful mission," Gibson said. "[If] we lost the very next one, I think the Congress would have said, 'Okay, that's the end, guys. We just don't need to do this again.' I think that just would have been the end of it."

Because STS-27 survived, as did other flights with dings here and there, a false sense of security cropped up somewhere along the line. Tiles were beat up a little bit? No problem. That much damage was a shame, but at the time, a few—or a few hundred—screwed-up tiles was not considered much of a safety issue. It would take some time to repair, so the concern was more due to its impact on the processing flow to get the shuttle back out the doors and to the pad for another flight. In many ways, a straight line could be drawn from the flight of STS-27, the damage it sustained, and the failure to do anything about it, straight to STS-107.

Milt Heflin was primed and ready to go.

This was his moment, where he would see if he could at the very least *begin* to live up to the standard of mentors like Gene Kranz and Christopher C. Kraft Jr. He had been a flight controller, and now he was lead flight director for the flight of STS-30 in May 1989 to deploy the *Magellan* probe on its way to study Venus. The book had been thrown at Heflin during training simulations many times over, multiple failures, one after the other, piled on top of each other. The work was grueling, and for the first hour or so of the flight of STS-30, Heflin sat in mission control expecting a full-fledged emergency to take place any minute.

None ever did. "You are there to do what you've been trained to do, and there ain't a damn thing you have to do," Heflin quipped. "The easiest part of my job as a flight director was being *in* mission control during the real mission. The hardest part was getting ready to go in there."

NASA's bread and butter had always been its ability to sort through a problem, and to do it very quickly. If there was time to mull over an issue endangering a mission objective from getting accomplished—say, something that might have been keeping *Magellan* from being released from the cargo bay—there were all sorts of avenues to pursue. Call in the mission management team, the backroom flight controllers, and maybe even other NASA centers from around the country. When those kinds of things happened, it was all hands on deck. But there was no authority higher than that of the lead flight controller when a decision needed to be made right now, in real time. There might have been higher-ups sitting nearby, but that was somewhere else. Not here. Not in the control room, where the word of "flight" was law. It could have been the president of the United States, but even that was not enough to usurp the omnipotent power of the lead flight director. "Even though you've got somebody behind you who would be perhaps higher than you are in the normal, everyday food chain, it doesn't make any difference," Heflin said firmly. "You are in charge. They are *not* there to approve or disapprove of your decisions."

Heflin was certainly in charge, but he was in charge of a group of capable men and women who knew their stuff and who were also not afraid to speak up if and when they felt the situation warranted it. Said Heflin:

I'm often asked, "How in the world did you stand that tension and stress?" I am surrounded by people, and they are not going to let me make a mistake. They're too damn good. They're not going to take me down a path, and if they think I'm headed down a path and I'm missing the point on something, they will get in my face and tell me. That's where trust comes in. A team is put together with people who if they don't know the answer, they will tell *you they don't know the answer. They will tell you what they're thinking or where they want to go, but they will not bullshit you. When they tell you something, you can take it to the bank. Lead flight director is far from a lonely position.*

Each of Heflin's first three missions as lead flight director—STS-30; STS-34, five months after his debut; and STS-41 in October 1990—released planetary probes to study our neighbors in the galaxy. STS-34 sent *Galileo* on its way to Jupiter, while *Ulysses* flew forth from STS-41 to begin its journey to the sun. At first, releasing such a probe on his rookie flight seemed not much more than the luck of the draw. Then, after building a rapport with NASA's Jet Propulsion Laboratory (JPL), it seemed only natural for him to follow up with the other two.

It was this kind of work within mission control and with JPL and various other NASA centers that began to foster what Heflin called one of his greatest strengths—team building. Over the course of his career, he came to be something of a peacemaker within those teams. Remembered Heflin: "This one particular flight director was probably one of the most technically capable we've ever had, extremely sharp. This particular person was assigned to some of the really, really, *really* tough sort of things we had to do for that reason. But he took no prisoners. There was a lot of carnage at the end of some tough, technical discussions, decisions or whatever. There were bodies along the way. It wasn't uncommon for me to come along after a little bit and smooth some of the roughness out."

Finally, it was important as a lead flight director to fully comprehend both the hardware and the science behind a payload such as *Magellan*. What kinds of sensors and appendages did it have? How would it fit in the payload bay? Once the probe was deployed, what could the orbiter do and not do in order to get away from it? "It was very typical to get the project manager for that particular spacecraft in our meetings to tell us, 'When you're playing with our payload out there, this is what you need to know about it,'" Heflin said. "The more we understood, we had a lot of sharp people who could help figure out some way to get around a problem."

That was Milt Heflin, a team player to the core, keeping everybody on the same page, playing the same notes.

The personal appearance was the kind of event every astronaut attended at one time or another. During this particular time in the barrel, John E. Blaha happened to be making a presentation before a group of engineers and their families in Sydney, Australia. Afterward, he took a few questions and someone asked him about the DoD flight of STS-33. He could not give details, he teased, but what Blaha did say left an air of suspense hanging in the room, just the kind of cliffhanger Hollywood loves. "It was the mission," Blaha said, "that ended the Cold War."

During ABC News's coverage of the 22 November 1989 launch of STS-33, the network was given just a few minutes' notice of when *Discovery* would actually lift off with Blaha, commander Frederick D. Gregory, and mission specialists Story Musgrave, Manley L. "Sonny" Carter, and Kathryn C. Thornton. Reporter Jim Slade announced that its payload was a six-thousand-pound satellite called *Magnum*. Slade said very matter-of-factly—the way all veteran reporters like to do, as if he had been in on the planning of the mission himself—that the spy satellite was being cast into geosynchronous orbit over the Indian Ocean so it could listen in on the military and diplomatic communications of the Soviet Union and China.

Whatever its payload might have been, if in fact it did end the Cold War, it did so with a broken toilet.

Imagine Robert M. Kelso on console during STS-33, shortly after beginning his very first shift as a flight director. Kelso was working the flight's graveyard shift, replanning for the next day's activities, while the crew was scheduled to be asleep. Normally, not a lot went on during the planning shift, making it a bit easier way to break in the rookie flight directors.

Chuck Shaw, leading the shift prior to Kelso's debut, put the crew to bed and asked Kelso to make sure they were not disturbed. Just watch over the ship and make sure nothing went wrong—that was all Kelso had to do. Shaw left the control room, and almost as soon as the door shut behind him, what seemed like a thousand caution and warning lights fired off almost at once. All heck was breaking loose. "In the control center, we were getting red lights all over the place," Kelso said. "On board, there were a bunch of klaxons going, 'OOOOOGA! OOOOOGA! OOOOOGA!'"

Information was suddenly flying at Kelso from every direction.

Flight! Only a couple of minutes until loss of signal!

That meant that *Discovery* was about to pass out of range of a TDRS that was enabling communications between the shuttle and Houston. It would be a few more minutes after that until another relay picked it up.

Flight! We're dumping a lot of gas out of the cabin.

The pressure inside the crew cabin had dropped, and it was getting lower. To compensate, relief valves were opening in the nitrogen and oxygen tanks and flowing air back into the volume at an extremely high rate.

Hey, flight! If that does not stop and we have to do an emergency de-orbit, we are off the range. The ground track is taking us away from a landing site in the continental United States.

Flight! We need to be looking at Africa if we need to land in the next hour or so.

Kelso could not help but wonder how things had gone so wrong so quickly. This had to be a simulation. He had not been on console for five minutes and now he was about to de-orbit his orbiter? This was not going to be a good day for Rob Kelso.

The orbiter and its crew went out of range of one communications relay satellite, and when it came within reach of the next, telemetry showed that the cabin had repressurized. The situation was beginning to return to normal, but then the whole process started anew, alarms and all. Kelso and a few others headed to a back room to make sense of it all. It did not take long to figure out the source of so much commotion. The culprit turned out to be a malfunctioning Waste Containment System—the broken toilet. "When the astronaut pulled up the handle to flush, it broke and it did not close the overboard vent line," Kelso explained. "He forced the handle forward, which opened the large lid that normally isolates the bowl of the potty between the cabin atmosphere and space. All of the air in the cabin is now rushing under this guy's rear end, through the gate valve, down into the bowl of the potty, and overboard the shuttle through this vent line. He's got his rear end, in all reality, exposed to space."

After managing to close the vent, the astronaut tried working the handle again, creating the second round of alarms. To make matters worse, two vents for the life-support system were in the Waste Containment System closet, right above the user. "The regulators for the huge nitrogen tanks and the big oxygen tanks were opened full wide to keep up with the loss of the air going out of the potty, and the life-support vent is blowing air right on top of his head," Kelso added helpfully. "So it sounds like a big, huge air dryer . . . SSSSSSSSSSH!"

If Kelso was having a hard time working the problem, just think of the crew's reaction to the sound and fury of what was taking place. "He opened the door to the potty, and looked into the mid-deck at the rest of the crew," Kelso described. "Everybody's eyes are *huge*. They're as wide as saucers. They're all going, 'We didn't touch a thing! We don't know what happened, but it was not us!'"

Although Kelso was unsure if the crew was fully aware of the source of the problem, there was still no air-to-ground communication because of the instructions from Shaw. Kelso was prepared to break the silence if and when another STS-33 crew member attempted to use the badly broken toilet. "I had the whole team looking for any signs of life that somebody might get up and go to the potty," he said. "Sure enough, early in the morning at the end of the sleep shift, somebody started turning on power for the coffee warmer. We called immediately and said, 'Hey, the potty's broken. Don't go near the potty!'" The fix was simple. For the rest of the flight, a pair of vice grips was used to manually open and close the check valve. Problem solved, but before then, during an eight-hour debut as flight director, Kelso had been sure he was watching his career literally go down the toilet. A flight rule stating, in essence, that "missions with broken toilets end immediately" was instead rewritten on the spot during a mission management team meeting the following morning allowing the flight to continue.

As humorous as the toilet episode might have seemed, the mission of STS-33 was one that had been touched by sadness. S. David Griggs was originally named as the flight's pilot, but he lost his life less than six months before launch when the World War II trainer plane he was flying crashed in Earle, Arkansas. Blaha took Griggs's place on the crew. Then, less than two years after STS-33, Carter was also killed in a plane crash, this one a Southeast Airlines twin-engine commuter plane that went down in Brunswick, Georgia.

Could there have been a covert EVA on this mission? Good question. Musgrave had already done one spacewalk, the program's very first shuttle-based EVA, during the April 1983 flight of STS-6. Thornton would go on to perform one spacewalk on STS-49 in May 1992, and two more to service the Hubble Space Telescope during STS-61 in December 1993. And at the time of his assignment to the crew of STS-33, Carter was the EVA representative to the astronaut office's mission development branch. After his death, JSC's Neutral Buoyancy Laboratory—the pool used for underwater EVA training—was named in Carter's honor. That was a lot of potential to have on board if none was planned.

Rob Kelso had an American flag that he displayed in the control center whenever he worked a flight. He was proud to fly it in honor of a country and an agency that was coming out of the Cold War and bouncing back from the *Challenger* accident. "Flights at the end of the eighties and into the early and mid-nineties were one of the best waves of flying the shuttle program ever did," Kelso said. "We were flying missions that we didn't really know how to fly. It was a time of creativity, innovation, and doing one-of-a-kind missions. The flights we did in that time period had never been done before or since."

Kelso, serving as the Orbit 2 flight director, was certain that STS-36 played into exactly that kind of "can-do" American attitude. Its payload was classified, but the buzz was that *Atlantis* and its five-person crew deployed hardware known as KH 11-10, which was, according to a NASA website, "an electro-optical reconnaissance satellite that was heavier than other KH-11 satellites and believed to include a signals intelligence payload. It had wider spectral band sensitivity, perhaps 'real time' television capability and other improvements." To get the payload where it needed to go, the orbiter traveled in a highly inclined orbit of sixty-two degrees from the earth's equator. That carried the flight path north of the Arctic Circle and as far south as the coast of Antarctica.

That is as much as most of the world would ever know about the accomplishments of STS-36, and while STS-33 might have brought the Cold War to a stop, the very next DoD flight was even more grand, according to Kelso:

STS-36, in some circles, is probably one of the most important shuttle missions we ever flew in the 135 we did. And yet it'll probably never get that recognition. It was just fascinating to do those DoD missions. The servicing flight where we went up and fixed the lens on Hubble gets a lot of praise as one of the most important missions. But out of all the ones I know of, from STS-1 to the conclusion, probably STS-36 other than Hubble stands out as the most important mission that we ever did, certainly one of the most important. NASA wouldn't just come up with the unique inclination and do all that analysis for abort modes and landing sites if it wasn't damn important. This the only flight we did before or since to an off-nominal inclination launching from the Cape.

Mike Mullane and David Hilmers were both on this flight, but neither cared to elaborate on Kelso's best-ever assertion. "I think that it is hard to rank missions by their importance," Hilmers said diplomatically. "It is kind of like beauty. It's in the eyes of the beholder. We were instructed not to comment specifically about DoD missions, but I was proud to be a part of two of them and of their role in enhancing the security of our nation."

As much as the DoD missions had accomplished—what with ending the Cold War and all—only three remained after STS-36, concluding with STS-53 in December 1992. In the wake of the *Challenger* calamity, the Department of Defense's effort was refocused on moving its initiatives to unmanned lift vehicles. Only those that could not be launched on the Titan IV rocket had remained on the shuttle's launch manifest.

Spaceflight experience was in relatively short supply during the October 1990 flight of STS-41, but Thomas D. Akers did not much care.

Richard N. Richards had flown once before, but this was his first outing as a commander. Bill Shepherd was also making a sophomore voyage into the heavens, and an initial shot at serving as the mission's flight engineer. Akers, pilot Robert D. Cabana, and fellow mission specialist Bruce E. Melnick were rookies. In a very real sense, they were learning the ropes together, each in a brand-new position. "Generally, I never felt being experienced was as big a plus as some people did. You aren't up there long enough to ever really get good at it," Akers joked. He continued:

The guys who go up and spend months in the space station, they get really good at adapting to living in space. I flew four times in my career, for a total of thirty-three days, and I learned something new on every mission, about every day I was up there. I always thought it was funny—you come back from space after your first flight, and now you're a, quote, flown astronaut. It was kind of like the eagle of knowledge had lit on your shoulders. It's not that way at all. You come back and you go, "Boy, I don't feel a whole lot different than I did two weeks ago, in terms of what I know." In fact, it kind of reinforces how little you know when you get up there and get to do it for real.

Akers had been a flight test engineer in the air force prior to joining NASA, so the flight's primary objective of deploying the *Ulysses* spacecraft to study the sun was not exactly in his wheelhouse. He and his crewmates were trained to deploy the craft, which first headed to Jupiter before using that huge planet's gravity to swing it into an orbit around the polar regions of the sun. The astronomy—determining what the data *Ulysses* was sending back actually meant—was left up to experts back on the ground. Rather than using a robotic arm to release *Ulysses* into space, it was simply released off its berth in *Discovery*'s payload bay.

The planetary probe was not the crew's only priority. The *Intelsat VI* communications satellite was hurled into space by an unmanned commercial Titan III rocket on 14 March 1990. However, the rocket's second stage was left attached to the satellite, and that, in turn, prevented the firing of a solid rocket motor that would have lifted it to its prescribed orbit. Although the second stage was eventually jettisoned and the satellite raised to a higher altitude, it was unusable and in need of a rescue.

During the flight of STS-41, Akers and his crewmates conducted a study to determine if the satellite was in any shape to be saved. "They were wondering how much damage was being done to *Intelsat VI* and its solar panels," Akers said. "We had some test panels on the end of the arm that we just basically stuck out into the breeze, to see how much damage was being done by those very small and very few particles as you go around the earth." As it turned out, such debris strikes were minimal. Plans were made for a future shuttle crew to capture and repair it while on orbit.

Akers was so busy with *Ulysses*, *Intelsat VI*, and a dizzying array of secondary experiments, he was never fully able to take time to soak in the experience of spaceflight. At just over four days on his first flight in particular, there was very little time to do so:

We deployed Ulysses *six hours after we got in space, so the main job was over the first day. We had some secondary payloads that we worked with for the next two days and then we got ready to come home. There were two days there doing secondary payloads that you had* some *time to look out the window. I tell people, I never got time while I was up there, even on four flights, to appreciate what I was getting to do. I'd tell myself every time after the first one, "I'm going to take some time this flight and just look out the window and enjoy." You're always so busy, you always feel like you're behind.*

Less than two years later, Akers would have even more time on orbit to *not* look outside. This time, he was going to be face-to-face with *Intelsat VI* itself.

Flying a high-profile mission like STS-26 had its fair share of perks for Mike Lounge, but it came with the price of intense attention from the media and an interested public.

Going back to the Mercury era, spending time in the barrel had always been part of the deal when it came to flying in space. In contrast, Lounge's next flight, STS-35 in December 1990, featured round-the-clock work with the Astro-1 package of four telescopes. Straightforward and dedicated to astronomy, the mission allowed Lounge and his crewmates the luxury of getting the job done without the superfluous fuss kicked up by STS-26. "We operated 24/7 an observatory that had a key role to play," Lounge said. "We were busy every minute of the flight, and it was a longer flight, nine days versus four for STS-26. There was a lot more going on on 35, I'll have to admit."

There had indeed been a lot going on with STS-35, even before its eventual launch. The flight was scheduled to launch 16 May 1990, but a series of hydrogen leaks forced several delays. On 8 October 1990, *Columbia* was moved from pad A to pad B so *Atlantis* could take the STS-38 DoD crew into orbit the next month. The very next day, however, a tropical storm forced a rollback to the VAB. STS-35 finally got going in the early morning hours of 2 December 1990. "It was a horrible seven months," admitted Jeffrey A. Hoffman, who—along with Robert A. R. Parker, Ronald A. Parise, and Samuel T. Durrance—was one of four astronomers assigned to the flight. "I was fortunate, in that I had already been assigned to STS-46, the first flight of the tethered satellite. So I at least had something challenging to do while we were waiting. But the poor guys on the rest of the crew, it was really kind of dispiriting."

Just ten hours into the mission, a computer data display unit used to help point Astro-1 appeared to shut down automatically. Almost immediately, there was a noticeable burning smell, and in space that is never, ever a good thing. "You'd get this smell of burnt plastic permeating the spacecraft, which is not a nice smell to wake up to in the morning," Hoffman said. Because this was the first time the computer had been used for celestial observations, a certain amount of difficulty in getting the software up and running was expected. But with one data display system down, the crew made do with a second until it, too, bit the dust four days later. A post-flight examination revealed that one of the filters on the outside of the computer had become clogged with lint and had not been properly cleaned on the ground.

Hoffman recalled, "We had to put on the breathing masks, and they asked us to basically make a fire in the spacecraft by turning on the computer. Of course, the computer shut right down again. The only good result of that was that we demonstrated that the breathing masks took so long to put on, that they would have been basically useless in a real emergency. That led to the procurement of the current version, which is a fast-donning, full-face mask. It was a very frustrating experience."

Ground teams at Marshall Space Flight Center were able to essentially uplink enough information to aim Astro-1's ultraviolet telescope, with, hopefully, some fine tuning by the crew. "I had never used big telescopes, so it was sort of a joke," Hoffman said. "The other three astronomers on board were all optical astronomers, whereas I'd been an X-ray astronomer, and I didn't use regular telescopes. They let me, once or twice, use the little paddle. To be honest, I was happy after that. I looked out the window and took lots of pictures."

The crew was experiencing at least one other frustrating problem. A waste water dump system got clogged, making it impossible to use the on-board urinal. It did not take long to fill up a huge plastic contingency bag, and when that happened, things *really* took a turn for the worse, Hoffman said: "We needed contingency urine collection devices. Normally, you're only supposed to use them once and throw them in the wet trash. But because we knew we were limited, they said, 'Would you mind? Just hang them up on the wall and use them as many times as you can.' The problem is that they have little pinholes in them. They were very old. I think they're from the Apollo era. So you get these yellow bubbles coming out. I'm not trying to gross you out. This is history. This is the way it was."

After the supply of male urine collection devices ran out, the all-male crew gave ones for females a shot. It did not work out so well. "They leaked a lot," Hoffman said. "We had socks up there to soak up the urine. It was pretty gross. When we finally landed, I remember the look on the face of the technician who first crawled on board, because it must have really reeked. We couldn't really smell it that much, because we were kind of inured to it. It was pretty bad, but we survived."

It was a somewhat inglorious end to Lounge's spaceflight career. Lounge, who died on 1 March 2011 due to complications from liver cancer, knew that STS-35 would probably be the capstone to his career in the astronaut office. When the mission was cut short a day due to an iffy weather forecast for the primary landing site at Edwards Air Force Base, Lounge took a few moments for himself as *Columbia* was prepared for reentry on 10 December 1990. "As we were packing up and spending that last orbit around the earth really just waiting, I spent a lot of time enjoying weightlessness and very slowly bouncing from the mid-deck up through the access hatch into the flight deck and back," Lounge said. "I remember thinking, 'This is probably the last time. I'd better soak it in.'"

The high-gain antenna on the Compton Gamma Ray Observatory would not budge, so the crew of STS-37 and teams on the ground went to work.

Commands to deploy the antenna were sent up from Goddard Space Flight Center in Greenbelt, Maryland, and time after time, the efforts failed.

Multiple firings of *Atlantis*'s Reaction Control System would not shake the antenna loose.

Wiggling the thirty-seven-thousand-pound observatory on the end of the robotic arm did not work.

Veteran spacewalker Jerry Ross and rookie Jay Apt were up next, and this was exactly the kind of issue they had trained for hours on end to address. Their contingency EVA was the first spacewalk since before the *Chal-*

lenger accident, and Ross had been in on the last one before the loss of the STS-51L crew, too. The satellite was a collection of four separate gamma-ray detection telescopes named after Dr. Arthur Holly Compton, a Nobel Prize winner in physics. It included several EVA-friendly capabilities, and one of them allowed Ross and Apt to deploy the antenna by hand and lock it into position. Still, would some other gremlin somehow prevent them from successfully completing the job they set out to do? This was far from a done deal. Although Ross and Apt had trained to handle a wide assortment of problems during their EVA, but not knowing exactly why this particular bugaboo was preventing the antenna from being deployed made the spacewalk's outcome all the more uncertain.

While Ross moved over to Compton, Apt went to work getting tools ready in case they were needed. After getting clearance from the ground, Ross gave the antenna a push. And then another, and another, and another. Finally, he was able to shake the antenna free.

Ross and Apt then went after the primary objectives of the one spacewalk they *did* have scheduled during the flight of STS-37, and that was to prepare for moving humans and equipment around a planned space station, at the time known as *Freedom*. "Everybody had pretty much decided that we were going to go build this space station, and they were starting to get some pretty grandiose plans which said that we were going to be building and maintaining this space station with spacewalks," Ross began. Despite such plans, NASA had lost much of its spacewalking expertise in the dark days following *Challenger*—astronauts, trainers, flight controllers, and engineers were leaving en masse. "There just weren't many people left that had done spacewalks," Ross said. "I was one of the very few left that had ever done one. Probably even more critical was the flight control, flight design, and hardware design people, and we needed a very robust team to be able to address the looming EVA wall that was coming at us to build the station."

Ross and Apt put together a group of tests in order to rebuild the agency's knowledge and experience base from within, and then picked a flight—STS-37—on which to place them. The gear they wanted to try out included Crew and Equipment Translation Aids (CETA)—carts on two segments of track running nearly the length of the payload bay, one movable and one fixed. Six weeks or so after getting the plan approved, both men were named to the crew. It was a proud moment for the veteran. "We got to help design and develop this test hardware, and then go test it in space," Ross said. "So I got to conceive of something, helped to get it onto a flight and then got to go test it. That was pretty cool."

Together on orbit, Ross and Apt tried a handful of concepts with the translation aid—one in which they pulled themselves along manually; another worked mechanically, much like a railroad hand cart; a third had an electrical generator to power the spacewalkers back and forth; and on yet another, the track would follow a tethered astronaut up and down the rail. Said Ross:

While it looked fairly rudimentary, it really did give us exactly what we were looking for. It helped us to build up expertise, and it gave us some good results which were incorporated back into the hardware. One of them was a cross-section of the handrails that we use all over the space station now, to give us better grip with less hand force. And, we have two CETA carts on the station, which kind of ended up being hood ornaments more than anything else. Initially, the carts were supposed to have been on their own track, as opposed to being on the same track with the mobile transporter, which gets in our way routinely when we're moving around.

The evolution of what would become spacewalks to construct the ISS had begun.

Brian Duffy and Byron K. Lichtenberg took very similar—and very different—paths before they found themselves together on the crew of STS-45 that flew in March and April of 1992.

Both men were fighter pilots in the air force—Lichtenberg flew 238 combat missions in the skies over Vietnam, while Duffy graduated from the United States Air Force Academy in Colorado Springs, Colorado. After making it through the military branch's test pilot school, Duffy went on to oversee testing of the F-15. That, however, is where their paths diverged.

Becoming an astronaut had long been an all-consuming passion for Lichtenberg, and he began laying the groundwork early. To do so, he knew that he should probably become a test pilot, and to do *that*, he would have to get an advanced degree—at the very least, a master's. He went to Brown University for aerospace engineering, and in the process he joined the air force's Reserve Officers' Training Corps (ROTC). He did a tour of duty in

Vietnam in 1972, just as the Apollo program was ending. Coming up was the Space Shuttle, and it was going to need several people "in the back," he said, to launch satellites and do spacewalks and science experiments.

Figuring it could only increase his chances of flying, Lichtenberg left active duty in the air force and went back to graduate school, this time to get his doctorate in biomedical engineering from the Massachusetts Institute of Technology (MIT). "I made a career path change and said, 'I'll be a fighter pilot with a PhD degree, and that should be pretty good credentials to get me into the space program,'" Lichtenberg recalled. "Everything was fine until about halfway through the PhD program, and I realized that it was not my life's passion to be an educator or researcher."

He might not have been an ivory-tower academic, but his seat on the shuttle was about to open up through his work at MIT:

I was in a lab group that was working on the vestibular system, which is your inner-ear organs of balance. We had submitted a proposal to NASA to do space experiments to try to understand how the vestibular system adapts to weightlessness and how space-motion sickness happens. It was a hot topic for NASA at that time, so they accepted the proposals. About a year later, they came back to each principal investigator on the experiment team and said, "We're going to interview people for this job called a 'payload specialist.' They're not going to be professional, career astronauts based in Houston, but rather scientists. We'll give them the training they need to live and work on the Space Shuttle, and then train them to do the experiment."

There were two payload-specialist spots opening up on STS-9, which would eventually fly on *Columbia* in late November–early December 1983—one for the European Space Agency (ESA), which designed, built, and paid for the Spacelab-1 module, and investigators from the United States were to use the other. Lichtenberg's boss at MIT, Laurence R. Young, nominated him for the American position. The scientists came up with a list of six names to be sent to Houston for what amounted to pass-fail astronaut physicals. "Other than that, NASA had no say in who was going to be selected to go fly as a payload specialist," Lichtenberg said.

Lichtenberg got the job, as did ESA's Ulf D. Merbold. As a payload specialist, his paycheck and benefits came from MIT, which in turn billed NASA for his services. Lichtenberg and Merbold were the first two payload specialists ever to fly, and the going was not always smooth. NASA had already selected a group of mission specialists to fly on the shuttle, and each came with a blue-chip pedigree. There were those within the agency's hierarchy who did not understand why anyone else was needed. Lichtenberg continued: "When the payload specialists came on board for this particular mission, NASA looked at it and went, 'Wait a minute. Why are flying these people? We've got these twenty mission specialists that we've already trained, plus the other scientist-astronauts. We don't need payload specialists.'"

The payload specialists were considered not much more than third-class citizens, behind the pilots and mission specialists. Remember, Lichtenberg had dreamed of becoming an astronaut for decades, just as long as, if not longer than, the "real" astronauts. "They pretty much tried to do what they could for a while to discourage us," Lichtenberg said. "For the first time, the folks that were running the astronaut office did not have the final say over who was going to fly on the mission. I think that really bothered them."

Nearly eight years would pass between Lichtenberg's first flight on STS-9 and his second, STS-45. In the meantime he cofounded Payload Systems Incorporated, which landed the first commercial payload on the Russian space station *Mir* and was a founding member of the Association of Space Explorers. Lichtenberg was again a payload specialist on STS-45, only this time his paychecks were coming from Lockheed Palo Alto Research Laboratory. The mission's flight plan included the debut of the Atmospheric Laboratory for Applications and Science (ATLAS-1), put together by the United States, France, Germany, Belgium, the United Kingdom, Switzerland, the Netherlands, and Japan. Twelve instruments within the ATLAS-1 platform performed fourteen investigations in four fields—atmospheric science, solar science, space plasma physics, and ultraviolet astronomy.

Attitudes had evolved, and Lichtenberg found much more acceptance while preparing for his second spaceflight:

Christa McAuliffe died and NASA saw the public's reaction to the fact that she wasn't selected as a career astronaut, but she was on Challenger *and she died just like the other crew members. They were all going to the same place, so she was a regular astronaut, too. John Young [Lichtenberg's STS-9 commander] actually said this to me, "You're not an astronaut until you've flown in space, and if you fly in space, then you're an astronaut." The NASA viewpoint changed tre-*

mendously. We were much more integrated into the crew. We did a lot of bonding sessions and some personality assessments. We had offices right there in the astronaut office building. We were much more accepted, so it was a lot different—and better.

Of the two, Duffy's was the more typical route to orbit—if there was such a thing during the shuttle era. Service academy. Test pilot school. Apply to the astronaut office. Done, done, and done. Duffy got a hole shot, too, because not only did he apply to the astronaut office, he was accepted on his very first application in June 1985—the last ASCAN class before the loss of *Challenger*.

The day of the accident, Duffy's wife asked him what was going to happen next. He told her that NASA would figure out what happened, fix it, and then get back to flying. He had lost friends before in aviation accidents, which sometimes meant that there were stand-downs to figure out what had happened. For better or worse, this was one of those times. "I didn't expect this to be any different, to tell you the truth," Duffy said. "My wife often says, 'It's not like this was the first dangerous job he's ever had. It's not like he was a doctor before and all of a sudden, now he's riding rockets.'"

Duffy was not a scientist like Lichtenberg, but he understood the importance of the flight's research and getting it across to the public at large. "People are interested to hear what it is that you're doing on a particular flight, and why it matters," Duffy said. "Whether it's collecting data in the unique, weightless environment; growing crystals; or trying to understand human physiology better. The important thing is to be able to translate it in a meaningful manner."

No matter how hard Pierre J. Thuot and Richard J. Hieb tried, the *Intelsat VI* communications satellite simply would not cooperate during the first two spacewalks of STS-49 in May 1992.

With a diameter of 11.7 feet and a height of 17.5, the contraption was never going to be easy to capture by hand. Thuot and Hieb were also supposed to attach a live rocket motor to the satellite to replace one that had failed. This, however, was getting just the slightest bit frustrating. Owned by 124 member nations of the International Telecommunications Satellite Consortium (Intelsat), the satellite was supposed to be a large part of a global telecommunications system. For the time being, however, it was not much more than a very large piece of space junk.

Bruce Melnick twice used the remote manipulator arm in an attempt to maneuver Thuot into position to grab the bulky mass. The plan was to attach a specially designed capture bar to the bottom of the satellite so it could be brought into the payload bay and repaired. After an initial attempt on the fourth day of *Endeavour*'s maiden voyage, it was left pushed away and wobbly, obviously not an ideal situation in such close proximity to the brand-new orbiter. "There was a lot of concern amongst our crew and I'm sure on the ground also," said Tom Akers, who was scheduled for his own spacewalk later in the flight to test construction techniques for the upcoming space station venture.

For a year and a half, the crew had trained for this moment: Thuot and Hieb outside; Melnick on the arm; Akers serving as IV—the intra-vehicular crew member helping out wherever he could; commander Daniel C. Brandenstein deftly holding *Endeavour* in formation with the satellite; pilot Kevin P. Chilton closely monitoring a fuel supply for the maneuvering jets that was starting to get dicey; and mission specialist Kathy Thornton operating the cameras. "Rick and Pierre went out and did it exactly the way we'd trained," Akers continued. "It was just slowly turning at less than one RPM [revolution per minute], but on the simulator, Pierre would put that bar on and it would never budge. Up in space, when that bar touched that moving satellite, it just started wobbling and basically got out of control really quick." If plan B was to make a second attempt, those involved had to all but drop back and punt when a second go at capture the very next day also fizzled when the satellite again ended up wobbling. Plan C was unprecedented. Melnick had an idea—what if three people went outside, Thuot and Hieb joined by Akers, to grab the satellite by hand?

Not everyone was on board at first, Akers remembered:

The ground team was reluctant to do that. They had thought about it and discarded the idea. I remember our commander, Dan Brandenstein, did a really good job of talking to the ground. Of course, when you talk to the ground, the whole world hears. So he was very diplomatic, but very firm, that we needed some margin, that we couldn't just go out and do what we'd done twice before and hope *that it would work. We needed something that ensured it would work. The ground team rethought the situation, decided, "Yep, that's the right answer," and agreed for us to go out and do it. It wasn't something when the first time somebody said it, everybody jumped on and said, "Yeah, that's what we ought to do."*

The theory behind sending the third spacewalker into the fray was understandable, in that it provided an extra set of hands. "The problem was that Pierre couldn't get the capture bar on the satellite because it was moving," Akers continued. "The idea was, 'Okay, we have to grab the satellite and stop it from rotating. Then, we can maneuver the bar up slowly, taking our time, and put it in place.' If two people grab it, it can still have one axis it can rotate. But if you put three people around it, then you can control all three axes. It's like a milk stool. If you've got a three-legged stool, it's more stable than a two-legged stool."

Once everyone agreed on the best course of action, planning kicked into overdrive on the sixth day of the flight. Astronauts jumped into the pool at the Neutral Buoyancy Lab to figure out where Thuot, Hieb, and Akers would need to be placed. Runs were made in the vacuum chamber. *Intelsat VI* had not been designed to be lugged this way and that by human hands, but by chance, it had three equally spaced rods near the bottom that the spacewalkers could safely grasp. Figuring out what to do dominated conversations on orbit. Chilton and Melnick came up with procedures for what would take place, even though neither would actually take part in the spacewalk itself. Nearly two decades later, Akers marveled at such a concerted effort. "We'd never talked about it, trained for it, or planned for it," he said. "My opinion, it's one of the best examples of teamwork I'd ever seen."

There was risk, Akers continued, but there was a certain amount of that on any EVA. The bottom of the satellite featured a thin layer of metal, almost like that of a tin roof back in his native Missouri. It might have been possible to slice open a glove on the metal layer, but Akers insisted that there was not much more risk than that, and certainly not as much as what the media portrayed.

Once the spacewalk actually began, it took nearly three hours to maneuver into position and firmly grasp *Intelsat VI* by hand and another five to get a kick motor successfully installed that eventually took it to its proper altitude of 22,300 miles above the earth. At a total of eight hours and twen-

8. STS-49 spacewalkers (from left to right) Rick Hieb, Tom Akers, and Pierre Thuot conduct the first and only three-person EVA to capture the wayward *Intelsat VI* satellite. Courtesy NASA.

ty-nine minutes, the EVA was the second longest in the history of human spaceflight and the longest to that point.

"Any spacewalk is hard work, especially if you're free floating," Akers said. "If you have your feet restrained, it's not as hard. We spent a lot of time in those foot restraints, holding the satellite and getting the bar on there. It was hard work, but we didn't come back in dead tired, let's put it that way." Pictures of the historic EVA are some of the most famous of the shuttle era, showing the three astronauts working side by side. The photographs were nice and all, but their crewmates could not resist taking a good-natured shot at the trio of spacewalkers. "The guys inside said they got bored taking pictures," Akers said. "Those pictures, if you look at the whole sequence, the earth in the background goes from the right to the left to the bottom. We made several trips around the earth while we were standing in that position. They said it was it was like watching paint dry, us out there putting that bar on."

Despite the lengthy EVA, Akers was headed right back outside the next day alongside Thornton to continue figuring out ways to construct a space station two hundred miles up. Because of the additional work with *Intelsat VI*, their two planned EVAs became just one. Originally, Akers and Thornton were to put together a truss pyramid that looked as if it were made of

Tinkertoy nodes and poles, with Thuot and Hieb taking it back down again on the flight's final spacewalk. That was the plan, but the reality of the situation turned into something much different, Akers said:

Kathy and I, they gave us a new plan the night before. We built about half of the structure we had planned, and then had to tear it down. In the middle of all that, the KU band antenna broke. We had to go do a pass to stow it. It was probably the hardest EVA I've ever done, just because nothing was in the order we had trained to do it in. We also had a whole lot more trouble putting up those poles and nodes than we had anticipated. We had trained in the water, and you had the viscosity of the water to help your control when you moved the poles. Well, you get up into space and you're moving one of these poles, you don't just quit pushing to make it stop like in the pool. You've got to pull it back the other way. We knew we had a long day of work ahead of us when we started, and it sure enough turned out that way.

The fourth and final EVA of STS-49 took seven hours and forty-four minutes to complete, more than an hour longer than expected. The dynamic duo—Thornton and Akers—would again team a little more than a year later, this time on one of the most critical shuttle missions NASA had ever faced.

They were not a lint-filled and burned-up computer or a used urine-collection device, but Jeff Hoffman still faced problems aplenty during his next shuttle flight—STS 46 in July-August 1992.

The first problem took place when the European Retrievable Carrier (EURECA) was released a day late due to extensive communications problems between *Atlantis* and the payload. After the satellite was deployed, an attempt to thrust it to an altitude of 310 miles was halted just seven minutes or so into an expected twenty-four-minute burn due to unexpected attitude data from the satellite. The problem was resolved using correct uplink information without having to recapture EURECA, and the satellite was subsequently boosted to its planned orbit on the next-to-last day of the flight. The satellite—carrying a package of fifteen experiments—was retrieved by the crew of STS-57 eleven months later.

EURECA's issues, in turn, delayed the first testing of the Tethered Satellite System that had been jointly developed by NASA and the Italian Space Agency. With *Atlantis* lowered to a relatively low orbit of 160 nautical miles, the plan was to release the satellite on a tether twelve and a half miles long and one-tenth of an inch in diameter. The hope was to begin learning how to use tethered satellites as a means of generating electricity and to study Earth's magnetic field and ionosphere. There was even a theory that a tethered system could one day be used to generate artificial gravity on a long-duration mission to Mars. The extension of a forty-foot boom from the orbiter's payload bay went well, but after that, it was a series of one malfunction after another. First, an umbilical at the top of the boom would not pull out. More than two hours later, a thruster firing from the orbiter freed the umbilical.

Hoffman, the flight's payload commander, had always found the tethered satellite more complicated than he felt its original designers fully appreciated:

They never designed an Attitude Control System for the pitch and roll of the satellite, because as long as there's tension in the tether, those are under control. I was in it from pretty much the beginning, and it's our job as a crew to try to think of all the different things that can go wrong. That was one of the first questions I asked: "Suppose you lose tension in the tether. Then what happens?" The answer was, "Oh, but that's never *going to happen." The deeper we got into it, the more you learned about all the potential instabilities that can happen, particularly when you're trying to reel in the tether. In the end, our project manager agreed to put in the money, to put in an Attitude Control System. Then, there was no way to bring the tether deployment to an easy stop. All you could do was slam on the brakes, which could cause tremendous instabilities. There was no way to start it up if it had been stopped. There was a whole litany of things which, during the last few years before the flight, we convinced project management to put in, to give the crew some control over what was going on. It turned out that every single improvement that we put into that system, we ended up using. It was a* very *crew-intensive mission.*

The issues were only just beginning. The first attempt to deploy the satellite ended when it began to move unexpectedly side to side, and ninety minutes later, another effort was stopped when the tether reel jammed after extending only 587 feet. Two more efforts to retract and then get a running start to free the tether ended up with only 830 feet of the twelve-and-a-half-mile anchoring line freed from its spool.

After a rest period, the crew tried one more time. The powered cord was retracted, but stopped at 733 feet, now unable to move either in or out. The implications were clear—with so much line freed and uncontrolled, there was a very real possibility that it could wrap itself around *Atlantis*. The entire crew—Hoffman along with commander Loren J. Shriver, pilot Andrew M. Allen, fellow mission specialists Franklin Chang Diaz, Marsha S. Ivins, Claude Nicollier, and Franco Malerba—leapt into action. "The satellite got over to just about forty-five degrees [from the boom, at which point the tether would have been cut]," remembered Hoffman, who was making the second of what would turn out to be a trio of flights within the span of just three years. "Loren was trying to fly the shuttle to get back underneath it. We were trying to control the attitude of the satellite to get it under control. That was pretty hairy. That was *very* hairy, actually."

Discretion being the better part of valor, the call was made to end the troublesome experiment. Hoffman and Chang Diaz suited up for a contingency EVA, and began a pre-breathe exercise in which they took in pure oxygen in order to purge their systems of nitrogen. Had they actually ventured outside, Hoffman would have climbed the mast and pulled the tether back in hand over hand, with Chang Diaz collecting the wayward line. Rather than add a risky spacewalk to a situation that was already tenuous, nearly a day after starting to deploy the satellite, the boom was lowered one panel at a time while the crew watched carefully for additional slack inside the structure. When there was none, it indicated that the issue was at the uppermost controlling mechanism. The boom was re-extended with the reel brakes engaged, clearing the jam. More than three hours after starting the operation, the satellite re-docked at the top of the boom and the whole thing was subsequently brought back into the safety of the cargo bay.

Despite the disappointment, Hoffman could at the very least find solace in the view. "We did go down to the low altitude, which was visually the most spectacular thing I've never done in the shuttle, because you're down where the atomic oxygen is rather thick," Hoffman said. "The entire shuttle was just glowing bright orange, just spectacularly beautiful. I don't know if you've ever seen Saint Elmo's fire. Saint Elmo's fire is more bluish. This was kind of an orange-white, just an ethereal glow. It was so bright, you could see it with your naked eyes. It was just spectacular."

The problem with the tether was discovered following an extensive post-flight investigation into the episode. One of a few extra bolts that had been added to the system after its original design caused the jam on orbit, and in all the starting and stopping, it was actually badly bent, which further exacerbated the problem. This had been a problem with NASA's reel mechanism—having been formed just four years before the flight, the brand-new Italian Space Agency's satellite itself worked perfectly.

As long as he was not blown to kingdom come in the next few seconds, Jerry Ross knew exactly what he was facing.

The countdown for STS-55 on 22 March 1993 had gone smoothly, and all three of *Columbia*'s main engines crackled to life right on time. Smoke billowed from the pad, a familiar signal that something amazing was about to happen. It did. At T-minus three seconds, the main engines shut down automatically when an incomplete ignition in the one on the lower right-hand side was detected.

The shuttle program had just experienced the third of what would eventually be five launch-pad aborts—after STS-41D in June 1984 and STS-51F in July 1985, with STS-51 to follow in August 1993 and STS-68 in August 1994. Making the fourth flight of his career, Ross knew from experience that he was up against a delay of about a month to allow time to replace all three main engines, and as it turned out, he was almost exactly right. It also meant that he was going to have to head back to Germany in the meantime and repeat the medical tests and blood draws that he had already been through in preparation for the flight of that country's Spacelab module. "I'd had a *long* period of training, and I just wanted to get the thing off the ground and get it over with," Ross could not help but admit.

Dan Brandenstein, serving at the time as the head of the astronaut office, had called Ross while he was in quarantine for the flight of STS-37 to ask if he would consider tackling the flight of the German Spacelab as its payload commander. It was going to be a life sciences mission, and there were doctors in the astronaut office. Surely, one of them could go instead, Ross wondered out loud. That was true, Brandenstein told Ross, but he also needed somebody who could stand up to the Germans and make the flight work. Ross laughed at the memory, and his reaction: "I said, 'Aww, man. I really don't want to do it, but you got me at a weak time. Here I am ready to go fly in space, and it's kind of hard to say no to going up again.' So I agreed to do it." Ross would not finish all his post-flight activities for STS-37 before venturing to Germany to begin planning for STS-55.

The mission was bought and paid for by the German Space Agency, but in order to recoup at least some of its costs, on-orbit usage was traded back to NASA and the European Space Agency. But rather than cutting back on its own expectations, the German agency simply piled the additional tasks on top of everything else it had planned. "By adding on another 25 percent of a mission for the Americans and 25 percent to ESA, they then had 150 percent of a mission to pack into just 100 percent," Ross said. The German mission manager was between a rock and a hard place. He did not want to ruffle any feathers at either NASA or ESA, so he could not—or would not—make the hard decisions on what to curtail or delete altogether. That put Ross in the uncomfortable position of being the bad guy:

It ended up being me making the arguments that hardware wasn't ready, it's not going to fly, all those kinds of things. It's not in my nature to be argumentative or dictatorial, but I didn't have a choice. I was not going to let them make the crew fail, so we worked through all that. After the mission, a lot of the Germans came up to me and thanked me and told me that if I hadn't done what I did, it would have been a mess. They understood, but their management would not do anything about it because of the awkward conditions they were put into.

STS-55 eventually featured eighty-eight different experiments developed by scientists from around the world, and most of them had never had hardware fly in space. "To try to get the hardware interfaces, the flight procedures and the timelines all do something that would work took the vast majority of our time," Ross said. "It was a *very* challenging mission." It had been a pill to iron out so many details, but because of that very fact, Ross came away with a keen sense of satisfaction over the amount of scientific research that was accomplished on the flight.

There was to be no spacewalk for Ross on STS-55. Steven R. Nagel, his commander on both this flight and STS-37, knew that the Crown Point, Indiana, native loved not only EVAs but also spending time looking out the windows and taking photographs of the earth. This flight, however, Ross was spending sixteen hours a day in the depths of a laboratory with no windows. In good humor, Nagel pestered Ross about his lack of opportunities for sightseeing and went so far as to set up a "gotcha" gag, just the kind that famed prankster and pioneer Mercury astronaut Walter M. Schirra Jr. would have loved. Ross related:

9. Steve Nagel took this photo of Jerry Ross during the flight of STS-37, and Nagel made sure that a copy of it tagged along when the two of them returned to orbit together on STS-55. Courtesy NASA.

On STS-37, you may have seen this one picture of me looking in the aft window after we'd successfully repaired the gamma ray observatory. Steve had taken that picture. He had taken three—the first one and the last one were terrible exposures, but the one in the middle got it just right. So Steve asked for some of the ground support people to get one of those pictures and cut it out to the exact dimensions of that aft window. He knew as soon as we got on orbit, I would be taking the covers off those windows, activating the lights in the payload bay, and turning on the cameras so we could open up the payload bay doors. So I got up onto the flight deck, and I pulled off the cover. There's this picture of me looking in the window at me. I just started laughing—and Steve actually forgot that he'd had the picture put in there. He was wondering what the heck I was laughing so loud about back there.

The picture's work was not done. Nagel took it out of the aft window and then placed it in the rear of the Spacelab, just so it would *look* like there was actually a window there.

When human spaceflight was in its infancy, rendezvous was a hurdle of monstrous proportions to clear. Putting two craft even within feet of each other took orbital mechanics, mathematical calculation and physics, literally, to new heights. Started during Gemini and perfected by Apollo, the art of bringing two bullets together became standard, if not exactly routine, fare during the shuttle era. There were satellites to retrieve and space stations with which to dock.

The EURECA satellite—containing fifteen experiments primarily devoted to material and life sciences, as well as radiobiology—had been deployed by the crew of STS-46, and now it was time for the craft to come home and be studied. All pilot Brian Duffy and the rest of the STS-57 crew had to do was go out and grab it during their June–July 1993 flight. With smart and talented people working out the equations, it did not matter if the target was as small as EURECA or as big as the space station *Mir* or its offspring, the ISS. Neither task was exactly easy—far from it, actually. The shuttle program just made it look that way. *Endeavour* had on-board radar to help locate EURECA, beginning at about 149,000 feet out. The ground was also tracking the satellite, keeping tabs on how far it was from the orbiter. Checklists made sure that the two came together at an appropriate time and position.

As it was in the case of most satellite retrievals, STS-57 commander Ronald J. Grabe brought the shuttle in from underneath the 9,424-pound EURECA during a daylight pass over the surface of the earth, to make it easier to grapple with the robotic arm. "The setup work is done from the ground and then basically from inside about a mile or so, the crew is pretty much primary from that point on," Duffy explained. "The responsibility kind of shifts from the ground, where they do all the work to get you in position, and then the crew takes over for the final portion for the grapple."

When neither of EURECA's antennas could be latched, it was berthed anyway in the payload bay. They were secured during a spacewalk by G. David Low and Peter J. K. "Jeff" Wisoff, just like that. It was just one more example of NASA making the difficult appear simple.

A shuttle orbiter, five crew members, two satellites to deploy, another launchpad abort, and shrapnel sprayed over the cargo bay once on orbit. STS-51 had it all.

One launch attempt for the flight was scrubbed on 17 July 1993 when the pyrotechnics for a vent arm and SRB hold-down bolts inexplicably charged prematurely, and another try a week later was halted at T-minus nineteen seconds due to an issue within the right SRB. Rescheduled for 4 August, it could not go then, either, because of concerns that the Perseid meteor shower might somehow affect *Discovery* while it was on orbit. The 12 August try was aborted at T-minus three seconds because of a faulty fuel-flow sensor on the left main engine. As on STS-55, the flight was delayed still longer in order to change out all three main engines. After yet another slip on 10 September, the launch finally took place two days later.

The problems were not over.

Nine and a half hours into the flight, the crew—commander Frank L. Culbertson Jr., pilot William F. Readdy, and rookie mission specialists James H. Newman, Daniel W. Bursch, and Carl E. Walz—deployed an Advanced Communications Technology Satellite, which was attached to what was known as a Transfer Orbit Stage (ACTS/TOS). It was freed by the detonation of the Super*Zip separation system—two cords inside a hollow tube attaching ACTS/TOS to its berthing platform in the payload bay. Only one of them was supposed to have fired—the primary cord, with the other designed to be used only as a backup. Both cords went off due to a wiring miscue, and the result was not good, remembered Rob Kelso, on console for the mission as the Orbit 2 flight director. "There was so much pressure from both detonation cords firing that it sent shrapnel around the payload bay," Kelso said. "That was pretty harrowing. A piece of the shrapnel had gone through the radiator. We could've had a significant Freon leak, where we lose a radiator. That could've resulted in coming down early. We even found where shrapnel was lodged into the forward bulkhead. It could have done some significant damage if it had hit at the right place."

Although debris strikes were later found in thirty-six different locations and less than half of the Super*Zip was left attached to the berthing platform, telemetry showed that *Discovery* was still up and running smoothly. As Newman and Walz headed outside for an EVA to test tools, tethers, and foot restraints for the upcoming flight of STS-61 to repair the Hubble Space Telescope, they took a look at the damaged platform but stayed as far away as possible to protect their suits.

STS-51 was also to deploy and then retrieve another payload: the Orbiting and Retrievable Far and Extreme Ultraviolet Spectrograph–Shuttle Pallet Satellite (ORFEUS-SPAS). That assignment went off without a hitch.

The goal of STS-58 was simple enough. The second dedicated Spacelab life sciences mission, launched on 18 October 1993, delved deeply into the mysteries of human adaptation to the weightless environment of space.

On board were fourteen experiments focusing on the cardiovascular, regulatory, neurovascular, and musculoskeletal systems of the body. David A. Wolf joined the medical sciences division at JSC in 1983, seven years before he moved into the astronaut office and a full decade before his first spaceflight, to study those very issues. So intimately familiar was he with the research and the team that put it together, landing a spot on STS-58 had a profound impact on Wolf. "I was part of their research communities, and I knew how hard they'd worked for many, many years," he said. He continued:

It was a real, genuine honor for me to get to go up. I felt like their ambassador in space, in this community I was so close with. To get to go up and conduct their research at the detailed, hands-on level in space, it made me take extra extreme care to precisely get our data. A large part of my life had been dedicated to this broad research team, dedicated to life-science research. I felt a great responsibility to them, and I think they also felt a great trust in me. I didn't feel like I was coming in from the outside to do someone else's work. I felt real ownership and membership in the teams involved in the work, which gave me a great sense of satisfaction. I did feel it was the perfect mission.

Essential to the flight's cardiovascular research were measurements of the pumping dynamics and chamber sizes of the heart. The initial plan was to acquire the data during the flight of STS-58 with ultrasound equipment in the Spacelab module that was able to take two-dimensional, real-time moving pictures of the heart. Late in pre-flight preparation, however, issues began cropping up. It was critical instrumentation, and the glitches worried mission managers. If it failed outright, a crucially large portion of the data would be lost. However, shortly before launch, the decision had been made to fly as a backup the very same American flight echocardiograph that Wolf had constructed with his own hands some eight years earlier. As it was, the prime unit did indeed fail during the activation of Spacelab. Wolf brought his echocardiograph out from beneath a floorboard and hoped for the best, because, quite simply, it had not been tried out in years. He was not disappointed. It was up and running in no time, and providing the critical cardiac imaging that otherwise would have been lost. "That was particularly satisfying being that involved in the recovery," Wolf said. "My machine was a more old-fashioned mechanical scanner, but it worked fine—something to be said for the 'old-fashioned' ways. The older technology I had built actually came out and salvaged a large and important part of our data. I used my own machine in space that I had built when I first came to NASA."

As a doctor himself, Wolf understood the impact of the shuttle on medical research. From just such a perspective, the sum total of the information gained by the shuttle program's life sciences flights laid the foundation for modern space medicine and physiology. The capacity of the cardiovascular system was diminished during spaceflight. Why? Muscle and bone density decrease. Why? The shift of body fluids affects renal and endocrine systems, and the way blood pumps throughout the body. Why? The balance and position-sensing organs of the neurovestibular system re-adapt in zero-g. How? That kind of research first gained momentum during NASA's three *Skylab* missions in the early 1970s, but shuttle astronauts took it to a much deeper and broader level.

Scott Parazynski's father, Ed, built a résumé that was every bit as impressive as his son's would one day become. After joining the Boeing Company as an engineer in 1962, he worked in conjunction with NASA to develop the Saturn IC booster for the Apollo program. An interest in all things space first gripped the younger Parazynski at the age of five, and it turned out to be a firm hold that would never be lost.

Every child wanted to be an astronaut in the sixties, but relatively few had a father who was so deeply immersed in the industry. There young Parazynski was, meeting one of the original German Pennemünde rocket scientists, touring the New Orleans Michoud plant where the booster was being assembled, attending the launch of *Apollo 9* and staring wide-eyed as *Apollo 11*'s Neil Armstrong and Buzz Aldrin collected the lunar samples that he himself would one day take to the summit of Mount Everest. The youngster built and flew model rockets, and had gathered space program posters and photos of his astronaut heroes. One day, he would orbit the earth alongside one of them.

When the Apollo program ended, Ed Parazynski moved into program management and business development positions with Boeing, in Washington DC and later overseas to Africa, the Mideast, and Europe. Moving so often might have been a "very exciting" opportunity, but as the younger Parazynski did so, the world was changing around him.

While he was in the eighth grade in Beirut, Lebanon, violence was pronounced, with one bombing taking place just around the corner from his family's home. Then, when he was a high school senior at the Tehran American School in Iran, protests started every day at sunset's curfew. Dissident Iranians took to the streets and were on the rooftops of mosques, chanting over loudspeakers. Occasionally, there was gunfire. The Iranian Revolution overthrew Shah Mohammad Reza Pahlavi on 11 February 1979, and eleven months later, a crisis of monumental proportions began when students overran and took sixty-six hostages at the American embassy in Tehran. Fifty-two of them would be held in captivity for the next 444 days.

Despite the chaotic atmosphere, the young man was not much bothered:

All young kids think that they're immortal and invincible, so I really didn't have a huge appreciation for life risk at that point. I thought it was a bit of an adventure. I wanted to go out and explore. But, there were other times that were a little bit scarier than others. I remember one time being out at a place called the St. George Hotel in Beirut. We were swimming in the pool there, and there was a sniper across the street who was taking potshots at people on the street. We had to kind of run for cover fairly quickly. That was the end of our swimming there.

Ed Parazynski eventually moved his family to Athens, Greece, during that tumultuous year in the Middle East. The younger Parazynski graduated from the American Community School in Athens in 1979, before entering California's Stanford University that fall. The groundwork for a career as an astronaut was being laid. Parazynski was accepted into Stanford's medical school, and in due course landed a coveted NASA Graduate Student Fellowship at the agency's Ames Research Center near Sunnyvale, California. There, he studied the fluid shifts that take place in the human body during spaceflight. It was a vital piece of the astronaut puzzle, for when he applied to the office for the first time while still in an emergency medicine residency in Denver, Colorado, he was selected as a member of "The Hogs" ASCAN class in March 1992. Parazynski would always figure that getting that long-awaited dream gig boiled down to somehow managing to stand out in the interview selection process, over and above the background and experience that any candidate brought to the table. Do not forget good old-fashioned luck, either. "Quite honestly, there are a lot of people who could do the job quite well," Parazynski said. "NASA selects astronauts from a really talented group of people, any number of whom could do the job well. So I do consider luck as a significant part of the equation. I remember my selection week and thinking that of the twenty-two people who were there the vast majority would've been just fantastic astronauts. You really knew why they had selected the people that they had selected for interviews."

As the Shuttle-*Mir* and ISS programs began to heat up, none of Parazynski's classmates had to wait more than four years to fly—and some much less than that. Steven L. Smith and Jerry M. Linenger were the first of the Hogs to be assigned to flights, and next up were Parazynski, Joseph R. Tanner, and Jean-François Clervoy of ESA for the November 1994 flight of STS-66.

Clervoy never forgot how he learned of his assignment. Hoot Gibson was on the phone, telling Clervoy to get to his office as soon as possible. Had Clervoy said something wrong in one of the many engineering meetings he had been attending while getting up to speed on operations of the robotic arm? Once he sat down with Gibson on that Friday, the chief of the astronaut office, the feeling persisted. Gibson let him off the hook, telling the Frenchman that an announcement for the crew of STS-66 was being made the following Monday.

Clervoy needed to be there because he was going to be a member of that team, along with Parazynski, Tanner, commander Donald R. McMonagle, pilot Curtis L. Brown, and payload commander Ellen L. Ochoa. "Ahhh . . . my heart started *boom*, *boom*, *boom*," Clervoy said. "I thought, 'Wow . . . this is becoming real.' There was no agreement committing NASA to ESA to provide me a flight. It was not a given. You think it's still a dream." There was only one catch. Clervoy could not say anything to anybody about the assignment over the weekend. His wife was back home in France, but he did not dare tell her over the phone:

If there is something in the media before the chief of the astronaut office has made the official announcement, it looks very bad. So I kept it for myself. I couldn't sleep. Maybe some people that weekend thought I had a special look, a special happy face. I was very, very excited, because from one minute to the next, something that was still kind of a dream was very concrete and real. I could feel it. I could feel I had already one foot in it. Until it happens to you, especially when you know what it means to go to space—all that energy, all that training, all the high level of performance and the competence that is required for you to be assigned—you still feel something in this business is a bit unreal, a bit a dream. From one minute to the next, boom, *it was there. It was concrete.*

The rookies of STS-66 were often like schoolkids passing notes and crib sheets during class. If one of them found a better way of performing or prepping for a task, he presented it to the rest of the Hogs. Clervoy explained:

Scott was the first of our class to train for spacewalks. Right after his first sessions, he sent to the whole class his notes to help us be more efficient on our future first spacewalk training class. Joe was MS-2, which means he was the flight engineer seated between the two pilots for ascent and entry. So he was the one in our crew to practice intensively cockpit procedures involving manipulating hundreds of switches. As I was seated next to him on entry, he helped me early in training to learn quicker than if I had had to learn all those switches on my own. In my case, I was the first one of the class to train on the remote manipulator system. I wrote also a quick note to summarize my findings which could help future trainees.

So intensive were Clervoy's note exchanges that they picked up his own nickname:

Scott and I were in the same blue shift on STS-66. We prepared simultaneously our own cheat sheets to help our work in the simulator, and regularly cross-checked our respective cue cards to take the best of each other's card. They eventually converged to what became quickly a reference in the astronaut office. I improved it a lot on my following mission training, and it is known today as the "Billy Bob card"! So I wouldn't say that we leaned on each other because we were all competent in our respective tasks, but we all helped each other with tidbits that each of us thought could be beneficial to others.

On STS-66, *Atlantis* took an ATLAS laboratory into orbit for a third research cycle on the sun's energy output, the middle atmosphere's chemical makeup, and how each of those impacted global ozone levels. The Cryogenic Infrared Spectrometers and Telescopes for the Atmosphere–Shuttle Pallet Satellite (CRISTA-SPAS) was also deployed and then retrieved eight days later to collect data on the earth's middle and lower thermospheres. When it was retrieved, McMonagle and Brown tested a technique called the R-bar approach, to save propellant and reduce the number of thruster firings that could potentially damage the Russian space station *Mir* during future dockings.

As always, the flight plan was packed with a mind-boggling assortment of secondary experiments, covering everything from the effect of microgravity on developing rats to protein growth and spacecraft thermal control systems. None of the STS-66 crew had spent a lifetime studying earth sciences, but Parazynski took to the challenge with relish. "In medical school, you're just deluged with lots and lots of different information, and you have to process it really quickly," he said. "It was like going back to school. It was a big chunk of new science that we had to master, but that's what I really liked about being an astronaut. It was sort of like going to school 365 days a year. You're always learning something new."

John M. Grunsfeld had dreamed of this moment for as long as he could possibly remember. He was an astronaut, yes, a member of the STS-67 crew on board *Endeavour* and waiting to be catapulted into the dark early-morning sky on 2 March 1995. That was hard enough to believe, but for Grunsfeld, there was still more reason to revel in the experience.

Not only was the Chicago-area native about to fly into space for the first time, he was to do so in order to help operate the Astro-2 observatory from its fixed position in the cargo bay. It was another goal he had nurtured since childhood. "That was my dream as a kid," Grunsfeld said. "I just sort of assumed, given *2001: A Space Odyssey* and other works of science fiction, that by the time I was a professional astronomer and astrophysicist that all astrophysicists would go to space the same way they go to mountaintops. Of course, as I got older, I got to see that we weren't making the same kind of progress that Arthur C. Clarke predicted. But in my case, it actually came true."

Better yet, this was not just another telescope. Astro-2's package of three ultraviolet telescopes—one from the University of Wisconsin, one from Goddard Space Flight Center, and one from Johns Hopkins University—

had first flown on STS-35 five years earlier. They enabled celestial observations from above Earth's atmosphere, which absorbs lower-energy infrared light and higher-energy ultraviolet light and X-rays. With that kind of distortion, Earth-bound astronomers and astrophysicists were getting only a minute portion of the picture. The STS-67 press kit put it this way: "Seeing celestial objects in visible light alone is like looking at a painting in only one color. To fully appreciate the meaning of the painting, viewers must see it in all of its colors. Getting above the atmosphere with space instruments like the Astro ultraviolet telescopes lets astronomers add some of these 'colors' to their view of stars and galaxies." Added Grunsfeld, "You have to go above the atmosphere to see ultraviolet light, and stars produce lots of ultraviolet light. We had three telescopes to study the physics of stars, galaxies, quasars, planets, and the moons of planets. Our job was to be telescope operators. For me, that was actually pretty amazing, because just three years prior, I was at Caltech [California Institute of Technology], observing at Mount Palomar, on a telescope not much different from what we had in the payload bay. We were up on the mountaintop, and then three years later, I'm on orbit doing the same thing."

Grunsfeld left that Palomar mountaintop and walked straight into the astronaut office, a member of the 1992 candidate class with Jean-François Clervoy and Scott Parazynski. With missions like this one bumping up against the oncoming Shuttle-*Mir* program, NASA needed astronauts and it needed them to be ready to fly as soon as humanly possible. It was, Grunsfeld admitted, "the sweet spot to arrive." He had made a career out of building experiments to go to space, and he was about to go along for the ride himself:

The whole language of the Space Shuttle, the language of the experiments that we fly and operate, was completely familiar to me. While people were struggling with trying to learn about the shuttle's communication system, I looked at the schematic and completely *understood how it all worked. There were no mysteries, and so when I went through training, I exhibited a very high proficiency and was able to go through the classes very quickly. It was recognized by the leadership that I was somebody who demonstrated a level of competency and capability. I didn't need five years of training to get to the point where I understood the technical nature of what we were doing. Plus, we had an astronomy mission on the manifest, Astro-2, that would benefit greatly from having somebody with my experience on board. So it all just kind of came together.*

Grunsfeld was on his way, his work on orbit with space-based telescopes just getting started. Four years later, he met up with the granddaddy of them all.

NASA administrator Daniel S. Goldin was not amused, and he let Rob Kelso know about it.

Northern Flickers—the woodpeckers, not some obscure semi-professional basketball team—left approximately two hundred holes in the foam insulation of the External Tank that was to have helped propel STS-70 off pad 39B on its scheduled launch date of 8 June 1995. Created over an extended Memorial Day holiday weekend, the damage was as small as a single peck and scratch marks to as big as some four inches in diameter. "They saw this big orange-looking thing out at the launch pad and decided that would be a great place to make nests, so they started to attack," said Kelso, slated to serve as the mission's lead flight director. "They made holes all over the dad-gum thing. They'd peck through the outer foam, but then they'd hit the hard aluminum-lithium tank. They couldn't go through any further, so they'd get frustrated and go to another part of the tank."

The shuttle-based deployment of another TDRS would have to wait. Rather than making its way to space on 8 June, the stack was instead that day rolled back to the VAB for repairs. Launch was rescheduled for more than a month later on 13 July, and in the meantime, the crew—consisting of commander Terence T. Henricks; pilot Kevin R. Kregel; and mission specialists Nancy J. Currie, Donald A. Thomas, and Mary Ellen Weber—visited the Universal Studios theme park in Orlando, Florida. There, they were presented with a huge Woody Woodpecker doll in "honor" of the issues with the External Tank. Amused, the STS-70 crew turned it over to Kelso, who took Woody back to Houston.

Not only did Kelso take the famed cartoon character to JSC, he also made him a member of the flight control team—complete with a badge and headset. If this had been in the 1960s, the only things missing would have been its white shirt, black tie, and pocket protector. Woody even got to work different consoles on different shifts, and when that was not enough, someone even found some woodpecker cut-outs at a local hardware store and attached them to several of the mission control workstations in the brand-new shuttle control room that made its debut during this flight. The ascent, TDRS display, and reentry were handled from the old Flight Control Room-1, while on-orbit operations were controlled from the new room in Building 30 at JSC.

All went well, right up until the moment that Woody was captured on television. Goldin was soon on the phone. "I get this call from Dan Goldin, and he is irate," Kelso said. "He goes, 'I want *every* woodpecker out of that control center. I don't want him sitting with anybody. I don't even want to *see* him. I don't think the holes in the tank are funny, and I don't think Woody sitting around in the control center is funny.' So we had to go hide all the woodpeckers."

STS-70 was able to proceed on 13 July 1995. Just six days earlier, STS-71 had returned to KSC from the first docking with the Russian space station *Mir*. The turnaround from the landing of one mission to the launch of another was the quickest in the history of the shuttle program, woodpeckers notwithstanding.

Brian Duffy had been looking forward to getting his shot as a leftseater, and this was it. He was a rookie commander in charge of a nearly all-rookie crew during the January 1996 flight of STS-72.

After his second flight as a pilot on STS-57, the air force had been in touch with Duffy, wanting him to come back as an active member of that military branch of service. He thanked them for the invitation, but no, he was not ready to return just yet. He wanted a shuttle command of his own, but knew that it was no sure thing. "There's no guarantee that once you're selected into the astronaut office, that you're going to fly," Duffy said. "Once you fly once, there's no guarantee you're going to fly another time. So, it was a little bit of a gamble in doing that. I thought I would be a good commander, so it was something I wanted to try to do."

Two objectives of STS-72—the retrieval of a microgravity research satellite that had been launched out of Japan ten months prior to the flight, as well as the deployment and recovery of another experiment platform—were similar in nature to the capture of the EURECA satellite that he had helped perform on STS-57. That kind of prior experience might have played into his assignment, but he could not say for sure. Duffy never asked. As long as it was his name next to those three magical little letters—CDR—he was happy for the opportunity and determined to show that it had not been a mistake.

Leadership was nothing new to Duffy. He had worked his way through the ranks at the air force academy, and then again as a fighter pilot. Duffy had once been Blue Four, the new guy, the lowest man on the totem pole of a group of four planes. Typically, pilots moved up a spot to element lead, then to flight lead of the quartet of planes, and then even larger than that, the lead of an entire strike package—which could be as many as eighty to a hundred planes. "Leadership is something you work your way into," Duffy said. "You have to demonstrate capability and proficiency as you progress—otherwise, you stop. Being a leader is something that comes with the territory." There was a "fair amount of difference," Duffy continued, between heading up a shuttle crew and management in the air force, and there happened to be a fairly obvious reason for that:

You don't have six people with you in a fighter cockpit. The interpersonal skills are more important in a shuttle crew, just because of the nature of it. In a military situation where you're going to go on some sort of a combat mission, the leader puts the plan together and then tells everyone what to do, when to do it, and how to do it and then expects that to happen. Those sorts of things still need to happen on a shuttle crew as well, but it's done in closer quarters. It's a little bit different. You're not directing people on the radio. You're elbow to elbow doing things.

The quarters were certainly closer, but there was another dynamic in the mix as well. How would the flight's four first-timers—pilot Brent W. Jett and mission specialists Daniel T. Barry, Winston E. Scott, and Koichi Wakata—fare once they hit MECO and first encountered the effects of weightlessness? It was the unanswerable question on *every* spaceflight. Lead spacewalker Leroy R. Chiao was the flight's only other veteran, so he and Duffy had a good idea of how their bodies were going to react on the all-important first couple of days on orbit. The rookies? Not so much. "Space adaptation syndrome is repeatable and predictable," Duffy said. "I knew how Leroy and I would do, but everyone else was questionable. I made sure that every task that was to be done by a rookie would have a well-trained backup, and I used myself and Leroy in that role as much as the timeline would permit. I wasn't worried about it, but it did require some extra planning."

Aside from the two satellite captures, two EVAs were conducted to test a portable work platform for use in the construction of the ISS. Chiao and Barry performed the first, while Chiao and Scott did the second. Only one of them would ever be voted off the island. After flying twice more on the shuttle and retiring from NASA, Barry was selected as a castaway on the hit ground-breaking American reality television series *Survivor: Panama—Exile Island.*

Try again. That's all NASA could do when it came to the failed test of the Tethered Satellite System on STS-46.

Plans for a re-flight on STS-75 quickly kicked into gear once STS-46 landed. Jeff Hoffman eventually found himself on the crew of the following flight, the fifth and final mission of his storied career and one on which he became the first astronaut to log a thousand hours of flight time on the shuttle. He was not the only STS-46 veteran on the mission. Along with him was Andrew Allen, who moved from pilot of STS-46 over to command of STS-75, and Nicollier, making a then-unprecedented third trip into space with Hoffman. Franklin Chang Diaz was also making a fifth flight, and when the landing was delayed a day due to weather conditions, he joined the thousand-hour shuttle club as well. Rounding out the crew were rookie pilot Scott J. Horowitz and freshman mission specialists Maurizio Cheli and Umberto Guidoni. The flight hardware included the United States Microgravity Payload to study various materials in zero-g conditions. That worked all but perfectly, although a second shot at the tethered satellite did not.

The flight itself did not get off to a particularly smooth start in the moments after its launch on 22 February 1996. Just six seconds into the ascent, a meter showed that the left main engine was performing at just 40 percent thrust, instead of the expected 104 percent prior to its throttle down. At the same instant, all four primary avionics computers showed a failure message on the same left main engine, a signal that was accompanied by an illuminated status light on the control panel. Downlinked telemetry, however, showed that the engine was performing exactly as it should and the ascent continued unimpeded.

Three days later, it was time to give the tethered satellite another try. The craft had been almost completely redesigned. Now in place was an automated Attitude Control System, and gone were the upper umbilical and bolt that had so plagued the previous deployment. This time, the operation went perfectly, right up to the point where the tether snapped only a quarter of a mile or so from its full twelve-and-a-half-mile length. Hoffman would always remember the triumphant moment turning sour so quickly:

The initial part of the deployment went beautifully. We had no trouble getting it going. It was really quite an extraordinary thing to watch as this went out. To see this tether go up into space and sort of disappear into nothingness, it was a little bit like Jack standing at the bottom of the beanstalk. Visually, it was extraordinary. But just a little bit before we reached full deployment, I was watching the tether very closely. All of a sudden, I saw the tether go slack. It was this feeling of déjà vu. I thought, "This can't *be happening again."*

If there was a sense of relief to be had in the midst of the incident, it was that the break took place near the bottom of the line. Had it been further up, the tether would have been cut, forcing Allen into an emergency evasive maneuver to miss the elastic rebound of what could have been miles of cord. Later, Hoffman used a high-powered telephoto lens on his camera to examine the charred and frayed end of the broken line. From that documentation, it was determined that an electrical short had caused the problem.

The tethered satellite would never again be tested on the flight of a Space Shuttle.

When STS-77 flew in May 1996, the ball was in N. Wayne Hale's court at long last.

Ever since the very first flight of the shuttle in 1981, when John Young and Robert L. Crippen rocketed off into the wild blue yonder and beyond, Hale had worked his way through mission control. He was on console during that inaugural shuttle ascent to monitor propulsive consumables and then did the same thing for the Orbital Maneuvering and Reaction Control Systems during *Columbia*'s reentry and landing. By STS-8, he was the lead propulsion officer, and not long after that, he was getting up to speed on instrumentation and spacecraft analysis. Hale bumped up the ladder still more, serving as flight director for the planning team of STS-28 and then as Orbit 2 flight director on STS-33.

He grew into each role, and when STS-77 rolled around, Hale was ready.

For the first time in his career, he was to serve as lead flight director for a mission that was to feature an unprecedented four rendezvous with two different payloads—the Inflatable Antenna Experiment and the Passive Aerodynamically Stabilized Magnetically Damped Satellite (PAMS). Also on board was the Spacehab-4 module, which carried three thousand pounds of experiments and support equipment for twelve commercial space product payloads.

There were a lot of things on the to-do list with that flight. They were very, very significant experiments, and it was a varied flight. The experiments were not all the same kind of thing. I was flight director on some of the life sciences flights, and it was like, "Okay, time for another blood draw. Time to pull the plants out of the locker and see how they've grown." You get kind of stuck in one thing. STS-77 had a variety of activities. Of course, I'm kind of the geeky engineer guy, so I'm really proud of the fact that it still holds the record for the most rendezvous for any shuttle flight. That was really exciting. It's just another one of those flights—it wasn't a Hubble deploy. It wasn't a mission to Mir. *We weren't building the International Space Station, so it's kind of been lost in the mix of all the 135 shuttle flights that we had. It's still my favorite flight.*

Hale was not the only one getting his feet wet on STS-77. The only rookie on a crew of six, Australian-born Andrew S. W. Thomas had become an American citizen in 1986 in hopes of landing his lifelong fantasy title—astronaut. With a doctorate in engineering from the University of Adelaide already in hand, Thomas went to work first for Lockheed Aeronautical Systems Company in Marietta, Georgia, and then at NASA's JPL in Pasadena. In 1992—yet another Hog—he got *the* job he had so desired for so long. Although landing a spot on the crew of STS-77 was certainly an honor, Thomas readily admitted that just joining the astronaut office was quite possibly an even bigger milestone:

Once you're selected as an astronaut, it's pretty likely that you're going to fly unless you have a major medical issue or an accident or something. The most exciting point was in 1992, when I actually got selected. For me, the significance of that was that it was the culmination of a dream of many, many years. And by most people's standards, it was a fairly unrealistic dream. What are the chances for anyone, let alone someone who was in South Australia when Gemini and Apollo were happening? I remember in 1992, finding myself actually at the space center in Houston, standing next to Alan Shepard. Thirty years before, I had been pretending I was he in the backyard, pretending I was in the capsule, the same as every kid, with the chair turned on its back to simulate being on the launch pad. There I was standing next to him. I'd just joined this office which had made all those flights happen. It was a personally rewarding moment, a little bizarre, I have to say, because it had a degree of unreality about it. But the most important thing was the personal sense of accomplishment of having worked toward what most people would have thought was an unobtainable goal and making it happen.

Twenty-five hours into the mission, the *Spartan-207* satellite and its attached inflatable antenna were released. Mounted on three struts ninety-two feet in length, the antenna was deployed to a full fifty feet in diameter, making it roughly the size of a tennis court. The theory was this: if successful, a deflated antenna would require less stowage space and maybe even a smaller launch vehicle than rigid structures. Less stowage space meant lower cost, up to ten and as much as one hundred times less. That was the idea, but reality was something else. It did not work well enough to fly again.

After jettisoning the inflatable antenna, commander John H. Casper brought *Endeavour* close enough to allow Canadian astronaut Marc Garneau to grasp *Spartan-207* with the robotic arm. Three more times later in the flight, the orbiter rendezvoused but did not recapture PAMS. One of four technology experiments for advancing missions in space research payloads on the flight, it was designed to demonstrate aerodynamic stabilization in low-Earth orbit.

If the operations sound simple enough, the fact is that they were anything but. Consider the release and retrieval of *Spartan-207*. Once deployed, Casper backed away to a distance of about 400 feet directly above the satellite. Holding steady for eighty minutes or so, the antenna was inflated and experiments began. After that, *Endeavour* backed away 900 feet to allow for the test package to be discarded. Another separation burn took the shuttle away from the area at a rate of about two and a half miles per orbit, and the following day, it was as much as forty to sixty nautical miles behind the satellite. To catch back up, the shuttle was moved to within eight miles of *Spartan-207*, at which point it began a final intercept burn. About 2,500 feet away, Casper took manual control of the shuttle and flew to within 35 feet of the satellite. Garneau locked it onto the end of the remote arm and stowed it safely back into *Endeavour*'s cargo bay. The very same processes were repeated during the trio of PAMS rendezvous, up to a distance of about 2,000–2,300 feet. Three times, the orbiter held steady for about six hours while video was captured of the satellite's on-orbit attitude.

Hale always marveled at the degree of difficulty this flight exhibited, in spite of the disappointing results with the inflatable antenna:

You go on the Internet sites and the chat rooms, and a lot of people talk about, "We ought to build this kind of space station," or "We ought to build that kind of structure in outer space." I just get the impression that most folks have no *idea how hard it is to do things in zero gravity, in vacuum. The inflatable antenna was a wonderful experiment, but it didn't work out. The folks up at Goddard spent a huge amount of time on a really great experiment, trying to build this large space structure in a very cheap and hopefully efficient way. It just didn't work. Some of the things we learned might mean that something in the future might work, but that one didn't.*

STS-77 was certainly a busy flight, but was it as difficult as an ISS assembly flight? Thomas was not quite as certain on that point. "STS-77 was quite ambitious for the time, no doubt about it," he concluded. "By the standards of the assembly flights of the space station, it was probably not as hard as those have been." He added, "You're carrying up large payloads on an assembly flight. You've got to do a rendezvous. You've got to do a complex docking. You've got to do a deployment of a large payload with the arm. You've got to mate it to the space station. Plus, you've got to transfer crew and a lot of equipment. You've got to do a suite of EVAS. We didn't have EVAS on STS-77."

When Story Musgrave got back from STS-61—the first servicing mission to the troubled Hubble Space Telescope—he saw absolutely no reason that he could not fly again.

Although he was fifty-eight years old, to Musgrave, it was just a number. He was a go-to guy within the astronaut corps for all things EVA, and his performance reports were spot on. Musgrave wanted another Hubble flight, so he went back to work as a CapCom and waited. And then he waited some more.

Crew seats on the second Hubble servicing mission went elsewhere. Other flights fell by the wayside as well, and they were just the kind of trips into space that Musgrave felt were tailor-made for him. A physician and physiologist, he would have loved a life-sciences mission. According to Musgrave, he got turned down for that, too. A proposed long-duration stay on *Mir* did not happen. He felt sure he knew the reason. "After the great project management of the sixties and going to the moon, when NASA became defensive and congressional staff started to run NASA, you got a lot of people riding in the chain that didn't care about spaceflight," Musgrave began. "They cared about themselves. Not getting another flight was 100 percent politics. It had nothing to do with anything else." As time wore on, Musgrave was ready to take matters into his own hands:

I was getting passed over. For what reason? For no reason at all. I was going to take some action, and it would not have been pretty. I would've won. You start out gentle. You start out telling them, "Hey, I've been recommended for this mission. What's the problem? Is there a problem with my performance? How come I'm not getting assigned to a flight?" You go and talk with them to start with and you escalate the pressure until you get some justice. Wherever you've got to go, you escalate the pressure, even if it gets to the courts. I would have done that.

Musgrave's fifth and final flight came in the form of STS-80, slated to fly in the fall of 1996. On board were two primary payloads—ORFEUS-SPAS II and Wake Shield Facility-3 (WSF-3). The WSF-3 was a twelve-foot-wide stainless steel disk designed to generate an "ultra-vacuum" wake ten thousand times greater than what was possible on Earth while being towed behind the orbiter. In that kind of environment, the goal was to grow thin semiconductor films for use in advanced electronics. It had been flown twice before, and twice before, the results had been less than spectacular.

Radio interference and control problems during the February 1994 flight of STS-60 forced the wake shield to be suspended at the tip of the robotic arm the first time out. On its next flight, STS-69 in September of the following year, it placed itself into an unexpected safe mode. When a subsequent attempt by the ground to trigger the flow of the thin film material went bad, the shield was shut down again. Ultimately, just four of seven film growth runs were completed successfully.

When Musgrave was named to the crew of STS-80, some might have been tempted to figure it was simply a way, at best, to mollify Musgrave, and at worst, to shut him up. According to Musgrave, there was at least an element of that in play. "I knew it was the end of the game," he concluded. "They said, 'We're not going to assign you this flight unless you'll go away. So, understand, if we assign you this flight, it's the end of the road.' That happens in organizations, especially when they're in failure."

If Musgrave's personality was sometimes hard to gauge, he had proven his abilities before, during, and after a spaceflight. This was a then-record sixth shuttle flight for Musgrave, and it also tied John Young—who had flown Gemini and Apollo in addition to the shuttle—for the most in any type of spacecraft. Musgrave was the only astronaut to fly each of the five different orbiters, and you could tell, he once said, "'cause there are ten fingernails imbedded in the instrument panel of every one of them!" Regardless of the feathers he may or may not have ruffled, Musgrave would not have been on the crew of STS-80 had he not been able to get the job done. "If you want to take a positive viewpoint of how that happened, somebody may have said, 'We've got to make that thing [the wake shield] happen, and we need someone who will get in there and attack the details,'" Musgrave said. "I'm not an expert, of course, on the manufacturing of electronic chips in space. What I am good at is doing details. I had a lot of nicknames, and 'Dr. Detail' was just one of them." When the flight finally got off the launch pad on the afternoon of 19 November 1996, Musgrave was exactly three months past his sixty-first birthday, making him the oldest astronaut ever to fly at that point. He saw himself as something very much like a fine wine:

I was getting much better. There's a huge art to spaceflight, and there's a huge art to pursuing those details. I can't say I was any better than I was on 51F. I was incredibly good on STS-51F. I was really on top of the total picture, so in a way, maybe I peaked then. I was the systems engineer, up and down. I did rendezvous. I was this and that, everything. After that, I was becoming more comfortable with the whole process—how you learn a flight, how you scope out what's demanded of you, how you scope out what's going to make a system fail, what's going to make it work. It's very complicated.

This time, the wake shield experiment worked almost perfectly, and seven thin films of semiconductor material—the maximum number expected—were grown. The updated ORFEUS-SPAS II telescope got twice as much data as on its previous shuttle flight, doing 422 observations of nearly 150 celestial objects, including the moon, nearby stars, distant Milky Way stars, stars in other galaxies, and Quasar 3C273—a very bright and active nucleus of a young galaxy.

The flight's one big glitch took place as Tamara E. Jernigan and Thomas D. Jones prepared for the first of two planned EVAS to evaluate equipment and procedures for upcoming construction of the ISS. They could not open the airlock hatch due to a small screw that had become stuck in a mechanism securing it in place.

The wake shield had been retrieved about halfway through the flight, and the ground was handling operations of ORFEUS-SPAS II. To Musgrave, the cancellation of the spacewalks was a "bummer," but he had never been one to have to look for things to do. He took photographic studies of the earth's surface. He did some zero-g experiments inside *Columbia*. He observed aurora, shooting stars, and purple lightning during night passes. He worked to make his final journey into space count, and he enjoyed every second of it. "It was, by an order of magnitude, the most relaxed flight plan I had ever flown," Musgrave said. "It was great to be able to have some time just to experience spaceflight, but especially Mother Earth and Earth photography."

Though free time was negated some by getting into and out of reentry mode and pumpkin suits, the originally scheduled landing on 5 November was delayed two straight days by bad weather around KSC. Despite his sometimes contentious relationship with the powers-that-be within NASA, he would years later laugh and wonder if maybe the wave-offs were for him. The few additional precious hours on orbit had been that much of a gift. At wheels stop, the flight of STS-80 lasted seventeen days, fifteen hours, fifty-three minutes, and eighteen seconds, making it the longest mission in the thirty-year history of the Space Shuttle.

In the minutes before that landing at KSC, Story Musgrave had not quite finished being Story Musgrave. He had no responsibilities during reentry, and preferred not to sit like "a lump on a log down in the mid-deck." Musgrave decided that he would be the first to personally observe the deadly plasma enveloping the orbiter as it tore back through the atmosphere. In so doing, he somehow managed to stand up, braced, throughout reentry, steep S-turns to bleed off speed, and landing, a feat of strength and endurance difficult enough for someone half his age. All the while, he used a small lipstick-tube-sized camcorder taped to a stick to record the beautiful show in an overhead window on the flight deck. He could see shock waves in the flames hurtling past the window in a mind-numbing array of colors. Although they had not expected Musgrave to remain in such a precarious position for so long, commander Kenneth D. "Taco" Cockrell, pilot Kent V. "Rommel" Rominger, and the rest of the crew seemed not to mind. They, too, were enjoying the sights on a small monitor wired into Musgrave's video camera. "Taco and Rommel, they were mature fliers," Musgrave said. "They didn't care if I was yelling and screaming about what I was seeing. They don't care. It's no distraction to them. They're focused on the flying and they enjoyed the plasma when they got a chance to enjoy it. Otherwise, it never would have happened. The downside of all that is that we should've had cameras in the overhead window for reentry, and all we do is hit the button to turn them on. I got the first recording, but I should not have."

On 2 September 1997, Story Musgrave's retirement from NASA was announced. "Throughout the shuttle program, from its earliest stages to the present, Story has been instrumental in developing the techniques crew members use to perform spacewalks," said David C. Leestma, then the director of flight crew operations, in a press release. "His knowledge, expertise and friendship will be sorely missed."

There was no other call to make. STS-83 would have to end on 8 April 1997, just four days into a flight that had been scheduled to last nearly sixteen days.

Columbia could fly and land with one of its fuel cells not functioning, but with a payload bay full of a Microgravity Science Laboratory that was packed with power-hungry experiments, there was simply too little margin for error in case anything else went wrong. The flight rules left no room for interpretation—if a fuel cell went down on orbit, the mission ended early.

Each orbiter was equipped with three fuel cells that used a reaction of liquid hydrogen and liquid oxygen to generate electricity, and each of those contained a couple of "substacks" made up of sixteen cells each. Data was erratic on fuel cell number two during pre-launch startup, but it was cleared for flight. Then, just two hours into the flight, the difference in output voltage between the two banks of cells began trending upward. If left unchecked, the mixture of liquid oxygen and liquid hydrogen could have led to an explosion.

Three purges of the balky fuel cell were attempted during the first thirty hours of the flight—one that lasted two minutes, another that took ten minutes to complete, and a final one of thirty minutes in duration. Nothing worked. "The kicker was that if it had happened on any other flight, it wouldn't have been such a big problem," said Rob Kelso, the lead flight director for STS-83. "That particular flight, the laboratory had large furnaces for experiments that would suck up a lot of power." The shuttle could easily fly on just two healthy fuel cells, but shutting down a third that had been dedicated largely to running the Microgravity Science Laboratory meant that much of the flight's research went out the window. The crew could make laps of the earth's surface and take in all its pretty scenery, but not much else.

Kelso got on a private air-to-ground loop and told STS-83 commander James D. Halsell Jr. what he probably already knew. "I called Jim and explained to him that we were going to have to pack up and de-orbit," Kelso remembered. "It was just something we had not seen in the fuel cell before, and he understood that." Before *Columbia* touched back down at 2:33 p.m. EDT on 8 April, there was already talk of a quick turnaround and re-flight of the same orbiter, crew, and payload. The Microgravity Science Laboratory would still be in the payload bay, and the crew—who, along with Halsell, also consisted of pilot Susan L. Still and mission specialists Janice E. Voss, Michael L. Gernhardt, Don Thomas, Roger K. Crouch, and Gregory T. Linteris—could be held together as a group. The bad fuel cell, of course, would need to be swapped out for a good one, and standard operating procedure after any liftoff called for the main engines to be replaced as well. Other than that, workers in the OPF would need to stick some more consumables on board and keep air in the landing-gear tires and *Columbia* would be ready to roll right back out to the pad. The crew, maybe they would have to do an ascent and entry sim or two, and they would be good to go, too.

Just three days after STS-83 landed, NASA issued a press release confirming that a re-flight was in the works. The shuttle launch manifest was tight for the rest of the year, but a mulligan was not out of the question. In fact, from the tone of the release, it seemed probable. "While shortening STS-83 was disappointing, we now are in a position to do everything possible to complete the mission as STS-83 Re-flight (STS-83R) with little or no impact to downstream flights," said Tommy Holloway, the shuttle program manager. "Also, it provides us with a unique opportunity to demonstrate our ability to respond to challenges such as this one." Exactly two weeks later, the flight of STS-94 was officially announced. Everything was staying the same, except for the mission patch. Its red border had been changed to blue, and the "83" replaced with a "94." The seven astronauts of STS-94 strapped back into *Columbia*'s crew module on 1 July 1997, just fifty-six days after they had landed. Just one mission—STS-84, the sixth to dock with the Russian space station *Mir*—had taken place in the interim.

STS-94 was all but flawless.

STS-85 did not dock with *Mir*, nor was it one of the famous servicing missions to the Hubble Space Telescope. On this flight, it was all science, all the time, and Stephen K. Robinson was fine with that.

Those kinds of flights got a lot of attention, certainly. Robinson, though, had first applied to the astronaut corps in 1983, twelve years before he was finally accepted. "I have no idea how many applications and updates were involved," Robinson once said. "I do know I have a nice collection of rejection letters!" Once he was finally ensconced in the astronaut office, if his bosses wanted him on STS-85—which flew in August 1997 with the CRISTA-SPAS-2 satellite on board to study Earth's middle atmosphere—that is exactly what he was going to do, no questions asked, no complaints made. "When it's your first flight, I'm not even sure that you're fully aware of the implications of different kinds of flights," Robinson said. "You've worked your whole life towards this one goal. You're not going to say, 'I'd rather have chocolate ice cream than vanilla ice cream.' That was my attitude then, and it kind of stayed my attitude, actually."

For Robinson, there was also a larger purpose in play. "I feel like spaceflight isn't *for* me," he continued. "I'm working for the spaceflight program, and when I get assigned, I'm going to be really happy that I get to be a part of that. It isn't about making this particular astronaut's desires come true."

The first seven astronauts, the legendary Mercury Seven, were rock stars in the early 1960s. Rugged and handsome, each one had the swagger of a baseball, hot dog, apple pie, and white-picket-fence American hero. They had been specifically selected to be strapped inside the pointy end of a ballistic missile and launched toward who knew what, hopefully before the dreaded Soviets. They were Big Al, Gus, Scott, Wally, Gordo, and Deke.

And John, it was impossible to forget John. Alan B. Shepard Jr. was the first American in space—if you do not count the monkeys who flew before him—and he was able to parlay the honor into great fame and fortune. The suborbital hop lasted barely fifteen minutes, and the tiny capsule splashed down just 302 miles from where it began. Two flights later, John H. Glenn Jr. became the country's first person to orbit the earth on 20 February 1962. In just under five hours, Glenn made three laps of the globe while crammed into the cabin of his *Friendship 7* capsule. The straight-as-an-arrow marine was greeted by another ticker-tape parade and trip to the White House to meet with President Kennedy, becoming in the process the poster child for good, old-fashioned American values. He became so famous and valuable an icon that his path back into space was blocked by the White House. Glenn asked to get back into the flight rotation every month or so, and every month or so, he was given the brush-off. It was not until years later that he discovered the reason. Glenn recalled:

President Kennedy had said that he would just as soon I wasn't used again for a while, and I wasn't aware of that at the time. I guess after my flight there had been such an outpouring of national attention. It's sort of hard to comprehend the attention we had and the—I guess "adulation" is the only word that comes to mind, but that's what it was. That's not very humble of me to say that, but that's exactly what it was. I don't know whether he was afraid of the political fallout or what would happen if I got bagged on another flight. I don't know what it was, but apparently he didn't want me used again right away.

Not content to stand idly by and watch as his colleagues were strapped into their Gemini capsules, Glenn left NASA two months after President Kennedy's assassination. In early 1964, he withdrew from his first political race after a slip in the shower left him with a badly damaged vestibular system in his inner ear. His career in politics facing the equivalent of a launch pad abort, Glenn accepted a management job with Royal Crown Cola and spent the rest of the sixties with the company. It was not a position that necessarily hinged on his celebrity status, although that obviously did not hurt. He continued making appearances on NASA's behalf and building his contacts within the Democratic party in his home state of Ohio. Glenn lost his first full political campaign, a 1970 primary for the United States Senate, to Howard Metzenbaum. Four years later, rather than run for lieutenant governor as requested by party elders, Glenn went after the Cleveland entrepreneur again. Metzenbaum hinted during the election that Glenn was not up to the task because he had spent most of his career as a government employee—it became known as the "You've Never Held a Job" issue. The former astronaut responded with a patriotic fury during a press conference:

I served twenty-three years in the United States Marine Corps. I was through two wars. I flew 149 missions. My plane was hit by anti-aircraft fire on twelve different occasions. I was in the space program. It wasn't my checkbook, it was my life that was on the line. This was not a nine-to-five job where I took time off to take the daily cash receipts to the bank. I ask you to go with me, as I went the other day, to Veterans Hospital and look those men with their mangled bodies in the eye and tell them they didn't hold a job. You go with me to any Gold Star mother, and look her in the eye and tell her that her son did not hold a job. You go with me to the space program, and you go as I have gone to the widows and the orphans of Ed White, Gus Grissom, and Roger Chaffee, and you look those kids in the eye and you tell them that their dad didn't hold a job. You go with me on Memorial Day coming up, and you stand on Arlington National Cemetery—where I have more friends than I like to remember—and you watch those waving flags, and you stand there, and you think about this nation, and you tell me that those people didn't have a job. I tell you, Howard Metzenbaum, you should be on your knees every day of your life thanking God that there are some men—some men—who held a job. And they required a dedication to purpose and a love of country and a dedication to duty that was more important than life itself. And their self-sacrifice is what made this country possible. I have held a job, Howard.

When Glenn trounced the Republican mayor of Cleveland for the senatorial seat, he was on his way. He was elected to three more six-year terms of office and, along the way, became an influential legislator in Washington. He ran unsuccessfully for the Democratic nomination for president in 1984, beaten back, ironically enough, by Walter Mondale, who had been a fierce critic of NASA and its human spaceflight program.

All along, the goal of returning to space remained despite his advancing age. It was not a pipe dream to Glenn, something to joke about at dinner parties. In 1996 he came up with a plan: rather than jockey for a sight-seeing experience on the shuttle, he worked the science angle. Astronauts in the normal thirty-to-fifty age range experienced body changes during spaceflight. Something somewhere along the line got tweaked in their cardiovascular and immune systems. Their bones weakened. The same thing happens to the elderly—Glenn had his pitch. He was more than seventy years old and had spaceflight experience. Why not send him up for an extensive battery of tests into the matter? He had no problem whatsoever with being a guinea pig, if it meant getting another flight. Glenn insisted:

This is not something just to give John Glenn another flight. This is something to really start a new area of research that I think can be very, very important. That's the reason it's so fascinating to me. Much as I'd like to go up again and just joyride around, we don't have the luxury in our spacecraft yet of just letting people go up just to get the view. This is an area of very, very proper research that has the potential to it of enormous benefit for—well—the graying of nations, they call it, all over the world. Our average age is getting older and older, and if we can learn some of the things by starting a program like this now, it could be tremendously beneficial, I think. That's what makes it more exciting even than just looking forward to going up again from a personal basis.

On 16 January 1998, NASA administrator Dan Goldin announced that Glenn would be getting his wish and flying again. Less than a month later, the rest of the STS-95 crew was outlined as well. Joining Glenn were commander Curt Brown, pilot Steven W. Lindsey, and mission specialists Scott Parazynski, Steve Robinson, ESA astronaut Pedro Duque, and fellow payload specialist Chiaki Mukai. Goldin had been impressed with the sheer force of Glenn's will. "Unbelievable," Goldin said. "I'd call him tenacious to the second power, to the third power, today. He just keeps moving forward. He's intense. He's deeply committed to this mission. He studies longer and harder than people I've met in my life." Goldin laid down the law before assigning Glenn to his coveted spot on the flight. He would not and could not fly if it was deemed not safe for him to do so, so that meant Glenn could not fly unless he passed the most rigorous of physicals. And, finally, he could not fly unless the National Institutes of Health found significant scientific merit in the work that Glenn would be doing. It was on this last issue that the endeavor found its greatest criticism. Surely, cynics gossiped, this was a political payback from American president William J. "Bill" Clinton, thanking Glenn for his long years of service—of *Democratic* service—in Congress. They said research was a secondary concern. Goldin responded to the criticism without reservation:

I invite the critics to go through the ten peer-reviewed science projects Senator Glenn is going to perform. He is performing ten experiments—depressed immune system, muscle loss, bone loss, sleep disorders. He's doing two of the three sleep disorder experiments. The next largest number of experiments will be done by Steve Robinson, six. There's a broad range. Think about immune system suppression. When you get older, stresses bring viruses in your body that have been suppressed for years out. Same thing may happen in space. We're going to study what happens to the Herpes virus, and maybe we'll learn a little bit more about what stress-induced activities do to the virus activity in the body. This is important. This is important.

What happened next was an interest in all things NASA in general, and in Glenn in particular, that far surpassed most anything else the agency had done in decades. This seemed to be a bigger deal to the average Joe on the street than Hubble or the newfound cooperation with the Russians on *Mir* and the upcoming International Space Station.

Some four thousand reporters were credentialed for the launch, which took place on 29 October 1998. President Clinton was there, as were seventy members of Congress. The incomparable longtime CBS News anchor Walter Cronkite came out of retirement to see his friend off on the adventure, and sat in on CNN's extensive coverage. That kind of avalanche of attention could have woefully impacted the mission long before it ever left the ground. Brown was not the sort to allow such a thing to happen, and neither was Glenn. As a whole, the group was shielded from much of the spotlight's glare by NASA's public affairs branch. "I think that public relations protected us quite well," Parazynski said. "Curt is a no-nonsense guy, so there were certain rules of engagement and certain limitations in terms of availability for doing those kinds of things." More than a decade later, Robinson remembered:

I think Curt was very professional about it. We had some very early discussions as a crew. He brought the crew together and said, "We're going to have a lot of undue attention, but we have a job to do, a difficult job to do, a dangerous job to do. It's part of our professional requirements to support the amount of attention. We need to have a professional bearing and reaction to the public eye. However, first and primary are the mission objectives. Every member of the crew, including Senator Glenn, has responsibilities pertaining to the mission objectives." I thought it was a very professional approach. In the end, it worked. We did not really allow ourselves to get distracted by any of that.

Glenn tried his best not to come onto the crew as the hero of the legendary Mercury Seven, one of America's original astronauts, the first American to orbit the earth. He was just John, as much as it was possible to be. He did not attempt to corner the market on the attention being lavished on the flight—in all honesty, that is the last thing John Glenn needed to do. At least in the beginning, there might have been some nervousness among the rest of the crew, some awe, maybe. When Robinson was a child, he had won second place in an art contest with a drawing of the Atlas rocket that took Glenn on his first journey into space. As payload commander, it was Robinson's job to sit down with Glenn and discuss his duties. "I was very nervous," Robinson admitted. "In a way, I was his supervisor. I had to make decisions about what he was going to do on orbit. I'm thinking, 'Me?!?' I kind of practiced what I was going to say to him." Robinson started. There was so much to do, with more than eighty experiments on board, every member of the crew would have to pull his or her own weight. That included Glenn, who would be assigned tasks with little or no say in the matter. "He basically said, 'Well, of course. That's the way it works,'" Robinson said. "His response in training was just fantastic. He took on whatever we asked him to do."

Once, Glenn came close to balking at taking on responsibility for a scientific payload that required a laptop computer to operate. He told Robinson that he usually left "the computer stuff" to younger people. "I said, 'Well, sir. I can't really give you a choice in this. You *have* to do this,'" Robinson said, smiling at the memory. "He said, 'Okay.' I don't think it was too long after that when I had the training people calling me up, saying, 'Would you get this guy to quit asking for more training?' He just went after it. By the time we flew, he knew more about that computer than I did."

As hard as Glenn trained, there was always another hand to shake, another question to be answered, another dinner to be interrupted. While at the

10. Astronaut-turned-senator-turned-astronaut-again John Glenn coaxes President Bill Clinton (left) into sampling the kind of food that would be taken on the flight of STS-95. Curt Brown, the second-best commander Glenn ever had during spaceflight, looks on. Courtesy NASA/Joe McNally, *National Geographic* photo.

Cape for the flight's Terminal Countdown Demonstration Test—its dress rehearsal, if you will—the crew headed to a local restaurant for dinner. Heads turned as Glenn walked in, and the whispers began. Soon, there was a line to meet him. "He was so gracious about it. He took time for every single person. It actually drove Curt Brown nuts, which made us all very happy. We thought his reaction was really funny," Parazynski said with a laugh. Through it all, Parazynski took from the experience an important lesson:

If people are interested enough to wait in line to meet you, to talk about the space program, America's leadership, the astronaut corps and whatever else, always take the time to spend with them. If they want an autograph or photograph, they're American taxpayers who are genuinely interested. I'm going to be genuinely interested in them, too. That's just one of many examples of John's grace. He was such a superstar—he still is—but at the time, it was in the news everywhere that he was going back into space. He was an ultra-celebrity, and he still managed to take time for people. That's a good lesson for all of us.

Glenn might have been a superstar, but his crewmates could not resist the temptation to gig him once in a while. As mission managers were dropped off at the Launch Control Center on the way to the pad, they asked for the crew's boarding passes. Six of the seven *STS*-95 crew members—everybody except Glenn—had a laminated card ready to present. One by one, out of the shoulder pockets of their pumpkin suits came the boarding passes. Glenn looked in his pocket . . . nothing. He looked in another . . . still nothing. It was only after they had made Glenn sweat for a moment or two that he was let in on the gag.

Then, during the flight, there came an appearance on *The Tonight Show with Jay Leno*. The big-chinned host asked Brown, "Does Senator Glenn keep telling you how tough it was in the old days, how cramped it was, how small it was, how lucky you young punks are?" The commander brought the house down with his perfectly deadpanned answer, "Well, Jay, actually, no, he doesn't *always* do that—only when he's awake." As they rode out the wave on their post-flight media blitz, the crew heard each other's talking points so often that the rest could almost repeat them back verbatim. At their last stop together, they did just that. Glenn got up and started in on a delivery he had made many times before. Soon, he was joined in his speech by the rest of the crew. "When it came Senator Glenn's turn to say his spiel, we all kind of butted in and said it all together, word for word," Robinson said. "He just sat back, smiled, and let us do it."

As it turned out, Glenn would not be bested. These young bucks might have thought they were funny, but Glenn was not about to take their ribbing sitting still. Parazynski and Robinson both could not help but chuckle over the time Glenn fired a shot across Brown's bow. It was simple enough, but spoke volumes.

"You know, Curt," Glenn began. "You're the *second* best commander I've ever had on a spaceflight."

Game, set, and match, John Glenn.

Not since the Soviet Union's Valentina Tereshkova and the darkest days of the Cold War had a woman been in command of a spaceflight.

In June 1963, Tereshkova made forty-eight orbits of the earth during a nearly three-day solo flight on board *Vostok 6*. It was a historic journey, but one that was not repeated by another female for almost two decades—Russian Svetlana Savitskaya flew in 1982 and two years later became the first of her gender to do an EVA. Not until 1983 and the flight of STS-7 did Sally K.

Ride break down the space barrier for American women. Others flew with distinction, but as mission specialists. Yet with the exception of a brief few minutes of capsule orientations by Tereshkova, a woman had never actually flown a spacecraft. On the shuttle, the two forward-most seats on the flight deck, those occupied by the commander and the pilot, had without fail gone to men. Eileen M. Collins's selection as a member of "The Hairballs" ASCAN class in 1990 signaled the beginning of the end for the unwritten policy.

Collins had served as an air force instructor on the sleek T-38, commanded crews on the gargantuan C-141 Starlifter, taught mathematics at the air force academy, and was in the process of graduating from the branch's test pilot school when she was named to the astronaut office. There was fanfare when she flew as pilot on two missions to the *Mir* space station, but nothing like what she encountered after the announcement that she would command the flight of STS-93 to deploy the Chandra X-ray Observatory. When she got word of the assignment from George Abbey and Jim Wetherbee, her commander on STS-63 who was serving at the time as JSC's deputy director, Collins could not possibly forget her first act as a shuttle commander. "They said, 'Oh, by the way, you have to go to the White House in a couple of days. The First Lady wants to announce it,'" Collins remembered. In Washington, she met with President Clinton and his wife, Hillary, in the Oval Office before heading across the hall to a formal press conference. For once in her life, the knees of this steely-eyed jet-jockey-turned-astronaut went wobbly:

When you leave the Oval Office and you go into the Roosevelt Room, there's a hallway. I was in that hallway, and this has never happened to me, but I panicked. This little, I guess, panic feeling came over me and I went, to myself, "What the heck am I doing here?!? I'm not kidding. What am I doing here?" Then, my more calm side said, "You can be Eileen later, but just go in right now and be the woman commander." I wanted to go home to my kid and be myself. It was just kind of strange. I learned from that incident that I could step in and out of being the woman commander. It's not something I have to be all the time.

Collins needed every bit of that commander persona during the launch attempts of STS-93, the first of which came on 20 July 1999, the fortieth anniversary of the *Apollo 11* moonwalk. Just as the sparkling igniters were going off at T-minus eight seconds, a primary hazardous gas monitor indicated a hydrogen concentration of more than twice the allowable limit in an aft engine compartment. A controller in KSC's Firing Room 1 manually halted the ground launch sequencer—a program that controls the final moments of the countdown—a tenth of a second before the command to start *Columbia*'s main engines. There had been no time to check a backup system, which did not show any sort of hydrogen spike. When primary readings returned to normal within seconds, it proved to the shuttle's technical community that there had not been a leak after all. "I'd never seen anything like it," Collins continued. "Normally, in the simulator when you have a scrub at T-minus nine seconds or whatever, you'll have a stop. In the shuttle, the clock got down to eight seconds, and it went T-minus eight, nine, eight, nine, eight, nine. It kept cycling back and forth. I'm sitting there going, 'What the heck is going on?'"

It came *that* close to becoming the sixth and final launch pad abort of the Space Shuttle era.

Because the three main engines had not actually roared to life, they did not need to be changed out, and managers called for just a two-day turnaround. After another scrub for poor weather on 22 July, STS-93 finally left pad B late the next night. The last-second stoppage a few days earlier had been the stuff of heart-pounding excitement, but it was far from over. About five seconds after the start of main engines and just before liftoff, a small deactivation pin ejected from a liquid oxygen main engine post and struck three hydrogen cooling tubes on the right main engine, rupturing them all. That caused a small, but visible, leak of about four and a half pounds per second. *Columbia* was about to run out of gas, its three main engines shutting down fifteen hundredths of a second early. STS-93 made it to orbit, but fifteen to sixteen feet per second slower than expected and about eight miles lower.

"Hydrogen leaked all the way uphill," Collins said. "The engines shut down early, and fortunately, it was just a fraction of a second, so we were able to make up our altitude difference with our orbital engines." Although Collins would later call it "quite serious," the crew did not learn of the issue until it received photos of the leak from the ground.

Problems plaguing Collins's first ascent as a shuttle commander did not end there. Five seconds *after* launch, a half-second short in an electrical bus caused the shutdown of the primary controller on the center main engine

11. The Chandra X-ray Observatory rests in *Columbia*'s payload bay prior to the launch of STS-93. The largest payload ever deployed by the Space Shuttle, Chandra had an eventful journey to orbit. Courtesy NASA.

and of the backup controller on the right. Had the backup not kicked in on the center engine, it would have failed outright. As it was, all of its data was lost during ascent, while the right engine was without control and redline protection. Rapid-fire audio from the mission control center in Houston left no doubt as to how serious the issue was. For Collins and pilot Jeff Ashby, it was as if they were doing just another intense training exercise in the simulators. Remarkably, the engines' actual performance was right on target. After STS-93 returned safely to Earth, the dramatic electrical short was traced to a wire rubbed raw by the head of a screw in a midbody bay. "It was a huge concern and a major malfunction," Collins admitted. "Fortunately, the shuttle's built with a lot of redundancy, so we didn't have an impact to the mission."

A little more than seven hours into the flight, Catherine G. "Cady" Coleman deployed the 50,162-pound Chandra X-ray Observatory, the largest payload ever carried to orbit by the shuttle. "The crew just forgot about the ascent problems," Collins concluded. "Actually, the hardest part of that mission was not deploying Chandra. That went perfect. The hardest part of that mission was doing all the secondary science experiments. There were just so many of them, it was a challenge."

With the exception of the STS-99 Shuttle Radar Topography Mission in February 2000, only one more flight devoted solely to research remained before the program focused its efforts on construction of the International Space Station and servicing missions to the Hubble Space Telescope—the ill-fated voyage of STS-107.

3. Hubble Huggers

To fully grasp the grand achievements of the Hubble Space Telescope, it is first important to comprehend just how closely the project came to being forevermore known as an abject failure.

For nearly two decades before its initial deployment, Hubble consumed many within the agency's science community. Measuring 43.5 feet tall and weighing in at twenty-four thousand pounds, the multi-billion-dollar instrument was monumental not only in sheer size but in potential as well. This was to be the creation that would help answer the questions of Creation: When did galaxies form and how did they evolve? What about black holes—do they actually exist? How old is the universe? How did all this begin in the first place? For decades, astronomers had discussed such mysterious phenomena as dark matter, quasar engines, and white dwarf stars. Now, the hope was that Hubble would allow them to actually see such things. Apollo took us to the moon, and Hubble was poised to take us light years beyond that.

Coming out of the darkness of the post-*Challenger* era, Hubble was precisely the shot in the arm NASA so desperately needed. A media circus turned up at KSC for the 24 April 1990 launch of STS-31 to deploy its bus-sized payload. Weeks later, that same media contingent was again present, but in a much darker kind of frenzy. Word had spread that Hubble was beaming photos back to Earth that were very nearly as fuzzy as those taken with a five-dollar disposable camera, and just about as useful. Any goodwill that had existed was instantly erased by a near-sighted boondoggle. Without a subsequent—and successful—servicing mission, Hubble would be not much more than NASA's orbital tombstone.

Long before either ever actually came into existence, the telescope had been developed with the shuttle in mind. Instruments were designed to last only two or three years, then to be replaced by the latest and greatest in astronomical technology. To accomplish these tasks, a suitable vehicle was required to transport a crew to capture, service, and then redeploy the mammoth instrument. Dr. Edward J. Weiler, at the time the chief scientist of the Hubble program, never fully recognized just how important the shuttle was in the grand scheme of things until the situation was at its most dire. "Hubble," he stated, "would've been canceled within a year or two after the launch if we didn't have the option of going up there and fixing that optical flaw. It would have been a billion-dollar embarrassment to the agency. It's not an overstatement to say the shuttle saved Hubble. Without the shuttle, there'd be no Hubble today. There would have been no Hubble in 1993."

Three and a half years after Hubble's ignominious outset, the crew of STS-61 saved the day and NASA's teetering reputation on the first of five servicing missions. There were plenty of other capability failures in later years, but almost all were well within the telescope's design tolerances. Anomalies cropped up during EVAs on the servicing missions, but, again, they allowed the astronauts and mission control to display the kinds of problem-solving skills the agency had demonstrated since its earliest days. Once past the horrific embarrassment of the initial screw-up, Hubble turned out to be one of the shuttle's most significant triumphs.

John Grunsfeld had nurtured a deep and abiding passion for all things to do with astronomy almost since birth, and he flew three of Hubble's five servicing missions. He became known as the "Chief Hubble Hugger," and he now calls the telescope "the most significant scientific instrument ever created by humans." To hear the conviction in his voice as he described Hubble's profound effect on the science community was quite convincing:

Its impact on the scientific fields that it was designed for . . . its impact on the general person-in-the-street's knowledge of the universe that we live in . . . as an icon for science . . . as an inspiration for students to study science and engineering . . . and also as a brand name for NASA . . . it has had the farthest reach of anything we've ever done. We've suddenly gone from these first views of what's out in the cosmos, to having a very detailed, data-driven picture of everything from a few hundred million years after the creation of the universe all the way to the present. We've learned about the formation of galaxies, stars, the chemical elements we're made of, and black holes. We've discovered dark energy, dark matter, how the planets work, and what our solar system looks like. The bulk of that knowledge has come in the last twenty years with the Hubble Space Telescope. The Hubble isn't just down the street or on a mountaintop. It took this fantastic space ship—the Space Shuttle—to enable all that.

The concept of a space-based telescope was first floated in 1923 by Austro-Hungarian-born German physicist and engineer Hermann Oberth, widely considered one of the founding fathers of rocketry and astronautics. More than twenty years later, American astrophysicist Lyman Spitzer Jr. began what would become a lifelong quest to make just such a thing reality. Never mind that any kind of a man-made object, much less an intricately designed telescope, in space was still more than a decade in the future. NASA approved the not-so-imaginatively named Large Space Telescope in 1969, and in 1974 designers came up with a plan that would one day drag the project back from the brink of oblivion—why not design the telescope so that its parts would be interchangeable by spacewalking astronauts who arrived via the shuttle? The European Space Agency came on board in 1975, and two years after that, Congress approved funding for the project. The project was named after Missouri-born astronomer Edwin Powell Hubble, who confirmed that our own Milky Way was just one of many galaxies in the universe.

After the 1975 flight of the Apollo-Soyuz Test Project, NASA was temporarily out of the human spaceflight business. Those that were left in the astronaut office turned their attention to what came next, the shuttle, and at virtually the same time, Hubble. Astronaut Bruce McCandless II had already taken part in NASA's grandest achievement to that point, serving as CapCom while Neil Armstrong and Buzz Aldrin left their footprints on the dusty plains of the Sea of Tranquility. Nearly fifteen years later, in February 1984, McCandless became the subject of the shuttle program's most iconic photograph. Snapped by STS-41B crewmate Hoot Gibson, the shot captured McCandless during the first untethered test of the Manned Maneuvering Unit (MMU) floating free some 320 feet from *Challenger*. In between chatting it up with Armstrong and Aldrin and his own first venture into the great unknown of space, McCandless was heavily involved in drawing up plans for the design and servicing of Hubble.

In August 1978 McCandless took a call from George Abbey. "He said something to the effect of, 'Those people up in Huntsville [at the Marshall Space Flight Center] are trying to build an EVA-repairable telescope. I want you to go up there and straighten them out,'" McCandless recalled. "I interpreted that as meaning, 'to help them out,' although it's possible in light of later conversations that what he really wanted me to do was go up there and tell them how stupid it was and help kill the whole thing." McCandless subsequently worked on Hubble off and on for more than a decade, until being named to the deployment crew of STS-31.

On trip after trip, astronauts ventured to Lockheed Missiles and Space Company in Sunnyvale, California, in order to gain hands-on experience with a Hubble mock-up. The challenge early on was basically two-fold. First, make as many integral pieces of the telescope as possible either replaceable or repairable, and second, the tools necessary to accomplish those tasks needed to be designed.

Design work on Hubble was very much trial and error, right down to the nuts and bolts that would hold everything together. Seven-sixteenths-inch, double-height bolts with hexagon-shaped heads were standard on Hubble. That may not sound like much, but it enabled the crew to work with a single-sized ratchet rather than fumble around with various sockets like so many millions of shade-tree mechanics back on Earth. One set of tools was devised, and then another. Finally, on the third try, engineers settled on a workable tool solution. Plug-in portable foot restraints were developed. Wing tabs were placed on electrical connectors. Boxes were labeled with inch-high lettering so EVA astronauts would not have to guess what was inside.

What McCandless called "a war" was fought with ESA and British Aerospace over Hubble's solar arrays, the telescope's twin wings that contained forty-eight thousand solar cells to convert the sun's energy into electricity for the instrument. Original specifications called for that crucial component's ability to be jettisoned, but not necessarily replaced. "You wouldn't believe the amount of blood, sweat, and tears over that before we finally prevailed," remembers McCandless. "I will, with a bit of hindsight, say luckily so, because we're on the third set of solar arrays right now. If we'd had to come home after the first set failed, it would have been a pretty short mission."

In some places on Hubble, clearances were a microscopic fifty-thousandths of an inch. "We had to pretty much jury-rig as we went along, on an individual basis, how these things would be held in place," said McCandless, who went on to add that in the end, "we were very proud of ourselves. When we got ready to launch, we figured that everything on the telescope could be replaced, repaired, or upgraded, except for the main wire harness, because it went everywhere, and the primary and secondary mirrors. Of course, you'd *never* have any problems with them."

Hubble's first projected launch date in 1983 turned out to be as unrealistic as the telescope's original $400 million price tag. Hubble's debut in space was subsequently pushed back to 1986, and even then its guidance software was going to be iffy at best. Officially announced on 19 September 1985 for a launch in August of the following year, the originally dubbed STS-61J deployment mission featured John Young as commander, with Charles F. Bolden Jr. as pilot and Kathryn D. Sullivan, Steven A. Hawley, and McCandless on board as mission specialists. The *Challenger* tragedy changed everything.

Hubble was placed in high-tech mothballs—a huge clean room at Lockheed, where it was regularly powered up and purged of nitrogen to the tune of some six million dollars per month. On the crew side, Young was reassigned in April 1987 to the post of special assistant to the center director for engineering, operations, and safety. Officially, according to a NASA press release, he remained "eligible to command future shuttle astronaut crews." Then, on 17 March 1989, came another NASA press release announcing the crew of STS-31 to deploy Hubble—with Young's name noticeably absent. Instead, Loren Shriver was appointed mission commander. Although Young would never again rocket into space, he maintained a strong presence within NASA's hallowed halls until his retirement from the space agency in late 2004. In fact, it was Young who called Shriver to ask if he was interested in taking over command of the Hubble crew.

Shriver's first spaceflight assignment had been as pilot on STS-51C, NASA's first DoD mission in January 1985. Preparations for that flight had been so secretive, crew members were initially not permitted to invite guests for the launch. There was also speculation that mission specialist Gary E. Payton—an air force officer on an air force DoD mission—had been added to the crew specifically to prevent others on the flight from gaining any more knowledge of the flight's payload than was absolutely necessary. Nobody could say anything about STS-51C, but NASA was very busily telling the whole world about Hubble. Shriver, to be sure, noticed a distinct difference between the two flights:

If there were ever two missions that were completely opposite in terms of the public attention that was given to them, it would be my first and second missions. The first one [was] a DoD [flight]. Not that there weren't some people interested in what I was doing, but I just couldn't say anything about it. Now along came the second with the Hubble Space Telescope deployment. Of course, it seems like sometimes everybody in the world was interested in that and what it would be and what it could do. There was a lot of publicity surrounding the mission as to the anticipation of the astronomy community for years and years to get a space-based telescope they could use, and finally it was about to happen.

Ed Weiler began work as chief scientist on the Hubble program in 1979, and he remembered a profound sense of excitement about the project as the launch date approached. This was going to be a huge feather in the agency's cap:

We knew we'd be taking a step in astronomy as big as the step that Galileo took in 1610. For the first time in human history, instead of using the human eye, Galileo put a telescope in front of his eye. And in one giant leap for mankind, so to speak, he increased the capability of the human eye by a factor of ten. Astronomy increased greatly after that, but they did it in little steps . . . a factor of 20 percent better, 30 percent better. For the first time since Galileo, astronomy would be taking another one of those great leaps, not just baby steps, of increasing the capability of ground-based telescopes by a factor of 10. As it turned out, it was more like a factor of 50, because Hubble performed much better than we ever thought it would.

After a scrub on 10 April 1990 due to a faulty valve in an Auxiliary Power Unit, STS-31 launched on a sunny Florida morning. On the morning of the flight's second day, the crew went to work on deploying their precious payload. Hawley gingerly lifted Hubble out of *Discovery*'s cargo bay, but in doing so, he encountered the somewhat unexpected:

For a massive payload, in particular one that's going to be in proximity to the shuttle, what you're worried about is a very, very remote failure case that the arm could fail in such a way that it drives by itself. The operator is always there watching, but, obviously, if it's close to structure, you may not have time to react. So they intentionally limit how fast the joints will drive so that they give you some time to react. Hubble being at the time the biggest payload we had ever deployed with the arm, the amount of motion you could command was limited, because it was designed to protect for this failure case, at least while you were close to the payload bay. So what we found as we started to lift Hubble out of the payload bay was that it didn't come straight up out of the bay. It seemed to wobble around a lot more than the simulator had predicted, and that was a bit of a challenge.

There had been considerable pre-launch debate concerning how high Hubble could—or should—be flown. After a 305-second burn by the Orbital Maneuvering System shortly after launch, *Discovery* was ultimately parked at an orbital altitude of nearly 311 miles. Hubble was released into its own station above Earth some 20 nautical miles higher than the carrier shuttle. By comparison, the highest altitude achieved by STS-32, two flights before, was "just" a little more than 178 miles. That's not all. Fifty percent of the flight's Orbital Maneuvering System propellant was spent getting into orbit, and another 49 percent would be needed for its de-orbit burn. "If you looked up at the key on-board shuttle fuel gauges, the moment you got there, they were reading 49 percent," Sullivan said. "You've still got five or six days to go, and you're already through half your propellant. Any indication of a leak and you're getting out of there fast, or we don't get to come home and talk to you about it."

It was time to get down to business. Hubble's first set of solar arrays unfurled without a problem, but the second unexpectedly stopped midstream. Computer software designed to sense binding in the unfurling mechanism was incorrectly stopping the process, and it left the mission in a quandary. Without the second power source, Hubble would be dead on arrival. After trying unsuccessfully to restart the array's motion, McCandless and Sullivan were sent downstairs to prepare for a contingency EVA in which they would have attempted to crank the array out by hand. They were suited up and the airlock had been depressurized down to five pounds per square inch when controllers at Goddard Space Flight Center came up with a work-around for the computer glitch. The would-be spacewalkers stood by, sealed in the airlock, as the software fix was implemented. If there was subsequent grumbling in some corners about leaving McCandless and Sullivan in the lurch down in the airlock, the end result was the same—the problem was solved and the mission to deploy Hubble was successful.

Flight director William D. "Bill" Reeves made the call and he stuck with it:

Every time I see Kathy she still points her finger at me and accuses me of locking them up in the airlock, but it's all in fun. She knows full well we did the right thing. The problem is when you're getting ready to send the crew out the door, you depress the airlock from cabin pressure down to five pounds per square inch pressure. Then you stop and do a check on the suits to make sure the suits are working. Then you continue to depress down to vacuum and then open the door and go out. They were at that five psi check when we stopped them. We just left it there. The reason I elected to do that was because once you go to vacuum and once you open that door, you have opened yourself up to all kinds of issues. One of the things you learn as a flight director early on—and you're taught that from all your predecessors and everything else—is don't do anything that can make your situation worse. You've got to be careful. It might sound like it's no big deal to go do it, but you got to be constantly thinking about the consequences of what you're about to do. Every decision you come up with, in the back of your mind is, "If I don't do this, what's the worst thing that can happen to me?" You've got to keep making those decisions in your head. You have to do the right thing. To this day, if I had to do it over, I'd do exactly what I did. I could have gone ahead and said, "Go to vacuum," had the crew go outside, and then we could have turned around and had a problem getting the hatch closed and getting the crew back in the airlock or getting the airlock repressed—a whole new set of problems.

Hubble was released on *Discovery*'s twentieth revolution of Earth, one orbit later than planned. After station keeping with Hubble from a distance of about forty miles for a couple of days, the crew of STS-31 returned home on 29 April 1990. The strange thing was, no one directly involved in the deployment flight itself could have done anything to prevent what was about to take place. They had done nothing wrong. Reeves would later rue the fact that Hubble's subsequent problems, and the resulting high-profile missions to repair them, took the bloom off the rose of a mission that had taken the telescope into space in the first place:

That really took a lot of the satisfaction away from the deploy team and the team that put it up there. As a matter of fact, all of the focus on Hubble has always been on the repair missions and the repair of the telescope and everything else ever since. I've always been a little put out about the fact that the deploy team kind of got left out of the hurrah, because we got it up there. We did our job. We got the thing deployed, and we did it the way it was supposed to be done. We didn't build the telescope. We weren't responsible for the mirror. It was just kind of sad it had to work out that way. It was a great accomplishment, and one of the highlights of my career was doing that.

Hubble was up there, all set to start snapping pictures of the universe. The first ones, hot off the press, were not pretty.

Ed Weiler could not wait.

For more than a decade, the Hubble Space Telescope had been the greater part of his life's work. He was not alone. As STS-31 left the launch pad, there was no way to oversell the science community's excitement. Astrophysicist Dr. John N. Bahcall of the National Academy of Science and newly elected president of the American Astronomical Society told the *New York Times* a couple of weeks before the liftoff, "I'll be drunk with excitement. We'll be looking at things people have never seen before. Who knows what surprises the deity has in store?" Yet almost as soon as Hubble was online and beaming pictures back to the ground, it was clearly evident that something was amiss with the blurry images. A certain dialing-in period was to be expected, but then hours turned into days and days turned into weeks with little to no improvement. Hubble was screwed up, and it had been long before it ever left the ground.

On 21 June 1990, a little more than two months after Hubble reached orbit, a press conference was held to announce to the world that NASA's highly touted piece of machinery was not working correctly because of a problem with its mirror. Overnight, Hubble became a national joke, if not an outright embarrassment. Pundits had a field day. Late-night talk show host David Letterman did one of his famous top-ten lists with the near-sighted telescope squarely in his sights. His excuses for the Hubble fiasco included:

10. The guy at Sears promised it would work fine.
9. Some kids on Earth must be fooling around with a garage door opener.

8. There's a little doohickey rubbing against the part that looks kind of like a cowboy hat.
7. See if you can think straight after twelve days of drinking Tang.
6. Bum with squeegee smeared lens at red light.
5. Blueprints drawn up by that "Hey, Vern!" guy.
4. Those damn raccoons!
3. Shouldn't have used G.E. parts.
2. Ran out of quarters.
1. Race of super-evolved galactic beings are screwing with us.

In some corners there was despair and in others a furious anger. Weiler fully admitted that his "depth of depression was unspeakable." He added, "I won't say it's quite like losing a child, but it's not much better. For most of us, we'd spent ten, twenty years of our lives on Hubble. The guy who thought up Hubble in 1946, Lyman Spitzer, that was already thirty-four years of his life. To go from being a national hero being involved with Hubble to walking around your neighborhood pushing your little girl in her stroller and having neighbors come up to you and say, 'Gee, we really feel sorry for you having to work on such a national disgrace,' that kind of gets to you."

If Weiler was a case study in the sadness that initially enveloped the Hubble debacle, Story Musgrave might best represent the rage. He had worked on the program since 1975, two years before funding was officially approved by Congress. He worked to develop not only EVA procedures for Hubble but also the tools with which to do the job. And here the telescope was, a dud. Musgrave charged, "It was just an egregious, negligent error. That's all it was . . . just a horrible error. I couldn't believe it. It's like *Challenger*. You weep, but then you just get unbelievably pissed off, just angry at the people—how could they possibly behave that way?"

Such grave mistakes were hard to comprehend on many different levels, and not just for Musgrave. *The Hubble Space Telescope Optical Systems Failure Report* was released in November 1990, and a little more than a year after that, two reporters from Connecticut's *Hartford Courant*, Eric Lipton and Robert S. Capers, wrote a series of Pulitzer Prize–winning stories on Hubble's stumbles. Both the official report and the journalists' investigative reporting offered searing indictments of what exactly had gone wrong.

For starters, NASA's report confirmed that "most, if not all[,] of the problem" rested with Hubble's 94.5-inch primary mirror. Manufactured by the Perkin-Elmer Corporation (later to be known as Hughes Danbury Optical Systems Incorporated), the mirror was essentially the wrong shape, flattened away from the mirror's center by a millionth of an inch. The difference was a tiny fraction of the width of a human hair or piece of paper, depending on who happened to be making the comparison. It was close, yes, but close only counts for horseshoes and hand grenades, as the old saying goes, but *not* for the Hubble Space Telescope's primary mirror. The error was no less than ten times larger than allowed tolerances.

Rough grinding on the glass that became Hubble's primary mirror began in December 1978 at the Perkin-Elmer plant in Wilton, Connecticut, and polishing was completed in April 1981 at a facility in nearby Danbury. Great precautions had been taken—a system to gauge the polishing progress was used only at night to prevent trucks passing the building from disturbing the process. Speed bumps were taken out of the parking lot and the plant's air conditioning was turned off. Yet it was the very measuring system that had been designed to correct any problems that doomed the project in the first place. During the construction of a barrel-sized Reflective Null Corrector (RNC), a measuring laser was bounced by chance off an area in which special nonreflecting paint had worn off. That, in turn, resulted in the RNC's field lens being installed 1.3 millimeters too far from its lower mirror.

Stunningly, technicians tried to correct the problem with three twenty-cent washers. The fix did not work and the delicate work continued, no one any the wiser. An Inverse Null Corrector, "designed to mimic the reflection from a perfect primary mirror," according to NASA's failure report, found fault with the RNC's measurements. A second null corrector, one that turned out to be making correct readings, also clearly showed an error in the mirror. Both measurements were ignored. Perhaps the most glaring error was that no outside testing of the mirror was ever conducted on the mirror prior to flight.

The study released by NASA concluded, "The most unfortunate aspect of this HST [Hubble Space Telescope] optical system failure, however, is that the data revealing these errors were available from time to time in the fabrications process, but were not recognized and fully investigated at the time.

Reviews were inadequate, both internally and externally, and the engineers and scientists who were qualified to analyze the test data did not do so in sufficient detail. Competitive, organizational, cost and schedule pressures were all factors in limiting full exposure of all the test information to qualified reviewers."

Such inattention to detail is precisely what ruined Musgrave's confidence in the process:

You really want me getting spun out? There was never any testing planned or asked for. It was never part of the plan. It was acceptable in the plan that the instrument that guided the polishing machines to put the curve in would be the same damn instrument which would do the final check, meaning you've not run a test, okay? You've not run any tests at all. The story is so horrible, because other instruments entered into the equation at times and showed them they had the wrong mirror. They didn't listen to those other instruments because they weren't part of the formal process. Doggone it, man, when the coarse ground block of glass came into the final lab for the final polish, it had the wrong curvature. They never asked the question—why would they ship us the wrong curvature? Well, the reason they received the wrong curvature was because they put washers in the [RNC] instrument. The only instrument they were relying totally on, they put washers in. Damn it, you don't behave that way. A high school kid would not have done that.

Hubble very quickly turned into a public-relations debacle for NASA. Just a year and a half or so removed from the resumption of flights following the *Challenger* disaster, such a widely derided blunder was the very last thing NASA needed. No one had died this time, but in a very real sense, the implications were very much the same—if this could not be fixed, it might very well have meant the end of the American human spaceflight program. The agency could have imploded at that point, lost in a war of finger pointing and backbiting.

Instead, what Weiler called a "full-court press" took place to repair Hubble. A strategy panel was formed that included members such as McCandless and Spitzer, who had dreamed of a space-based telescope since the 1940s. Meetings were held between August and October 1990 at the Space Telescope Science Institute (STSCI) in Baltimore, Maryland, and in Garching, Germany, near Munich. A number of options were discussed, including bringing

Hubble back home and installing a perfect backup mirror that had been manufactured by Eastman Kodak Company. Sending an astronaut into the depths of Hubble to add material directly to the primary mirror was mentioned, but planners decided against that as well. In the end the call was made for Hubble's near-sightedness to be remedied by the Wide Field and Planetary Camera 2 (WF/PC2, in NASA-speak pronounced "whiff-pick") and the Corrective Optics Space Telescope Axial Replacement (COSTAR), two instruments that amounted to very expensive pairs of spectacles.

Story Musgrave was announced as the payload commander for fix-it mission STS-61 on 16 March 1992, a full five months before any other crew member. The mission had drawn the plum, if somewhat intimidating, assignment of setting NASA in general and Hubble in particular back on course. It would be every bit as high profile as the earlier deployment flight of STS-31, though for starkly different reasons. When it came to going outside to remedy what ailed Hubble, this was what truly defined the content of Musgrave's character—he was not just interested in playing the music, he wanted to help write the notes. He explained it this way:

I like the art of developing the flight. I like choreographing the dance. If you have a very "mature" flight, "mature" means all you do is learn the checklist. You don't learn anything else. The checklist is what items to enter into a computer or what switch to throw or what knob to turn. You don't learn much else. I'm far more excited about a complex mission where I have to write *the checklist and I have to develop the mission itself. I'd been your lead spacewalker for decades. I picked it up from [*Apollo 9 *lunar module pilot] Rusty Schweickart in 1972. I was the point of contact for all spacewalking issues, hardware, and the whole thing. I don't know if that played into it. I just think when it got time to assign the payload commander, they wanted to get someone who could do the job. I'm a farm kid, so they knew I could do the job. I'm a farm kid . . . I know how to fix stuff.*

Joining Musgrave on the crew were commander Dick Covey and pilot Kenneth D. Bowersox. Jeff Hoffman was designated the flight's lead spacewalker and partnered with Musgrave on three of the mission's EVAs, while Tom Akers and Kathy Thornton took over for the remaining two. Swiss robotic-arm operator Claude Nicollier of the European Space Agency was tapped to handle the heavy maneuvering of Hubble, as well as the delicate touches required to move the flight's spacewalkers about during their EVAS. Each crew member was a veteran in his or her respective position, although from Akers's perspective, he did not think the requirement necessary, even if it cost him a spot on the flight. "The powers-to-be at NASA at the time decided that all of the crew members that went to work on the Hubble telescope needed to be experienced," Akers said. "They didn't want any rookies. Kathy and I both didn't feel that was necessary, and told people that. None of the four of us on STS-49 had ever done a spacewalk, and we went up and probably did one of the toughest jobs that had been done, without even training. We hadn't planned to do it that way, which was a good case for the adaptation of human beings in space."

From the top down, each crew member knew full well exactly what was riding on the flight of STS-61. On his fourth and final spaceflight and second as a commander, Covey was at the controls of a venture that could very well make or break NASA. He had been in a similar position during the flight of STS-26, the first to fly after the loss of *Challenger*, but many of the demands that weighed upon that flight, he said, were over as soon as the crew made it to orbit. This time around, not only was there was a huge telescope waiting to be repaired, but the world was watching and ready to pounce. "The pressure started early on, just in the design of the mission and in going through the training and the preparation for it," Covey remembered. "Then, it continued through the entire eleven days of the mission, going through the precise execution of all the spacewalks that we had. Until we landed, I didn't feel a relief of *any* pressure on STS-61."

So intense was the focus on this flight that for the first time during the shuttle program, an official backup EVA crew member was named on 9 March 1993. Gregory J. Harbaugh was less than two months removed from his own spacewalk during the flight of STS-54, while subsequent knee surgery had left him barely able to get into a training spacesuit. Nevertheless, he went on standby "for some insurance in the event of the unavailability late in the training cycle of any of the four prime EVA crew members," according to a NASA press release.

No one could have known then, but discussions to replace Musgrave were, in fact, held following his brush with disaster on 28 May 1993. During a thirteen-hour equipment test in a JSC vacuum chamber that day, temperatures reached as low as minus 170 degrees Fahrenheit. At one point Musgrave reported that his hands were feeling cold, but about eight hours into the session he mentioned that the pain seemed to have gone away. He elected to continue. At the conclusion of the test, he removed his gloves and, to his alarm, found that seven fingers had been blackened by what a NASA press release deemed "a mild case of frostbite." Although the announcement went on to state that Musgrave's condition was expected to improve quickly "with little impact to training and no impact to the mission," some NASA insiders used the incident to call the veteran astronaut's judgment into question. Many, including Covey, felt that Musgrave's distinctive tell-it-like-it-is personality rubbed some the wrong way and fueled debates concerning his place on the STS-61 crew. Covey's reaction as the mission's commander was two-fold:

It was a combination of things. I felt great sorrow for Story. He had injured himself and he certainly never intended to do that. He was injured and he had to recover from that, and so that was not a fun time. At the same time, I was a little bit frustrated with the idea that, here's Story Musgrave, one of the smartest men I know and one of the most studied in EVA and spacesuits, and he had not recognized [the onset of the frostbite] or aborted the run that he was in at a time when he could have precluded the injury. [Hearing talk of replacing Musgrave] was very difficult. The reasons for replacing him, in my mind, were personal and did not have anything to do with Story's technical ability or his ability to recover from the injury. I resisted any idea that there should be a change.

Musgrave would later call his case of frostbite "insidious," because he had felt comfortable at first and then had absolutely no feeling at all at the very same time his fingers were almost literally freezing off. He said he asked for temperature breaks to warm up, but for whatever reason, that did not happen. In hindsight, Musgrave would later admit that the test should not have proceeded in such drastic temperatures. As a result of Musgrave's predicament, heaters and insulation were subsequently placed in EVA gloves, lessening the risk of such a problem taking place during future flights.

Although Harbaugh did much of the mission's subsequent underwater training, Musgrave was determined not to lose his slot on the crew. He went to Alaska to consult with William J. Mills, a noted frostbite expert. He paid visits to various NASA centers involved in the Hubble reboot. It was a long ten weeks or so before Musgrave passed cold tests that allowed him to remain on the flight. According to Musgrave's biography *Story: The Way of Water*, however, it was another five years before his hands, particularly his right hand, were fully recovered to the extent where he could pick a pen up off a table. Musgrave had been aware of his rumored removal from the mission, and, true to his nature, he did not back down:

It's the same thing as the Challenger *accident. The people upstairs don't know the technology, they don't know what caused the problem, they don't know the fix for the problem, they've never done a spacewalk, they don't know* anything *about spacewalking. In a situation like that, you* hope *that the managers upstairs don't get involved in the process, because they don't understand the process. They don't understand the technology. They're not doctors, either, that know about frostbite. They should not be messing with the process, because they don't know* anything *about it. We sort of heard rumblings. We were afraid that micromanagement might creep in and make a wrong decision. I'm not very replaceable six months out, not with an eighteen-year history. I didn't care if the fingers fell off, didn't bother me. If they fall off, they fall off. I don't care. I'll shorten the gloves. I had no choice. It's May, and I've got a mission at the end of November–early December. I got back in the suit as soon as I got back from Alaska.*

The drama surrounding Musgrave's flight status did little to ease the intense pressure that most everyone within the NASA community was experiencing at the time. Covey knew that if the mission did not succeed in repairing Hubble, "it would have been perilous for the agency and for any of its other programs to survive." The Shuttle-*Mir* program was just over the horizon, and after that, the International Space Station. The U.S. Congress—for once—actually had a valid point. If NASA could not rebound from the bungle that was Hubble's mirror, why fund a cooperative effort with the Russians, or anything else for that matter? If a spacewalk to repair the telescope would not work, then it stood to reason that EVAs to construct the ISS were in serious doubt as well. The domino effect was at full throttle. The Shuttle-*Mir* program would be first to fall, then ISS, and quite possibly NASA itself after that. "I'm not overstating it," Weiler said. "That's the world I was living in. I think if that mission had been a miserable failure, the future of NASA space science would have been in doubt."

Intercenter rivalries could have brought the process to a standstill, but

12. Jeff Hoffman works in the payload of *Endeavour* as Story Musgrave draws close to Hubble on the robotic arm. If the work done by the crew of STS-61 had not been successful, NASA would have been in deep trouble. Fortunately, the flight's efforts paid off handsomely. Courtesy NASA.

they did not. Not this time, not with so much at stake. Beginning in the summer of 1990, the Hubble team came together as a cohesive and determined group. Johnson Space Center worked with Goddard Space Flight Center . . . Goddard worked with Marshall Space Flight Center . . . Marshall worked with NASA headquarters. . . . NASA worked with STSCI. Contractors were involved.

The gravity of the situation went to the highest levels of the United States government. NASA administrator Dan Goldin was called to the White House at one point to discuss the repair mission with President Clinton, and Goldin in turn made sure the repair crew of STS-61 knew exactly what was riding on the mission. During one training session at Goddard, Goldin brought them down for a chat at NASA headquarters in Washington and, in effect, told them that the agency's future was in their hands. No pressure, no pressure whatsoever. Hoffman and his crewmates fully realized what their jobs meant. "It didn't really affect our training or our operations, but we certainly recognized there had been such a furor over this disastrous mistake," Hoffman said. "We knew how important it was to get it right. Having worked on five different flights, we didn't work less hard in training for any of my other flights than we did for Hubble. The difference was that we got so much help. That was the wonderful thing about working on Hubble. Everybody was pulling in the same direction."

"The fact that there was so much negative reaction and being called a national disgrace, that was a real catalyst for something to happen that seldom happens at NASA," Weiler added. "[The] team that formed to put together the fix for Hubble was so united, I've never worked on a better team. Everybody had one clear motivation, and that was vindication." Such solidarity was critical, "and that's an understatement," Weiler concluded. The rallying cry around NASA quickly became "It was Hubble."

It was Hubble, so if the EVA crew needed a vacuum chamber run, they got it.

It was Hubble, so if centers needed to put aside their competitive jealousies to get this thing up and running finally, so be it.

It was Hubble . . .

It was Hubble . . .

It was Hubble, period.

Milt Heflin, who served as lead flight director for STS-61, felt the pressure as well. One day in October, a couple of months before liftoff, he got back to his office in Houston and found a single sheet of paper. Someone—he would never discover who—had copied a page out of the congressional record that said, in effect, NASA's future funding was in jeopardy if the agency was unable to pull off the Hubble repair. He was fine with it, because he knew the team he had assembled on the ground was ready to roll. "I wasn't cocky yet, but I was getting close to it," said Heflin. "I felt we were really prepared." Still, the note got his attention. "It made me think, 'Holy cow! We're doing something kind of important here,'" he added with a laugh.

The rules Heflin established for the flight were important ones. There would be no working problems in the hallways, in the background. They would be solved just as they had in training, through normal processes, in the control room. No one would try to reinvent the wheel when problems cropped up. For instance, before STS-61 ever left the ground, it had already been decided that if one of Hubble's solar arrays would not retract, it would be thrown away. When that actually happened during the flight, there was virtually no discussion of what to do. The solar array was discarded, simple as that. Heflin wanted to establish a cadence to the flight early on, a rhythm. With five EVAs planned, it was critical to get out of the starting blocks strong. Said Heflin:

We were able to do that, and that was probably one of the things that I was most proud of for the team. That was so cool. I mean, we're talking about the Goddard Space Flight Center, the team here in Houston, the astronauts, all the engineering folks. Everybody recognized with all the work we had to do, we had to be smart about how we used our time. I did not want the team to get behind. Going into the mission, when I showed up on console, seriously, I was cocky. I was cocky on behalf of the team. We were loaded for bear. We had trained so damn hard. We had "what-iffed" everything. We were ready to go.

SM-1—the first Hubble servicing mission, otherwise known as STS-61—launched from pad 39B in the early morning hours of 2 December 1993 with shuttle *Endeavour* carrying some 14,400 pounds of servicing gear, including more than 200 separate tools and crew aids. On flight day three, Nicollier grappled Hubble into an upright position in *Endeavour*'s cargo bay. The next day, Musgrave and Hoffman headed out the airlock an hour ahead of the timeline to begin the first of the flight's five EVAs.

After they first installed a cover on a low-gain antenna to avoid damaging it, Hoffman very carefully fed Musgrave into the depths of Hubble to replace two Rate Sensor Units, each of which contained a pair of gyroscopes. The maneuver saved time, because doing so meant they did not to have to remove Hubble's sunshade. In nearly forty hours of EVA time during the flight of STS-61, Musgrave remembers "almost no surprises," although there were some memorable exceptions:

Installing the rate gyros, I came up with a procedure that would save an hour by going inside the telescope. Jeff would grab me by the boots and feed me in.

I'd get in position with arms overhead before I got in. Over eighteen years, I developed procedures to remove components to be able to get access to the rate gyros. In all my practice, I was able to see these holes where the pins for the gyros would go into. When I got there for the real thing, I couldn't see them. I didn't have the visibility. To this day, I don't understand why . . . whether the helmet visors are different in the water tank. Were the suits different? It's not a flight suit that you wear in the water tank. That was a little disconcerting, because I couldn't see the holes. I had to feel for them.

The biggest surprise came at the end of the first EVA, when Hoffman attempted to shut Hubble's ten-foot-tall doors. Of all the things that could have gone wrong, latches at the top, middle, and bottom of the doors simply would not catch at the same time. If it seemed to be a minor inconvenience, it most definitely was not. It was the smallest of problems—what should have been a "no-brainer," Hoffman remembered—that turned out to be one of the flight's biggest headaches. Without securing the opening, thermal control would quickly have become an issue, allowing stray light to distort the telescope's imagery even more than it already was. This was a job that had to get done, and the quicker the better. "I thought to myself, 'Oh, damn . . . the very first task and already we've run into a major problem,'" Hoffman said. When it became apparent that Hoffman would not be able to work the glitch on his own, Musgrave free-floated over and the two went to work. Photographs were sent to the ground, and the two astronauts described as best they could the situation at hand.

At times, a sense of frustration came into play for Hoffman as he and Musgrave relayed what information they could to mission control in Houston. "It was clear they didn't see it in the same way that we did, because they came up with several suggestions which we didn't think were going to work," Hoffman said. The potential fixes were tried, but each came up short. Eventually, Musgrave and Hoffman suggested use of a ratcheting strap that would exert just enough pressure to allow the door latches to catch. The most obvious concern was that tightening the strap—which was capable of up to two thousand pounds of force—would crush Hubble like a soda pop can.

Heflin considered the information he was receiving from the EVA crew members as well as controllers back on the ground, and after doing so, gave the go-ahead to give the strap a try. Hoffman remembers the back-and-forth discussion:

We came up with the idea of using the tie-down strap, but in a very different way than it was originally designed to be used. That had a lot of the telescope people worried. They were uncomfortable with us doing this. It wasn't a shouting match, but we had what I'd call a good-natured discussion about the pros and cons of doing this. Milt Heflin made the decision that we were on the spot and we were in the best position to know what should or shouldn't be done. We did it, and it worked. That essentially saved the telescope.

The best compliment Heflin could have received? Story Musgrave approved of his decision-making process throughout the incident:

Mission control was perfect. Milt Heflin, he checked with the Hubble backroom, "Is this okay?" It's deferring to expertise. The come-alongs, you can break the telescope if you squeeze twelve hundred pounds on it. I did my best to persuade them that I only needed a pound or two. The mission controllers were right on. They're always right on. There was never a problem between mission control and us. Now, Milt may not have gotten approval from the backroom. They may have still been working it, but Milt's got to get the job done. Milt had faith in our judgment not to use force. Mission control had total faith in us. It was just gorgeous. I always mumble a little bit during spacewalks. I say where I am, how it goes now, and where do I go next because I want them on board with me, so they can help me. There's no such thing as an attitude that says, "Leave me alone. I'm doing my thing." Oh, no sir. You want them over your shoulder. The lead flight director always used those words, "They are there. Trust them. It is they that are there, and it is they that are doing the job."

So "hell-bent" was Heflin on staying on the timeline, he simply trusted in Musgrave's judgment. He knew that if the door had been left even partially open, it had the potential of consuming his team's attention for the next spacewalk. It was time to move on. "I knew Story," Heflin concluded. "I completely trusted the crew, and there were some on the ground who wanted to talk about it more. So I made a decision, and the decision was, 'We're going to do what Story wants to do.'"

During the flight's second spacewalk, one of Hubble's solar arrays would not retract due to a kink in its framework. Thornton released the entire assembly completely unfurled just as had been planned before the flight. As soon as she let go, Nicollier moved her out of harm's way on the robotic arm while Bowersox fired thrusters to back the shuttle itself away. It made for a striking sight, the twin solar panels flapping away, the sunlight catching them just right, so that they looked like some sort of very strange space-going, prehistoric bird.

Musgrave and Hoffman then made quick work of installing WF/PC2 during EVA number three. Expected to take four hours or so, the WF/PC2 swap took all of about forty minutes, which in turn allowed the two astronauts to move to the top of Hubble and install a pair of magnetometers. That EVA, along with the COSTAR installation the next day by Akers and Thornton, was the ballgame in the eyes of the public. Both instruments would hopefully be able to correct the flaw in Hubble's primary mirror, but they had to work in tandem with several other critical swaps, Akers reflected:

The WF/PC really is the thing that gave us all the pretty pictures of galaxies and things. COSTAR corrected the optics for all the other spectrometers and instruments that they use to look at the universe. You might ask why we changed solar arrays before we did the COSTAR, because COSTAR was a higher priority, right? You'd think you would do the high-priority things the first day, so in case you didn't get to go back out for some reason, you'd have the primary things done. You have to choreograph things to where you can maximize what you can get done on the mission. There are a lot of variables that go into that. From a point of view of science and getting pictures, those two instruments were definitely the highest priority. But even if you have a perfect mirror, if you don't have good gyros to maintain attitude and solar arrays to have power, it doesn't do you much good.

The spacewalk was the fourth and final one of Akers's astronaut career, giving him a total of twenty-nine hours, thirty-nine minutes in EVA time. It broke a NASA record that *Apollo 17* moonwalker Eugene A. Cernan had held for twenty-one years. Akers was not particularly impressed. "It's a very dubious record," he began. "It's kind of like having a record for being the slowest at something. There had been lots of astronauts who had done four spacewalks. It just took me longer to do my four than the other people. A day or so later up in space, I get an e-mail from Gene Cernan congratulating me. I didn't even know about the record, and I didn't keep it very long. It's been surpassed many times by the guys now working on space station."

Prior to the last spacewalk of STS-61, Hubble was boosted to an altitude of 321 miles. Musgrave and Hoffman installed a drive unit for the solar arrays, the Goddard High Resolution Spectrograph Redundancy Kit, and two protective covers over the two magnetometers near the top of Hubble. Nearly two decades later, the spectacular view during that last EVA was still vivid for Hoffman:

Although we were only basically fifty feet away from the orbiter, it looked really far away. It's not like being on an MMU and flying hundreds of yards away from it, but we were definitely out there. It's a very different perspective. Except for that, you're either in the payload bay or you're close up to Hubble. You're always near something. But when the arm was sort of drawn back away from Hubble, we were just sort of hanging out there in naked space, so to speak. That day, I was free-floating. The tether that was connecting me was loose, so there was no force pulling on me except the earth's gravity. I was essentially an independent satellite. It was just an extraordinary feeling. When I turned my back to the shuttle, it was almost like being lost in space. I could've spent hours out there. Every time we did a certain activity, the ground would ask us to stand by for five minutes or so, so they could turn on the power and make sure that everything was connected properly before we left and went on to do something else. So that did give us a little bit of free time now and then where you could sort of enjoy the whole experience of just being up there. It'd be a shame to spend all that time out in space, and be working so hard that you never really experience the magnificence of the environment that you are so privileged to be working in.

As the spacewalks progressed and it became clear that their efforts were paying off, Heflin could sense a feeling of relief spreading throughout his control room. "It was interesting to see," Heflin remembered. "You looked around that first spacewalk, and I mean, talk about game faces. There wasn't a single grin or smile on anybody's face. By EVA number three or so, you could see those folks grinning and smiling because we were well on our way to doing the job. For me personally, I think that's the most satisfaction I've had in the business, being a part of that team, watching that team work, seeing the locked jaws and seeing the stoic faces break into little smiles. That was really fun."

After the last EVA, there was one last bit of troubleshooting to do, due to erratic telemetry data from Hubble's subsystems monitor. After all that, it was re-released into space several hours late, on flight day nine. The crew could once again breathe. It had seemed rather ironic to be carrying the weight of the world on their shoulders in the zero-g of space. The crew received in-flight congratulations from President Clinton and Vice President Albert A. Gore Jr. Still, for Musgrave following the redeploy, there "was melancholy and sadness. I lost my baby, that's all."

STS-61 landed on KSC's runway 33 a few minutes past midnight on 13 December 1993. From the very first days of human spaceflight, astronauts and cosmonauts had always received the lion's share of attention. They were the superstars whose iconic words survived the ages—call it the "One Small Step" Syndrome, if you will. Yet it was the people back on the ground who truly made things tick, Tom Akers asserted:

It's not just the crew. When you come back and you get lots of attention for a mission, we always tried to give the credit where it's really due. That's to the thousands of people on the ground team that come up with the things for us to do up there. They actually did the real fixes on the Hubble telescope that we were just fortunate enough to get to install. We tried to pass that credit and attention around. It's uncomfortable, to be honest, dealing with the attention when you know that you're just one little cog in a big wheel.

On New Year's Eve, two weeks after *Endeavour* touched down at the completion of STS-61, Hoffman received a call from a friend at the STSCI that all was well. The repairs had worked and Hubble was producing stunning glimpses into the universe the way it had always been intended.

Mission accomplished.

For their efforts on the Hubble repair, the ground and flight teams were honored with the Robert J. Collier Trophy, the nation's highest aviation award. Not only that, but Heflin was also named *Countdown* magazine's "Astronaut of the Year," although he had flown not a single second in space. He and the crew of STS-61 were on a tour of Europe when they got word of the recognition. They gave him a good-natured hard time about it, but Heflin did not care.

They were laughing the laughs of victors.

There is no way to sufficiently pinpoint the origin of the invocation, but it had been around for ages. Legend has it that Alan Shepard uttered one of its more colorful variations moments before becoming the first American in space. It was simple enough, but packed a lot of meaning.

Please do not let me screw this up.

For the crew of STS-82, there was that kind of pressure. The first servicing mission had been one of the most-watched spaceflights since *Apollo 11*, but that had been because Hubble was broken. This time, it was not. Bowersox had served as pilot of STS-61, and for SM-2, he moved over to the commander's seat. "Sox" was joined by pilot Scott. J. "Doc" Horowitz; payload commander Mark C. Lee; Hawley, the robotic-arm operator who had deployed Hubble back in 1990; Harbaugh, who had trained as an EVA backup for the previous servicing flight; Steven L. Smith; and Joe Tanner. This time, their assignment was to upgrade Hubble, not to repair it.

Lead-up to the mission was so smooth, its launch actually took place two days earlier than originally scheduled in order to allow for more range opportunities. With no delays, *Discovery* leapt off pad 39A in the wee morning hours of 11 February 1997. Almost exactly two days later, at 3:34 a.m. Eastern Standard Time on 13 February, Hawley once again came into contact with Hubble through what must have been almost an extension of himself by that point—the robotic arm. Less than thirty minutes later, he positioned Hubble in the payload bay. "When I deployed Hubble, it wasn't moving," Hawley remarked. "I picked it up out of the payload bay and let it go. It's a little different to go capture a free-floating object, and then to berth it and to move EVA guys around on the end of the arm. So that's about as challenging as it gets in the robotic-arm world."

The next day, while Lee and Smith prepared to begin the flight's first EVA, air from the venting of the airlock rustled Hubble's winglike solar arrays. Although Hawley was sure the anomaly would cause controllers on the ground to cancel the EVA, it proceeded nonetheless. Two older astronomy instruments were replaced with new ones during the six-hour, forty-two-minute spacewalk. The Space Telescope Imaging Spectrograph (STIS) could sample five hundred points of a celestial object—be they regions in a planet's atmosphere or stars within a galaxy—in just one exposure. The old spectrograph could examine only place at a time.

During the flight's second spacewalk, Tanner and Harbaugh replaced a degraded Fine Guidance Sensor and a failed Engineering and Science Tape Recorder. Also installed was a device to further enhance the capabilities of the guidance sensor. Almost from the time *Discovery* caught up to Hubble, its astronauts had noticed a definite cracking and wear to Hubble's insulation on the side facing the sun and in the direction of travel. The worn insulation was found to be a fairly serious problem on further inspection by the flight's EVA astronauts. The concern was essentially two-fold: first and most obviously, ragged insulation could have compromised Hubble's thermal control, and second was the potential for debris to somehow damage its mirror.

While decisions were being made on what to do about the worn insulation, there were still another couple of EVAs to complete. Lee and Smith removed and replaced a Data Interface Unit and installed a Solid State Recorder to allow for simultaneous recording and playback. One of four Reaction Wheel Assembly (RWA) units, a large beach-ball-sized device located in one of the main equipment bays that was used to make large-scale changes in the telescope's attitude, was also replaced on that third time outside.

Afterward, mission managers opted to add a fifth spacewalk to repair the troublesome spots of insulation. When it did so, the ground went to work considering what was on board that could be used to cover them. Ultimately, spare multilayer insulation intended for contingencies was used precisely for that—the contingency created by a worn spot or two on Hubble. Harbaugh and Tanner started work with the insulation on two areas near the light shield just below the top of the observatory during the fourth and final scheduled spacewalk. Lee and Smith took over for the added assignment outside, and they attached several blankets to three compartments again near the top of Hubble during their five-hour, seventeen-minute outing.

When *Discovery* took its leave of Hubble, the telescope was in the best shape it had ever been. Less than three years later, however, the telescope was all but on its deathbed.

Three of Hubble's six gyroscopes were already down when a fourth failed on 13 November 1999, forcing it into a safe mode, unable to point precisely at its distant astronomical targets. By then, the STS-103 mission originally scheduled for launch in June 2000 had already been split into two flights—SM-3A and B—in order to repair the three gyros that had already gone sour. SM-3A was bumped up to 14 October 1999, with the other flight to follow in 2001. "Hubble had basically stopped doing science," John Grunsfeld reflected. "That was midway through our training, and with a lot of consternation, management decided that if they had one additional failure, it could put Hubble at risk of not surviving. They didn't want to take that chance."

On SM-3A's original flight plan, six planned EVAS would have carried out what amounted to a complete makeover on Hubble. Those surgical maneuvers had to wait. First, the telescope's life had to be saved by the crew of STS-103—commander Curt Brown, making his sixth spaceflight; pilot Scott J. Kelly; payload commander Steve Smith, back for a second straight Hubble flight; C. Michael Foale, on his fifth trip into space; Grunsfeld, for the first of his three Hubble servicing missions; Claude Nicollier, also making a second consecutive visit to the telescope, but this time moving outside to join the EVA crew; and ESA's Jean-François Clervoy, born and raised in France and nicknamed "Billy Bob" because of his decidedly non-Texan dialect.

The mission ultimately faced nine delays and scrubs. The first and most serious issue took place after STS-93 encountered substantial problems going uphill during launch on 23 July 1999. Approximately three weeks later, NASA ordered extended wiring inspections and maintenance across its fleet of orbiters. Launch of STS-103 was moved at first to no earlier than 28 October, and then to 19 November because of the amount of required wiring repairs. Then, an inch-long drill bit was found lodged in *Discovery*'s right main engine. More damaged wiring was discovered in an umbilical leading to and from the shuttle and its External Tank. The Thanksgiving holiday was observed. There was a dented main propulsion system line. A suspect weld was detected on the External Tank's pressure lines. And, of course, there were two very nearly prerequisite weather delays. Following the first of the weather scrubs, the crew members were drenched by a Florida rainstorm as they exited the shuttle on pad 39B.

The interminable setbacks put the flight up against an issue that seemed almost quaint in the years to come—the Y2K deadline. No one knew how computers would react to the ninety-nine to double-zero numerical rollback at the turn of the year, so rather than delay once more until after New Year's Day, the flight was cut from ten days to eight. "As it turns out, because the main thing we lost by cutting off a day was putting insulation on the upper end of the telescope, it was no big deal," Grunsfeld offered. "In the end, we never put that insulation on and the telescope's fine without it. But we'd already cut off 70 percent of the mission content to go up and replace these gyros quicker. For me, it was exactly the same number of EVAS. It was the team of Mike Foale and Claude Nicollier who lost an EVA, and so I felt bad for them."

In September, NASA lost communication with the Mars Climate Orbiter, which disintegrated as it incorrectly entered the planet's upper atmosphere. As he did with the first Hubble servicing mission back in 1993, NASA administrator Dan Goldin told the crew exactly what was required of them just before they hit the pad. "Dan Goldin visited us in quarantine at KSC, and he told us, 'We just lost a probe on Mars, so I count on you to brighten the image of NASA with a successful mission to repair the Hubble Space Telescope,'" Clervoy said. Then, Goldin addressed the Frenchman directly.

And you, Mr. Clervoy, do not forget you have six billion dollars in your hands!

The mission finally left the pad at precisely 7:50 p.m. Eastern on 19 December 1999. Although Clervoy had been in space twice, he was struck by the view from 330 nautical miles, a full 115 miles higher than he had ever been before. "It's exciting," Clervoy said. "It's quite fun to see the earth from so much higher, because you see more of Earth in one view. For example, I could see the east coast and the west coast of the U.S. by just swinging my head from left to right. If I am given the choice, I would rather prefer a high-inclination orbit, because you cover more visually. You cover more of the earth, and you can see beautiful things. You can see Greenland, the glaciers of Alaska, the Himalayas, China. You can see very well close-ups of the South Pole."

Clervoy had used the robotic arm to deploy an atmospheric research satellite during the flight of STS-66 in November 1994. On STS-103, he would be at the controls of the arm during all three EVAS. He was ready to go:

I had no fear about the robotic tasks. I was feeling very at ease with these. I was very comfortable and very intuitive, which joint to select and how to approach Hubble. It was like an extension of my body and my brain, so it was not like, "I hope *I will succeed because sometimes in the simulator I failed." No. I was just feeling in my proper role in that crew and in that mission. I liked this job very much. I felt, for me, it was not very difficult. I felt confident in my role, and so I felt it was really a mission that was tailored for me.*

During training, Clervoy developed three steadfast rules as a robotic-arm operator:

13. Jean-François Clervoy is a study in concentration during his work as a robotics operator for the STS-103 Hubble servicing mission. Courtesy NASA.

Never be responsible for a second of time being lost. If at all possible, he would *never* ask one of his EVA colleagues to wait while he worked on something at his station. He would be ready to serve and to move them any time they wanted to move.

Never touch the telescope with the arm or, worse yet, with one of the astronauts on the end of the arm.

Never initiate a move that is not *exactly* what had been requested. If Grunsfeld asked to be moved two inches to the left, pitched down ten degrees, and rolled left twenty degrees, that was what he would get. Nothing more, nothing less. Complicating matters even more was the fact that Clervoy would have to do almost continuous calculations in his head to correctly make the precise movements.

At Clervoy's workstation in front of *Discovery*'s aft windows were two independent hand controllers, one for rotation (RHC) and the other for translation (THC). On the side of the RHC was a switch that determined one of two rates at which the arm could be controlled—vernier and coarse. The vernier rate was used for very precise, very slow motions, just the kind of thing needed while working with a multi-billion-dollar piece of machinery like Hubble. A full deflection of either stick in coarse rate meant hammer-down, full-tilt operations that could, Clervoy remarked in jest, "scare those watching." The coarse rate was typically forbidden when the arm came within five feet of Hubble, but Clervoy said he was allowed to use the quicker means of operation in close proximity to the structure. He elaborated:

I was flying my colleagues around the telescope, inside the telescope, just a few inches away from structure of the telescope using coarse rate, just to be fast. I didn't want to be the cause of any delay. It worked well, very, very well. When you do a transfer from like the gyroscope spare-part location to the electronic bay where they're to be replaced, it's several meters. If I do that in coarse rate, I can do that just in ten, fifteen seconds. But if I select the slower rate, it can take two or three minutes to do the same translation from the forward end of the cargo bay to the aft end. You can save a lot of time. Normally, it's forbidden by the flight rules, so you need a special agreement from the flight director.

Another means by which Clervoy saved time was the use of a push-to-talk button he could trigger with his toe. That way, he could talk and fly the arm at the same time. Brown was skeptical at first. "My commander was very scared about this," Clervoy admitted. "When he saw me fixing this radio button with the gray tape, he told me, 'Billy Bob, you'd better not mess up with this. As soon as I see anything wrong there, you take all this off and you use the standard method.' Actually, it worked very well. Curt was amazed. He told me, 'Billy Bob, I never saw an arm operator like you.'"

Two days and eighteen hours into the flight, Smith and Grunsfeld began a grueling first EVA. They replaced three rate sensors and their all-important total of six gyroscopes, and installed six Voltage/Temperature Improvement Kits between the telescope's solar panels and its six ten-year-old batteries. Minor difficulties, including the removal of one of the old RSUs and a cap on the Near Infrared Camera and Multi-Object Spectrometer, lengthened the spacewalk to a tiring eight hours, fifteen minutes—the third longest in shuttle history.

A few minor objectives were left undone. In order to save time, mission managers also opted not to switch Grunsfeld, who started out on the arm, and Smith, who was free-floating, following the RSU swap. Although they had cross-trained many times, the decision resulted in Grunsfeld having to do a major task—installing the cell-phone-sized battery improvement kits—that had been meant for Smith to accomplish. Said Grunsfeld:

The gyros are really hard to install. They're in a very delicate place, with lots of stuff where you can hurt things. We had trained a lot, and we understood well the subtleties of the RSU *mounts inside the telescope. It was much better than they did on the first servicing mission, but still, they were tough to put in. I was never really concerned. I didn't pay a whole lot of attention to the time throughout the whole* EVA*. I never really looked at the clock on my* EMU *[Extravehicular Mobility Unit, otherwise known as a spacesuit]. I just figured we would work until we were done, and then we'd come in. We accomplished everything we were supposed to accomplish. The first day is always challenging. It sets the tone for the rest of the mission, but in addition to doing the tasks, we also had to set up the telescope to be serviced. There's a fair amount of overhead work just to get started.*

At eight hours and ten minutes, the mission's second EVA by Foale and Nicollier was not much shorter than the first. Together, they installed a new computer twenty times faster than the old one, as well as a 550-pound Fine Guidance Sensor. All that was left for Smith and Grunsfeld to do on the flight's third and final eight-hour spacewalk was to install a transmitter to send data to the ground, plus a Solid State Digital Recorder to replace a reel-to-reel device.

Like Grunsfeld, Clervoy paid no attention to time while his crewmates were outside:

You don't have a sense of time if you don't look at the watch, and I think that's especially so for those that are outside. Sometimes, we would ask colleagues outside, "Do you know how much time you have been out?" They would say only one hour and a half, when it was already three hours and a half or four hours. When you have your mind focused on a very precise goal, you don't think about eating or relaxing. I'm not saying other people cannot do it, but any people who train for space missions develop this ability to focus continuously for many, many hours in a row.

The telescope was released on Christmas Day. "We had to wait for them to test all the boxes we had replaced," Clervoy said. "When the ground told us they all worked perfectly, that was the best gift we could have hoped for on Christmas." Added Grunsfeld, "I can say that's been the highlight-holiday of my life. Christmas Eve, I was out doing a spacewalk on the Hubble and then on Christmas Day, we gave the science community and the world this amazing telescope. I can't think of a better gift to humanity than a working, operating Hubble Space Telescope."

The first three servicing missions featured two veterans from the preceding Hubble flights, but on STS-109, Grunsfeld was the only STS-103 carry-over. Experienced crew members gave new Hubble fliers a been-there, done-that confidence they might not otherwise have had. And, obviously, the assignment sat well with astronomer Grunsfeld:

It's not a requirement, but there was a desire by the Hubble team at Goddard Space Flight Center that somebody from a previous mission fly on the next. Steve Smith was able to hand over a lot of knowledge about how everything worked, and that was hugely valuable. The decision was made to do that again, and I'd like to think I did a good job showing leadership and creativity, understanding the hardware and contributing to the development effort. They decided I was the one who should carry the torch for the next mission. Right after we landed on STS-103, I started putting together the engineering development for the next mission and bringing new people in to try and learn the Hubble techniques. When it came time to assign the mission, I was the natural choice. I think that was deliberate. They wanted me to continue working the engineering and train other people, with the hopes that I would emerge as the leader for the next mission.

Emerge as a leader is *exactly* what Grunsfeld did. As payload commander for STS-109, he served alongside commander Scott D. "Scooter" Altman; freshman pilot Duane D. Carey; and mission specialists Nancy Currie and Jim Newman, who had previously flown together on STS-88, the first ISS construction mission. Also on board were Richard M. Linnehan, who had flown with Altman on STS-90, and rookie Michael J. Massimino. Newman, Linnehan, Massimino, and Grunsfeld were officially named to the flight on 28 September 2000. Altman, Carey, and Currie would not come on board until later.

Before the flight had a commander and full crew, Grunsfeld had been hard at work on the STS-109 flight plan:

Being the payload commander was an enormous difference in workload. On STS-103, I tried to spend as much time learning from Steve Smith as I could.

If somebody else is doing the work, I'm not one to sit back and say, "Well, I can just take it easy." I always try to pitch in when I can. Steve was very good about delegating work. Since I was the only one who was essentially assigned to the flight early on, I had to work not only all of the spacewalking activities—the engineering work for the instruments and the techniques and designing the EVAs*—I also had to do all of the work of what the commander would do until eventually a commander was assigned. So I had to do all the payload manifesting and the orbiter manifesting, and details like stowage issues, what camera is going to go on which spot, and what kind of fuel loads we would have for ascent and entry. You name it, and I was the representative from the crew for the flight to decide all of those things. Eventually, Scott Altman was named commander, but that wasn't until just a little over a year from flight . . . or so we thought, when we first were assigned. It was really a great experience for me to learn all these things, but it was quite a lot of workload.*

Columbia launched the seven-person crew near dawn on 1 March 2002. By the time they caught up to Hubble, Grunsfeld considered it a visit with a long-lost acquaintance. "I had dug in deep to learn many of the idiosyncrasies and oddities of the telescope, such that when I got there, it was like seeing an old friend," Grunsfeld began. "From my perspective, I was shooting for nothing less than 100 percent success. I really felt like we had to achieve everything that we had on our plate. I felt that way about every mission, to the maximum extent possible."

The mission's first EVA had a simple enough plan of attack—during their seven-hour spacewalk, Grunsfeld and Linnehan replaced Hubble's starboard solar array and then stored the old one in the payload bay for a later evaluation of its nine-year performance. Newman and Massimino were up for the next venture outside, and they swapped Hubble's port array before moving on to bolting in a new Reaction Wheel Assembly—the device used to make changes to the telescope's attitude. They also installed a thermal blanket, door stop extensions, and foot restraints to help prepare for the third EVA. There was still another task before going back inside. Newman and Massimino tested two bolts on the aft shroud doors and found that they, too, needed replacements.

So far, most everything was going according to plan. Massimino would remember years later:

Going into it, I wasn't sure how everything was going to work, but I felt well prepared. There really weren't too many surprises. The biggest thing is just the view around you. The beauty of the workplace is incredible. I enjoyed doing the mechanical work, and was happy that all worked out okay. You get prepared very well for that during training, but there's no training that can prepare you for the view that you see around you as you're working. I felt totally prepared for the work, but no way could I have anticipated how beautiful it was going to be up there spacewalking.

In June 2010, while promoting the IMAX film *Hubble 3D*, Massimino made an appearance on David Letterman's talk show. The normally ribald and irreverent host and the studio audience were left nearly speechless by Massimino's account of his spacewalking adventures on the flight. Massimino told Letterman:

When you first get to space and you look out the window of the shuttle, it's magnificent what you see. But it's kind of like looking at an aquarium. You're looking through a window. You might be eating something. If you need to go to the bathroom, you can. You're in regular clothes. But when you go out and spacewalk and see the planet, I found that to be a different experience. You can look anywhere. My first spacewalk, I didn't look around too much. I didn't want to get distracted. My second spacewalk, I really had a chance to soak it in. When I first looked, I almost couldn't stand to look at it. At the Hubble altitude, you can see the planet in its entirety. You can see the curvature, like you see in the movie. It's like a big ball. I just turned my head. The thought that went through my head was, "This is something I'm not supposed to look at. It's a secret. People aren't supposed to see this." There are no words to describe its beauty.

On a show like Letterman's, Massimino could not resist getting in a bit of a joke of his own: "I looked again, and the second time I looked, I started to get a little emotional. I've got to catch myself, Dave, because I was afraid I might start to tear up and that I would get some water in my spacesuit that could cause a problem. Then, there'd be an investigation and I'd have to admit I was crying! So I quickly got myself under control."

Turning serious again, Massimino continued describing the experience to a host and studio audience whose silence was almost palpable: "I looked a third time, and that third time that I looked, the thought that went through my head was, 'If you're in Heaven, this is what you'd see. This is the view from Heaven.' Then, it was immediately replaced by another thought, which was, 'Nah . . . it's more beautiful than that. This is what Heaven looks like.' That's what I felt looking at the planet. It's fragile. It's beautiful. It's perfection, and we need to take care of it."

While suiting up for the next EVA, Newman felt water on Grunsfeld's Primary Life Support System (PLSS) backpack. And that was, Grunsfeld knew, "a very bad thing." It would later be determined that when the suit was charged from *Columbia*'s airlock power supply, a voltage spike inadvertently traveled through the suit and flipped open a water valve that was not supposed to do so unless in the vacuum of space. A fix for just such a problem had been worked up but never implemented. The crew jumped into action and scrambled for replacements, eventually swapping out the PLSS and torso portion of Grunsfeld's EMU.

The situation confirmed for Grunsfeld a lesson he would never forget:

One of my leadership lessons for life is that mistakes are going to happen. Problems are going to happen. You need to learn to quickly trap those mistakes or problems. Identify them, limit their extent, fix them, and then move on and not think about them anymore. This was one of those situations. For an hour and thirty minutes, we pulled together as a whole crew, switched out all the suit parts, got me back in my suit and back in the airlock. We were able to go out two hours late. From that moment on, I totally forgot about it. I never really thought about it until the end of the EVA.

Grunsfeld would long remember his exchange with Altman, asking, "How long have we been out?" The STS-109 commander responded that it had been six and a half hours. Because planners had figured the EVA would take as long as eight and a half hours to complete, Grunsfeld took a shot at doing some mop-up work before making his way back into the airlock.

"You got out two hours late. You probably ought to come back in," Altman responded.

"Oh, yeah, I forgot about that."

The EVA to change out the original, twelve-year-old Power Control Unit (PCU) was a critical one. Due to a small manufacturing defect in the PCU, Hubble's main power distribution source, there was a high probability that the telescope would cease to operate within six to eight years if the task was not fully accomplished. If the crew tried to make the change and was unsuccessful, Hubble could be lost right away, Grunsfeld said:

It was a level of complexity and difficulty that nobody had ever tried in space before. So we had the combination of a critical component and a difficulty that pushed the envelope for spacewalking that made everybody very uncomfortable. This was one of those things where everybody on the ground was very skeptical it would work as scripted. It was just too complicated and too much work. It was believed that it would take longer than our STS-109 EVAs to complete. I signed up to do that task, and then I had to prove that it would work. One of my favorite spacewalking tools is one that I designed for that task, a connector wrench. We did that. We built really good mockups. I relentlessly trained, night after night. On my way home, for months and months, I would stop in our trainer building, go inside, and do that task until I knew it better than tying my own shoes. Then, we went to flight, and it worked like clockwork. Rick Linnehan and I went out there. We had practiced a lot, *and we executed it in a six-and-a-half-hour EVA.*

To make the changeover, Hubble was powered down for the first time in orbit. Linnehan went to work removing thirty of thirty-six connectors on the old PCU. He then began preparing the new unit while Grunsfeld finished with the remaining six connectors. Grunsfeld gently removed the old PCU from its slot on Hubble and replaced it with another one. The connectors were mated ninety minutes later, and an hour after that, the new PCU was confirmed to be up and running. Grunsfeld would later call the PCU switch the hardest thing NASA had ever done in space to that point, going so far as to liken the work to performing heart surgery.

There were two more spacewalks to go, and they were challenging as well. Newman and Massimino installed the Advanced Camera for Surveys, which became Hubble's primary imaging instrument. Massimino then took care of installing the Electronic Support Module, the first part of an experimental cooling system that would be put in place on the flight's last spacewalk. The cryo-cooler for the Near Infrared Camera and Multi-Spectrometer (NICMOS) was placed inside Hubble's aft shroud by Grunsfeld and Linnehan, while a Cooling System Radiator (CSR) went on the outside of the telescope. Working in tandem, Linnehan threaded wires from the CSR through the bottom of the telescope to his partner. As he did so, Grunsfeld in turn connected the wiring between NICMOS and its new cooling source.

On 9 March 2002, Hubble was redeployed. The flight's four spacewalkers had been outside for thirty-five hours and fifty-five minutes. That bettered by twenty-seven minutes the record for a single flight previously held by the astronauts of STS-61. At 4:33:05 a.m. on 12 March, Altman brought the orbiter to a stop on runway 33 at KSC, after completing a journey of nearly eleven days and 3.9 million miles.

Columbia would never again make a successful landing.

On 16 January 2004, a year to the day after *Columbia* left the launch pad on her final voyage, NASA administrator Sean O'Keefe met with approximately one hundred members of the Hubble team in a conference room at Goddard. Present were Grunsfeld, who by that time had taken over as the chief scientist for NASA; Ed Weiler; representatives from NASA headquarters in Washington, the STScI, and Goddard; and various contractors. What O'Keefe told those assembled left them in disbelief.

Due to a number of difficult factors, O'Keefe reported that he had decided to cancel plans for the fifth and final Hubble servicing mission. Steven V. W. Beckwith, then the director of the STScI, wrote in a memo later that day that the "mood in the room was decidedly somber." Weiler had insisted that funds were available to cover the cost of the mission, and that it had not been his recommendation to cancel it. Beckwith also noted that O'Keefe had determined the flight would extend Hubble's lifespan by only a brief time, perhaps only three years. "[In] that he is mistaken, but a short period (e.g. three years) obviously played a role in his thinking," Beckwith wrote. "Therefore, he thought the extra effort required to mount a single mission, SM-4, was not worth the scientific return. That drove the decision."

Grunsfeld had no trouble remembering his reaction to O'Keefe's pronouncement:

As I've said many times, I was stunned by that decision. We were a year after the tragic loss of Columbia *and the crew, but that's a decision an administrator gets to make, for better or for worse. At the time, I actually had already been assigned as the payload commander for SM-4. So it was not only a cancellation of the mission, but it was also a cancellation of my next flight assignment. So it sort of had a double whammy. It wasn't in pen, but it had been written in pencil that I would come back and lead that mission. I knew Hubble was too impor-*

tant, and I actually in my heart at the time believed we'd get back to Hubble one way or another, either with a Space Shuttle mission or with a robotic mission.

Just two days before the meeting, President George W. Bush had visited NASA headquarters to announce his Vision for U.S. Space Exploration, which included a return to the moon, crewed flights to Mars, and the eventual phase-out of the Space Shuttle program itself. It was against this backdrop that O'Keefe took considerable heat over his Hubble death sentence. Members of Congress—Senator Barbara Mikulski of Maryland and Representative Mark Udall of Colorado in particular—took up Hubble's cause. A website claimed to have collected twenty-nine thousand signatures in favor of saving the telescope. Children sent their school money to NASA.

In September O'Keefe hosted the Risk and Exploration symposium at the Naval Postgraduate School in Monterey, California. Mount Everest conqueror Ed Viesturs spoke, as did James Cameron, director of the blockbuster film *Titanic*. Moonwalker Harrison H. Schmitt was there, as was fellow Apollo astronaut Thomas K. (Ken) Mattingly II. Grunsfeld, Foale, and Shannon W. Lucid represented the shuttle era. Through it all, O'Keefe remained unmoved from his position. In a 10 March 2004 budget outline submitted to Christopher S. Bond, chairman of the U.S. Senate Subcommittee on Independent Agencies, O'Keefe gave his reasoning for the cancellation:

After much deliberation and consultation with shuttle experts regarding safety and risk considerations, I recently made the difficult decision to cancel the final Hubble Servicing Mission [SM-4]. The decision had to balance the world-class science that HST has produced, and would continue to produce, against the risks and complexity of preparing two shuttle missions in support of HST and unproven rescue techniques. The decision was not made with regard to budget considerations or any question as to the significance of the science returns of the HST, but rather was based on our assessment of what NASA must do to comply with the recommendations of the Columbia *Accident Investigation Board for developing on-orbit inspection, repair, and contingency rescue requirements for every shuttle flight.*

Grunsfeld was not going to let Hubble go down the tubes without a fight. Not Grunsfeld, not the man who had once been a kid who gloried in the small Sears telescope his grandmother had purchased for him. If he could see what he saw with that gift, the young Grunsfeld marveled even back then, just imagine what wonders there were to discover with a space-based telescope. With Hubble on the line, Grunsfeld became "part of a bigger effort" to save it:

It was very critical, in terms of the longevity of the telescope and some of the discoveries it was capable of making. There are several aspects, the biggest of which is that we needed to get the new Wide Field Camera 3 [WFC3], which was crucial for astrophysics, to bring Hubble up to state of the art. Changing out the scientific instrument computer was very important, or you couldn't get any data down to Earth. We wanted to install the Cosmic Origins Spectrograph. To give Hubble long life, we also had to change the batteries and the gyros. It all really worked together quite well to update Hubble to state of the art, bringing the new instruments online, making old ones come back to life and then providing infrastructure in thermal protection, gyros, batteries and computers to be able to give Hubble a long life.

In the end, astronomy won out. O'Keefe resigned as NASA's administrator in February 2005 to become chancellor at Louisiana State University. On 13 April 2005, physicist and aerospace engineer Michael Griffin was named as his replacement. Eighteen months later, Grunsfeld got his wish. Hubble would be serviced by a shuttle-based crew one more time. "We have conducted a detailed analysis of the performance and procedures necessary to carry out a successful Hubble repair mission over the course of the last three shuttle missions," Griffin said in making the announcement. "What we have learned has convinced us that we are able to conduct a safe and effective servicing mission to Hubble. While there is an inherent risk in all spaceflight activities, the desire to preserve a truly international asset like the Hubble Space Telescope makes doing this mission the right course of action."

The flight was tentatively set to launch in the spring or fall of 2008. Altman would return to Hubble as commander of STS-125, while Grunsfeld would be making his third Hubble flight and Massimino his second. Joining them aboard *Atlantis* were first-time spacefliers Gregory C. Johnson, the flight's pilot, along with mission specialists Andrew J. Feustel, Michael T. Good, and K. Megan McArthur. "I think NASA management decided that this, no kidding, would be the last time we would go to Hubble," Massimino said. "Because of that, I think they came up with the idea that they wanted to get as much experience as possible on that flight, with a mix of new people."

The plan had been to launch STS-125 in October 2008, but when Hubble's Science Instrument Command and Data Handling Unit (SIC&DHU) malfunctioned, the flight was delayed so that a backup could be prepared. *Atlantis* finally left pad 39A at KSC on 11 May 2009 with *Endeavour* standing by on pad 39B in case a rescue was needed. After everything it had taken to get this mission on the flight manifest, the pressure was on. If the work on Hubble's Power Control Unit during the previous servicing mission was a dicey operation, Grunsfeld knew the tasks to be accomplished on its Advanced Camera for Surveys on STS-125 would be very much like brain surgery. "We went *way* further than people had ever done, even including previous Hubble work," Grunsfeld said.

The crew desperately wanted to cross every single repair off the checklist. This would, after all, be the last time any human being came into contact with the telescope. Grunsfeld said, "There was an added degree of pressure, just that we *had* to accomplish everything on our plate. There wasn't a chance for somebody to finish up later on. On the International Space Station, if somebody on a given mission doesn't finish something, very likely, people inside can come out in a week or two and finish stuff. Or the next shuttle crew can do it. But for our work, there was *no* chance that any time in the near future that somebody would be able to come back and finish it."

The flight had its share of issues to overcome. Grunsfeld and Feustel were on tap for the first EVA, but when Feustel tried to loosen a single bolt to remove the old Wide Field Camera, it would not budge. After much deliberation, he was given the go-ahead to remove a torque-limiting device from his specially designed power tool, and go at the bolt with full force. Just before he tried the fix, Feustel asked, "What are the implications if I overtorque and break the bolt?" Massimino, supporting the spacewalkers from inside *Atlantis*, deadpanned in reply, "You sure you want to know?" With that, Feustel tried one more time. The bolt came loose at last. WFC2 was removed and a new one was installed in its place. Grunsfeld and Feustel then replaced the malfunctioning SIC&DHU. After a couple of other housekeeping tasks, their seven-hour, twenty-minute spacewalk was over.

On the flight's second EVA, Good and Massimino replaced three Rate-Sensing Units, their all-important gyroscopes, and a battery module. On the next, Grunsfeld and Feustel replaced the existing COSTAR with the Cosmic Origins Spectrograph, which allowed Hubble to peer further into the universe in the near- and far-ultraviolet ranges. After that, to repair the Advanced Camera for Surveys, they had to delicately remove thirty-two small screws from an access panel in order to replace four circuit boards and a new power supply. It was arduous and finicky work, and even though the flight plan called for the exacting task to be undertaken during two spacewalks, but Grunsfeld successfully completed it in just one.

Another frustrating issue cropped up during the next-to-last spacewalk, when Massimino attempted to remove a handrail that prevented placement of a device designed to catch the screws from a plate covering the imaging spectrograph. Removing the handrail had always been all but a given in training—all it took was loosening four large bolts and, boom, it was on to the real task at hand. That had been in the tank, during training. This was the real deal and in space, with no time-outs available, the head of the last bolt was stripped, making it impossible to remove. Again, the ground leapt into action and came up with a number of suggestions. Massimino made his way from Hubble to a supply container near the front of the cargo bay so he could pick up any number of odds and ends. "They were making me get a vice grip and they were making me get tape," Massimino would later say. "I couldn't believe I had to get tape. What are we going to use the tape for? What about staples and paper clips? You want me to get those, too? It was like I was going to the stationery store." Massimino could see Feustel through *Atlantis*'s aft windows overlooking the cargo bay, and at one point they made eye contact. They did not say anything, but they did not need to. "I could see his face, and I could see that he was just about as low as he could get," Feustel remembered. "Immediately, I started smiling and looking back at him.

"Neither of us wanted to say anything out loud so people would hear us, but we could see each other very clearly," Massimino added. "He might have been lying, I don't know, but he was just smiling and giving me a thumbs-up, like, 'What are you worried about? It's going to be fine. We can get through this.'"

Then came word to try the simplest fix of all. James Cooper, an engineer at Goddard Space Flight Center, suggested that Massimino rip the handrail off Hubble. As this was going to be the last servicing mission to the telescope, who else was going to need it?

I was shocked. I could not believe it. I was pretty surprised and upset about it. I had thought that the bolt head was stripped. I thought that the tool head was still in good shape, so I figured that no matter what I did to that bolt, it was not *coming out. I was just sick over it. All these other things that we had talked about would have multiple effects further down on what we were going to do. The thing to do was to just bust it off. Once I heard that, I was like, "Okay, we have a chance. Let's not screw this up. We have an opportunity to just start fresh."*

Massimino's primary concern was not so much that a quick rebound of the rail would somehow puncture his suit. His worry was that debris would manage to find its way into Hubble and damage its delicate instruments. Tape was placed around where the handrail would be broken to catch any wayward pieces, and Good was also standing by with a bag in which to immediately place its remains. The handrail broke right off, finally allowing the spacewalkers to get down to the business at hand. With this job done, on went the Fastener Capture Plate, out came the screws, and in went a new power supply board for the imaging spectrograph to complete the eight-hour, two-minute effort. The seat-of-the-pants fix had worked, leaving just one more Hubble EVA to be completed.

The crew woke up early to give themselves plenty of time to complete the final spacewalk. After trading out a battery module and replacing the Fine Guidance Sensor, Grunsfeld and Feustel were finished. After releasing Hubble for the final time, Grunsfeld briefly patted the telescope and gave it a salute. He was the last person to touch Hubble. "We were sending Hubble off in better shape than it had ever been before," he recalled. "So I gave Hubble a last pat and a salute and said, 'Hubble, you're The Man. Go do your stuff.'"

The two astronauts then slowly made their way back inside *Atlantis*, filled with the satisfaction of a job well done. The five STS-125 spacewalks had consumed 36 hours, 56 seconds, and over the course of the five servicing missions, sixteen astronauts spent a total of 166 hours, 1 minute working in and around Hubble. John Grunsfeld looked back at all that had been accomplished with understandable pride:

14. Hubble Hugger John Grunsfeld, who made three separate visits to the famous telescope, says good-bye to his on-orbit friend. Courtesy NASA.

In the end, Hubble is a scientific satellite. Hubble's never discovered anything. People using *Hubble have discovered things. Nevertheless, this was my third trip to Hubble. I spent the better part of twelve years of my life working on these Hubble missions. I did not try to anticipate how I would feel when all the work was done. On STS-109, I actually did feel a little bit of sadness thinking it was the last time I was ever going to see Hubble, and then this time, I wondered if I would feel the same way. And, in fact, I didn't. We were given a series of tasks on STS-125 that many people said would be impossible for us to have 100 percent mission accomplishment. So at the end of that last, fifth EVA, coming back into the payload bay having finished everything, I thought, "How can I at all be sad? We have given Hubble a new lease on life. We've brought its instruments up to state of the art. Our team on board, on the ground, and in the trenches had all accomplished something that people said was impossible."*

The final servicing mission extended Hubble's lifetime until at least 2013, if not later. Regardless, its legacy had been established. We cannot go to the stars, Ed Weiler concluded. Not yet. Hubble, however, has brought them to us. "Despite the world of *Star Trek* and *Star Wars*, it's going to be a long time before we develop the physics to be able to travel among the stars and other galaxies," Weiler said. "We dream of traveling in the universe. Our bodies may never do that traveling, but what Hubble has brought us is the ability of the human mind and imagination to go where our bodies can't. Hubble has brought the universe to us. Most importantly, it has brought science to America's kids. And, boy, do we need more kids to be scientists and engineers."

4. Sleeping with the Enemy

The photographs taken by U-2 spy plane pilot Richard Heyser did not lie. Soviet nuclear weaponry was on the ground in western Cuba just ninety miles or so from the Florida Keys, and preparations were under way for many more medium- and long-range missiles to come. The warheads could easily have reached most of the continental United States, information that chilled the American public to its very core. The Cold War had long been simmering, and over the course of two weeks in October 1962, it came the closest it had ever come to boiling over into an all-out nuclear holocaust.

The two nations stared each other down for nearly fifty years, hoping the other would blink first. In the early 1950s, Senator Joseph McCarthy became one of the most divisive figures in the history of American politics when his blacklists found communists around every corner. A decade and a half before Vietnam, the United States was embroiled in the Korean War in an attempt to stem the red tide. Bomb shelters were built in suburban backyards, and drills on what to do in the event of a nuclear bombing were conducted in school classrooms. Worse yet, such fear was not wholly unwarranted. The superpowers possessed more than enough weapons to destroy each other many times over, and all it would have taken was an itchy trigger finger here or there to send the world over the brink.

It was against this kind of backdrop that the earliest days of the space race were played out. Instead of lobbing nuclear missiles at each other, the United States and the Soviet Union vied for dominance of the last great frontier—space—beginning in the late 1950s and continuing throughout most of the following three decades. For years, the Russians held the upper hand. *Sputnik* was the first man-made object in space. Yuri Gagarin was both the first human in space and the first to orbit the earth. Valentina Tereshkova was the first female in space. *Voskhod 1* cosmonauts Vladimir Komarov, Konstantin Feoktistov, and Boris Yegorov were the first multi-person crew. The repeated body blows, one right after the other, hammered at the American psyche.

Something had to be done, and something had to be done *now*. The moon. That was it. The moon. The finish line to the space race was right up there in the sky for one and all to see. The race to the moon became a virtual war in that lives were lost on both sides, and while one country's goal of beating the other to the lunar surface was not necessarily an officially stated goal, it was one that was certainly well understood in and around the halls of NASA and its astronaut office. For Americans, the concept of living under a red moon was positively unthinkable.

Another concept would have been every bit as unsettling during that time frame. Working in conjunction with the Russians toward a common goal in space seemed pure fantasy, given the era's political climate. The Apollo-Soyuz Test Project in the early to mid-1970s showed that it could be done, but that appeared to be a one-time effort until the fall of the Soviet Union in the early 1990s. With it came a new openness between the two nations. While the joint venture known as the Shuttle-*Mir* program opened the door to further cooperation on the International Space Station, the fact that it ever happened at all is nothing short of remarkable. Frank Culbertson, the astronaut who served as the program manager for Shuttle-*Mir*, explained: "With the world and the politicians constantly looking over their shoulders, men and women worked through their problems face to face, building trust in each other and in each other's goals. This was often done at very personal levels and with high stakes, both physical and emotional, to achieve exactly what the participants set out to do and more: to execute a joint program of scientific achievement and space exploration by partners who had been archenemies less than ten years before."

Before the cooperation could stand any chance of success, however, both countries needed to clear some age-old hurdles.

On 17 June 1992, less than six months after the formal dissolution of the Soviet Union, the *Joint Statement on Cooperation in Space* was issued by the United States and the new Russian Federation. The agreement stipulated that the two countries essentially swap spacefarers—in 1993 a cosmonaut was to fly on board the shuttle, and a Soyuz capsule would take NASA astronauts to the Russian space station *Mir*, which had been crewed off and on since March 1986. Better yet, plans called for the shuttle to dock with the orbital platform by 1994 or 1995. The undertaking was expanded in 1993, when up to ten rendezvous flights and a space station were added to the mix. Ultimately, the goals of Shuttle-*Mir*—known also as Phase 1 of the International Space Station Program—were straightforward:

1. The two nations had to learn to work together. They had to learn respective languages and cultures, as well as the ways in which they did business on orbit.
2. Risk for the ISS had to be mitigated by testing hardware and procedures on *Mir*, and then react accordingly when the unexpected took place. And on *Mir*, the unexpected came to be almost expected.
3. The United States had not had a long-duration presence in space since the days of *Skylab*, so it had to get up to speed on living and working on orbit for lengthy periods of time.
4. Finally, a program of scientific research was to be conducted. Culbertson sometimes took heat from the scientific community for its interests being so far down the list of priorities, but he in turn reminded them that they were not last, they were fourth. If NASA and the Russians could not come to grips with the first three, science was going to be left behind altogether.

Within months of their space détente, contingents of NASA officials and astronauts were standing in places that only the famous fictional spy James Bond once could have breached—in Red Square, and inside the walls of the Kremlin and Star City, at one time the ultra-secretive home of Russian space operations. In theory, the situation was ideal if for no other reason than it was a concrete sign that the Cold War was beginning to thaw. McCarthy was dead and gone, but the possibility of disaster had long remained all too intact. The Soviets invaded Afghanistan in 1979 and then shot down Korean Air Lines Flight 007 on 1 September 1983, killing all 269 passengers—including Georgia congressman Lawrence McDonald. The United States planned a Strategic Defense Initiative global shield known as "Star Wars," and the Soviet Union was not happy about *that*. So divided were the two sides that many Americans would always consider the country's "Miracle on Ice" hockey victory over the Soviet Union in the 1980 Winter Olympics to be the greatest sports story of all time because of the backdrop against which the game was played.

Although American and Russian forces still had nuclear weapons pointed at each other, the big-picture fact of the matter was that tensions had eased exponentially with the fall of the Soviet Union. Working with the Russians, the only other space power in the world, made sense for many reasons and at the forefront was the country's dire circumstances. "They had a lot of capability, but they were severely underfunded," Culbertson said. "I mean, their budget had been cut to the bone. They were barely able to maintain *Mir* operations and had cut every other program they had going." Keeping the Russian presence in space alive served an ulterior motive as well, Culbertson continued:

There was obviously a feeling that if we helped them get through this period in space, that they would put more of their emphasis in that area than maybe in weapons or exported weapons. I understand that rationale, but I think preserving what they had done for the previous thirty years was also very important, because without all that, that all would have been lost to history. It would have been an interesting and incredible series of accomplishments, but nothing following if we'd let the whole thing collapse. A lot of talented people, a lot of hardware, a lot of experience would have just gone down the drain.

While the United States worked to get its shuttle program up and running during the 1970s and 1980s, Russia abandoned its own *Buran* reusable winged orbiter after a lone unmanned test flight in 1988 to instead concentrate on long-duration stays in space. Dick Covey, who helped lead discussions with the space agency on the other side of the pond, felt the partnership could work from the very beginning:

The Russians had gone down their path and we had gone down our path. Even though they flew the Buran *and showed that they could do it, they never really had any use for it because it didn't fit into their program. They had matured their space station and were keeping it manned all the time. We were flying the Space Shuttle routinely, being very successful with it and getting ready to build the space station which they became part of. By the time we got around to talking about exchanging crew members, [discussion] was more along the lines of, "Gee, let us show you what we've done and we'd like to see what you've done."*

It was not in a competitive sense, but in very much a cooperative sense. That set the groundwork for the International Space Station.

There were issues to overcome, and they were not minor. Most obvious was the language difference, and, at least in the beginning, planners attempted to compress what would normally have been a two-year course of language study into just six months. When some questioned the tactic, they were told that these were astronauts—they were smarter than most people, so they could handle it. John Blaha, the third American to do a long-duration stay on *Mir*, was not shy in expressing his opinion that the decision had been the wrong one to make:

I've told every senior person at NASA this. I started telling them this when I was in Star City and I kept telling them this when I returned. One of the largest failures by NASA management, in my view, was to not adequately prepare people in the Russian language before they went through the gates of Star City and took the cosmonaut training course in the Russian language. The human beings who paid for that in spades were the seven human beings who had to use the Russian language media to take the courses, learn the information, and take course exams. In my view, this management error made everything for each crew member three, four, five times as difficult as it needed to be.

The language barrier extended beyond normal conversation—Russian computer programming was written in Cyrillic, NASA's in English. Russian measurements were metric, NASA's were not. Issues on down to cabling connectors, wiring, and electrical grounding had to be discussed and worked through. That was not all. *Mir*'s orbit of 51.6 degrees to the equator meant that the shuttle would have to launch northward, rather than due east, to meet a vehicle that originated out of Russia. As a result, the shuttle could not receive nearly as much of a boost from the speed of the earth's rotation, which greatly reduced its lifting capacity. The station's orbit also made for a very small launch window, typically only about five minutes each day.

Those were issues for engineers and mathematicians, where common ground was easier to find. Friction on both sides eased, yes, but long-held animosities were not forgotten outright. When jet-jockeys-turned-astronauts-and-cosmonauts got together as the program got under way, there was an inevitable sizing up of the "other" side. And, after enjoying so many firsts for so many years, it was difficult for some in the Russian spaceflight community to accept help at a time when its budget had been cut to almost nil. "There were some people who were very realistic and understood this was the only way that they were going to continue in the business," Culbertson said. "There were also people who did resent the fact that they couldn't continue doing this on their own, and I don't blame them. It's kind of human nature."

Of course, those on the American side of the fence were not immune from anxiety over working with a former foe. Dave Leestma was the director of flight crew operations during the Shuttle-*Mir* era, and he remembered the reactions of some of his fellow astronauts when the subject was broached. "I had to take people and say, 'I want you to fly in a Russian rocket to a Russian space station,'" Leestma said. "They'd go, 'I didn't come here to do *that*. I came to fly in the shuttle.'" When it was announced in the fall of 1996 that delivery of the Russian *Zvezda* service module for the upcoming ISS program would be delayed due to continued budgetary woes, nerves over the partnership were not eased in the slightest.

Charlie Bolden graduated from the United States Naval Academy in 1968, during the darkest days of the country's involvement in the Vietnam War—its latest, greatest, and, ultimately, most futile attempt at stemming the tide of communism. After earning his pilot's wings, the newly minted marine flew more than one hundred combat sorties in Southeast Asia. By the time he retired from active duty in the marines in August 2004, he had reached the rank of major general. He was a steadfast career military man—he was all about the corps to his core, if you will. That experience might very well have played into his initial reluctance to work with the Russians. "I honestly did not feel that they had been equal partners over the decades," Bolden declared. "I'd been led to believe that we had given them information from our spaceflights, and that we had received little or no information from them in return. Taking quite a naïve and immature approach, I didn't think that was fair. I wasn't ready to sign up for them learning all about our systems."

The first tangible piece of the Shuttle-*Mir* puzzle was put in place when cosmonaut Sergei Krikalev was named to the crew of STS-60, which flew in February 1994—the second and last mission commanded by Bolden. Already a veteran of two separate extended stays in space, Krikalev was actually on *Mir* when the Soviet government crumpled. By the end of his six flights—which included two on the shuttle—Krikalev had been in space for 803 days, 9 hours, 39 minutes. Once he had the chance to work with Krikalev and Vladimir Titov, who became the second Russian to fly on the shuttle during a subsequent mission, Bolden became a believer. They were pilots who had just so happened to be on opposite sides of a global confrontation none of them had started. Bolden marveled, "After that flight . . . I had no question of the value and importance of our cooperative exploration efforts. The Russian government brought the whole crew over for some post-flight events, and we had an opportunity to observe the way the Russian people live. When you went out to the test facilities and saw a brand-new engine being fueled through a line that was rusting and corroding, you had to come away with a tremendous amount of respect and admiration for the people in the Russian space program, who were able to accomplish all that they accomplished."

Norman E. Thagard became the first American to launch from Baikonur Cosmodrome in Kazakstan a year later, and by the time the program was completed in June 1998, seven American citizens had spent a total of 831 days on *Mir*. Dave Wolf, who took the ride up on STS-86 for a stay on *Mir* and back down again on STS-89 more than three months later, remembered that most were able to rise above "us against them" feelings in time. "Some of the older generation of cosmonauts, just some, just a few—most were fully happy to have the U.S. astronauts involved—but some saw it as their country selling out their superiority in space and had second thoughts about the joint program," Wolf said. "It played out more as a coolness than anything else. I really can't think of any [issues] that didn't get overcome over the years."

Wolf worked hard to clear any nationalistic hurdles he faced:

I think people don't quite realize that the six or seven of us that [flew on Mir*] actually became cosmonauts and had a class of cosmonauts that we joined and graduated from. So we had camaraderie amongst ourselves as any class of astronauts would have, and became very close. I spent special effort to really live the life with them and join their program fully. I did not consider myself an American going in a Russian vehicle. I considered myself—in cosmonaut school, and then as a cosmonaut—genuinely joining the Russian pro-*

gram. They sensed that. They sensed my commitment and the genuineness of it, and treated me as such. I generated true friendships. I always felt that the basis of the Russian-U.S. program was based on mutual trust and respect that was built person by person, as opposed to an agreement on paper. I saw it as my duty to add to that.

The reality of it was that it sometimes took rough-and-tumble humor and sports to break down what remained of the Iron Curtain. Wolf had been trained as an air force "backseater," a navigation and weapons officer in the F-4 Phantom. Aviators on both sides often joked about how their governments had pitted them against each other. It was not all that long ago, after all, that they might very well have been called upon to kill each other. Maybe, just maybe, if they proved that cooperation between the nations in space could succeed, the same kind of objectives stood a real chance of working back on Earth.

The competitive natures of both sides played out in other ways. One game they played was technically soccer, but it was not the sport known by millions of schoolchildren and so-called soccer moms in American suburbia. This contest was played on an indoor court, and from the sound of Wolf's description, it seemed to be an extreme combination of soccer and rugby. "There were walls around the court, so it never stopped," Wolf recalled. "There was no out of bounds or time-outs. And it was *full* contact. It was rough, real rough. I liked it, and they knew I liked it. They invited me to play three times a week. Not many of the American cosmonauts actively engaged in that."

It was the most rugged of competitions, but it was hands down a better alternative than a salvo of bullets and bombs.

Many scenarios had been thrown Eileen Collins's way as an air force pilot, but nothing quite like this.

Signs of upheaval were apparent all around her during her two-week stay in Russia during training for the February 1995 flight of STS-63, which was to rendezvous, but not actually dock, with *Mir*. There were broken windows and deteriorating buildings in Moscow, to the extent that a couple of Russian women told Collins that they were embarrassed to have her see their country in such squalor. In Star City there were no lights in some classrooms. The pool used for spacewalk training was decommissioned and so dirty that it was impossible to see for more than a couple of feet. Still, Collins did not immediately recognize a lingering Cold War mentality:

I did not see that in the cosmonauts, and I did not see it in the engineers or the instructors that I worked with. But I'm sure that attitude was there, and it was probably in areas that I didn't interact with. I'm making this sound probably more rosy than what it was, but I saw people that were passionate about spaceflight. They were passionate about spaceflight, their history, and what they were doing. I saw some people with tears in their eyes about the fact that it wasn't progressing as fast as they would have wanted during that period of time.

Not that Collins was complaining about the conditions. She was a rookie preparing for a historic mission, the first in twenty years to come so close to a Russian spacecraft on orbit. *Discovery* was not scheduled to dock with *Mir* but instead close to within a few feet of the station. The distance had shrunk during negotiations, first from 1,000 feet, then to 300, before narrowing still more to 100. Finally, the sides settled on a distance of just 10 meters—or 32.8 feet. The approach and subsequent fly-around of the station would take some fancy flying by commander Jim Wetherbee, with Collins standing by with checklists, monitoring radar and keeping track of the "big picture." And that was just one part of an eight-day flight plan that also included twenty Spacehab-3 experiments, the *Spartan-204* free-flying research platform and an EVA by Mike Foale and Bernard Harris to test modifications on their suits designed to keep astronauts warm in the extreme cold of space.

Collins also faced a more subjective aspect of the mission—she was to be the first female pilot in NASA's history. The predicament was an odd one in which to find herself. Yes, she was breaking down a barrier that had existed for more than forty years. Thirteen women passed the same physical examinations as the agency's original seven male astronauts, but none of the Mercury Thirteen ever made it to space. Several American women had flown on the shuttle, but they did so as mission specialists. Collins was a pilot at the controls and just a step away from a command of her own. That she was a *female* pilot was completely beside the point to Collins. Nobody ever asked Alan Shepard what it felt like to be a male astronaut. When the attention she received appeared ready to get out of hand, she always refocused on the tasks at hand:

I never let that stuff bother me. I'd always, always come back to, "Why are we here? In general, we're here for human space exploration. But more specifically, what are we doing on STS-63? We're going to do the first rendezvous with the Mir *space station. We're going to deploy a science satellite. We're going to do a spacewalk." So if someone asked me a silly question, or one that might even be either intimidating or insulting, I always came back to the mission. If women—or men, actually—let those questions get to them, that's a waste of energy. Energy is better spent on the mission, versus worrying about some peripheral question.*

Eight of the surviving Mercury Thirteen attended the launch of STS-63 on 3 February 1995 as Collins's guests, and at least some of them would be there for all of her departures. Nine hours later, once *Discovery* was on orbit, the crew began a series of Reaction Control System (RCS) thruster burns to bring the orbiter in line with *Mir* in preparation for the momentous rendezvous. Two of the forty-four maneuvering jets developed oxidizer leaks, and one, an aft primary thruster, was pointed directly at the station.

Murphy's Law states that if anything can go wrong, it will, and this was a good example of an addendum to that rule—Kelso's Corollary. Anything that could go haywire still did, but Rob Kelso, the Orbit 1 flight director on STS-63, figured that when things went wrong, they tended to do so at the worst possible moment. This jet pointed in this direction, straight at the station, was just that. "It was the worst jet to have a leak on the worst flight, our first rendezvous with *Mir*," Kelso said. "If it had been any other leaking jet, a different direction, or another flight, it probably wouldn't have been a big deal." When the Russians discovered what was taking place, their initial reaction was understandable.

Nyet.

The concern was that fuel from the leaky jet might contaminate *Mir* in general and, in particular, that one of the leaky jets was aimed right at the periscope of the Soyuz capsule. The incident began an extensive series of negotiations and technical exchanges between the two space agencies, and the solution turned out to be a try-it-and-hope-that-it-works type of fix. Remembered Kelso, "I decided with my propulsion officer that we'd open up an isolation valve and allow a flow of propellant, to see if the force of the propellant hitting against the backside of the thruster valve would get it to close and stop the leak. Sure enough, it did."

Plans for the rendezvous were back on track. Collins marveled, "Our NASA flight directors and their support teams put together a plan to shut off fuel to that thruster. They convinced the Russians that we had our act together, and that no matter what happened, we were going to be safe. When mission control called to tell us the Russians had agreed to the close rendezvous, we said, 'Ha! We already know!' Vladimir Titov, who was on board the shuttle with us, had gotten on the radio and called his cosmonaut friends on *Mir*, who had already heard the news from Russian mission control."

The problem was exactly the kind of thing the sides needed to test their partnership, to see just how far they were willing to go to make it work. "It did force exchange between the NASA and Russian control centers," Kelso said. "We *had* to talk, and we had to come to an agreement on how we were going to deal with the problem." When all was said and done, Wetherbee performed a station-keeping operation for fifteen minutes at a distance of four hundred feet, before closing to a distance of a mere thirty-seven feet from *Mir*.

So close were the two spacecraft that Bernard A. Harris, the payload commander on STS-63, captured one of the most memorable images of the shuttle program. From a distance of just a few yards, Harris snapped a remarkable photograph of cosmonaut Valeri V. Polyakov peering through one of the station's windows at the visiting shuttle. "We could see that great big window," Collins said. "Bernard took a photograph of Dr. Polyakov in the window, and it made the cover of *Aviation Week* magazine. It was really, really neat." Then, with a bit of amusement in her voice, she added, "It was the only one he took that was focused. I went and pulled the negatives, and everything before and everything after was out of focus. That one picture was in focus, and it was fabulous."

STS-63 proved that the shuttle could move near *Mir*. Now it was time to close the deal and dock, and the job fell to *Atlantis* and the crew of STS-71 in late June–early July 1995.

Commander Hoot Gibson was on the fifth and final flight of his storied astronaut career, one that could very well have been derailed just five years earlier. In a 9 July 1990 press release, it was announced that Gibson and David Walker had both been removed from command of shuttle flights for "violations of Johnson Space Center, Houston, flight crew operations guidelines." According to an article in the *New York Times*, Walker was grounded from flight status on the T-38 for six months "for a number of infractions of NASA aircraft operating guidelines," which included a May 1989 midair close call between Walker's T-38 and a Pan Am Airbus 310. Walker returned to flight status on schedule and subsequently commanded two shuttle flights—STS-53 in December 1992 and STS-69 in September 1995.

For Gibson, who had made contact with another plane during a forbidden air-show race just two days prior to the announcement, the consequences were more immediate. He landed safely, but the sixty-nine-year-old pilot of the other plane, Henry W. Jones Jr., was killed when his plane crashed into a cornfield.

Barred from flight status on the T-38 for a full year, Gibson bounced back from the incident apparently none the worse for wear. He commanded STS-47 in September 1992, and then was named chief of the astronaut office shortly thereafter. He left that role to begin training for STS-71. How ironic it must have been, for just seven years earlier, Gibson had commanded the STS-27 DoD flight that deployed a payload that was likely used in some shape, form, or fashion against the Soviet Union.

This was the start of an entirely new reality. Americans and Russians were no longer blood enemies—they were not exactly bosom buddies, not yet at least, but the fighter pilots among them were not staring each other down in the skies over East and West Germany, either. Gibson's pilot for this flight was Charles J. Precourt, who had been in one of those air force F-15s. The Americans, Russians, British, and French grudgingly held sway over sectors of Berlin at the time, and all had representatives in the control tower at the airport in the city's Tempelhof Airport. Before the start of Shuttle-*Mir*, the former Soviet MiG jockey who served as his country's representative in the Berlin control tower was the only Russian Precourt had ever met.

That was changing, and quickly. There had been a few tentative and awkward exchanges with Krikalev and Titov here and there. Once the crew of STS-71 was named and the flight plan announced—Anatoly Solovyev and Nikolai Budarin would take *Atlantis* up to the station and Thagard, Vladimir Dezhurov, and Gennady Strekalov back down again, marking the first on-orbit exchange of crew members—the relationship evolved all the more. Precourt remembered:

When they first came over here, I had already started to learn some Russian. We were going to take them to the [Houston] Astros [Major League Baseball] game and down to Galveston, and I took Nikolai in my little pickup. We had a forty-five-minute ride down to Galveston. He didn't speak much English and I didn't speak much Russian, but we were able to, with a dictionary, talk about a few things on the way down there. You think about how the relationships started there and where they've evolved to today, it's pretty incredible because we couldn't even talk to each other back then and now we do everything together.

Atlantis ultimately made the first seven trips to *Mir* because it was the only orbiter fitted with a docking mechanism at the time, and the winged craft headed to the station for the first time after hurtling off pad 39A on 27 June 1995. It was the 100th crewed mission of the American space program, and it would mark the beginning of one of its most important post-Apollo undertakings. For the final fifteen years of the Space Shuttle era, its legacy would forever be linked with the Russians and the stations—*Mir* and ISS. This had better work—or else. Gibson brought *Atlantis* in on the new R-bar approach from directly underneath *Mir*, which not only allowed natural forces to slow its closure more than a standard advance from in front, but also reduced the need to fire maneuvering jets. *Atlantis* locked onto the *Buran* docking port on *Mir*'s *Kristall* module at 9 a.m. EDT on 29 June, and at a total mass of about 225 tons, they formed the largest spacecraft ever to that point.

Once the hatches were opened, Precourt was surprised by what he found. The *Mir* mockups back on the ground in Star City had been relatively spacious, which turned out to be almost nothing like the real thing. "What immediately struck me was how used it was and how packed it was," Precourt admitted. "It was literally like going into your garage and trying to clean it out after you've been packing stuff in there for twelve years, one of those garages that you no longer park your car in, you just pack stuff in. That's the way it was in the *Mir*."

Much would later be made of the conditions aboard the station, but Precourt balked at the notion that they were substandard:

I'd kind of take issue with people who are saying it's old and run down, it's about to fall apart, and so on and so forth. Yes, there were a lot of problems with what was going on inside the base block and the Kvant, *with some leaking fluid lines and things of that nature, but these are the kinds of problems that we all expe-*

15. During a brief fly-around of the structure, Russian cosmonauts captured this historic image of *Atlantis* docked with *Mir* during the flight of STS-71. Courtesy NASA.

*rience in our homes. I'm right now in the process of changing the coil to my air conditioner on the second floor. My family's without air conditioning. These are normal, everyday life events. What I hope the American public can glean from the education of the Phase 1 Shuttle-*Mir *program is that all hardware breaks down. We have to learn to take our hardware to space and not bring it home in a hurry like an airliner that might be flying home that has a problem, but learn to repair it out there. If we don't learn to do it out there, we won't ever be able to stay there very long, because hardware does fail. Our ultimate goal is to be able to go to the moon and Mars and put bases there for scientific research and for exploration purposes and stay there and survive. So these lessons are very valuable to us. The fact that the* Mir *went through ups and downs and we were able to live through that, I think, is a great testament to what we were able to do together. It should make people think twice when they try to criticize the Russians and their system for what it is or is not capable of doing.*

The two craft separated on 4 July—Independence Day in America—with a complement of eight crew members on board *Atlantis*. Along with

Gibson, Precourt, Thagard, Dezhurov, and Strekalov were STS-71 mission specialists Ellen S. Baker, Greg Harbaugh, and Bonnie J. Dunbar. After spending more than one hundred days on orbit, Thagard and his Russian colleagues were laid out flat in custom-built seats for the descent on 7 July 1995. Despite the precaution, the American was up and out of his seat on *Atlantis*'s mid-deck without a problem after landing.

The scenario posed by the instructor at the United States Military Academy at West Point, New York, had a chilling way of cutting right to the chase. Thousands of Soviet tanks were pouring through the Fulda Gap in Germany. There might or might not have been enough troops, tanks, and aircraft available to stop the attack. The clock was ticking with each and every German hamlet that fell by the wayside. Nuclear weapons were on hand. When and where would they be used? Cadet William S. McArthur had to figure out a way to stop World War III.

McArthur had been asked from his very first day at West Point if he was somehow related to Douglas McArthur, the iconic I-shall-return general from the Pacific theater of World War II. He was not, but even if he had been, it would not have mattered if the Soviet Union invaded West Germany and, from there, the rest of Europe. If such a nightmare had ever actually happened, the event would have changed the course of human history—if it did not end it altogether in an atomic holocaust. The Soviet Union was the enemy, and there was no way McArthur could have imagined the communist nation as anything but.

Then came the November 1995 flight of STS-74, the second shuttle mission to dock with *Mir*.

McArthur was not alone in what must have felt like something straight out of *The Twilight Zone*. Commander Kenneth D. Cameron was a marine colonel; pilot Jim Halsell and mission specialist Jerry Ross were both air force colonels; and Chris A. Hadfield was a pilot in the Canadian air force. For all of them, going to Russia for the first time during training was surreal—but rewarding, all the same. They were visiting the Kremlin and strolling right through the middle of Red Square—places that Ross knew full well would have been ground zero in a nuclear confrontation with the United States. Others felt at least some resentment on the part of the Russians, but McArthur insisted that he did not. There was one very good reason for that—physics.

"The physics required to get into space are not unique to a single country," McArthur explained. "The math and the science are universal. Newton's laws of motion don't apply just in England. The physical challenges of launching people into low-Earth orbit and bringing them back safely are identical in America and in Russia. It takes 17,500 mph to be in low-Earth orbit, and it doesn't matter whether you say it in Russian, English, French, or German. You still need the same velocity. At the cosmonaut and astronaut level, we had much more in common than we had differences."

Common sense dictated that the two sides respect the accomplishments of the other. The Soviets put the first human in space and launched the first space station, and no nation other than America had ever put a man on the surface of the moon. By the time McArthur finished his spaceflight career, he had flown once to *Mir* and twice to the ISS—one time as a member of the STS-92 crew and the second and final time on board the *Soyuz TMA-7* capsule for the start of a long-duration mission. Over the twelve years or so that he spent going back and forth to his once-rivaled country, McArthur watched as his working relationship evolved and matured from guest to colleague.

Ross, however, recognized that there were hostilities simmering just below the surface on both sides of the fledgling partnership. "We didn't trust each other," he admitted. "We'd been sworn enemies. I think most of us probably didn't even like the idea of working with the other guys all that well." In the long term, obviously, it was a good idea. *All* of mankind should take part, Ross said, and not just a couple of superpowers. Many obstacles—mistrust, differences in language, measuring units, distances, time zones, cultures, food, and so forth—were hard to overcome. "It takes time," Ross concluded. "I think the earlier we got started, the better, so we could get over some of those things and start to develop levels of understanding and trust."

The first docking mission with *Mir* ferried cosmonauts and one American astronaut back and forth, but the primary objective of STS-74 was to deliver a Russian-built docking tunnel to the station. To get the module ready for *Mir*, Hadfield used *Atlantis*'s robotic arm to lift it out of the cargo bay, and he then rotated it into a vertical position above the shuttle's own Orbiter Docking System. The two modules were positioned a few inches apart, at which point the fun began. With the robotic arm holding the Russian segment relaxed, Cameron fired the shuttle's maneuvering jets to ram the seg-

16. A module used to dock with *Mir* was the primary payload on STS-74; here, it is about to be berthed in *Atlantis*'s payload bay. If the two had not firmly latched, astronauts Jerry Ross and Bill McArthur were prepared—if not hoping—to go outside and strap them together. Courtesy NASA.

ments ran into a connection. "We hoped we'd be able to have enough velocity and contact that all the latches would engage," McArthur said. That was the primary plan. No spacewalks were planned for the flight, but Ross and McArthur had trained for a contingency EVA in which they would use a set of straps to ratchet the two docking mechanisms together.

Air pressure in the cabin of the shuttle was lowered to prepare for a potential spacewalk, and Ross and McArthur did a pre-breathe exercise of pure oxygen. They were not yet suited, but that was next. The thruster firing worked, however, and the American and Russian docking modules came together just fine. The twin tunnels were used to dock with *Mir* on 15 November 1995. Ross and McArthur did not get to do their EVA. Asked if he had wanted to go outside, even if it meant something had not gone according to plan, McArthur could not help but laugh.

Absolutely.

Only a few insiders knew at the time, but all was not well with astronaut M. Richard Clifford when he conducted a six-hour spacewalk during the

Shuttle-*Mir* flight of STS-76. Nearly two years earlier, he had been diagnosed with Parkinson's disease, a degenerative disorder of the central nervous system characterized by shaking; slowness of movement; stiffness or rigidity of the arms, legs, or trunk; and trouble with balance and falls as it progresses.

A 1974 graduate of West Point and a former test pilot, Clifford publicly disclosed his battle with the disease in an article he wrote for the National Parkinson Foundation.

In all likelihood, Clifford already had Parkinson's when he first flew in space during STS-53 in December 1992. Then, about six months after his second shuttle mission on STS-59 in April 1994, he went in for a routine flight physical and asked in passing if he could have an orthopedic surgeon check out his right shoulder, which he thought he had injured while playing racquetball. One exam led to another, and the news came out of nowhere—Clifford did in fact have Parkinson's. "I had never heard of this disease and as someone who considered himself to be in excellent physical condition, I naturally assumed it was something I could conquer," Clifford wrote. "In fact, my response was something like, 'Okay, fix it so I can get back to my racquetball!' Then, reality hit me. Hard!" Clifford continued:

It seemed impossible. I didn't want to believe it, and for a while, I refused to believe it. So many things went through my head when I began to learn more about my condition, but I was resolute and determined not to let it affect my outlook. The medical community respected my privacy and only those senior NASA managers with a need to know were informed. They asked me what I wanted to do, and my response was quick: I wanted to remain on flight status and remain in the cue for a future space flight. I wanted to remain an astronaut.

While NASA flight surgeons kept a careful watch on Clifford, he remained eligible to fly again. No one outside a tightly guarded circle of agency doctors, upper management, and family knew of the diagnosis, and he was determined to keep it that way—too many reporters could ask too many potentially embarrassing questions.

He landed a spot on the STS-76 crew, but despite testing much of the hardware and operational procedures for the experiment packages planned for installation outside *Mir*, Clifford was not initially assigned to take part in the spacewalk itself. "I informed management that I wanted to do the EVA and that I didn't know there were limitations imposed on my capabilities," Clifford continued in his personal account. "I think they were actually surprised by my desire to perform the spacewalk, despite my condition."

After STS-76, Clifford left the astronaut corps in 1997 to join Boeing and become the deputy program manager of the company's Space Shuttle efforts.

The Russian people impressed Tom Akers during his training for STS-79. The secretary in Star City's crew quarters had a husband who was a lieutenant colonel in the Russian military, and they lived with their children and an in-law or two in a small, two- or three-room apartment. The cramped conditions did not appear to matter to the woman. "She was happy and bubbly," Akers recalled. "Most of the people I ran into were happy with a whole lot less than most of us Americans would have been. They're just regular people. They worry about the same things."

And then there was the way Russia went about its business in space. In some ways, he said, it was more efficient—and, yes, better—than the American approach. "If something works, they don't change it," Akers continued. "It seems to me like here in America—and I've seen it personally—you start to do something and get a bunch of new engineers, and they want to redesign everything and reinvent the wheel rather than using existing technology that works." The end result, for Akers at least, was not so much an us-against-them remnant of the Cold War as it was a sense of finally being on the same team. His mind was on spaceflight, and not on things of the military and world politics. Rather than considering himself a warrior in enemy territory, Akers was instead an engineer working to resolve complicated issues with a group of coworkers.

On each of his three previous flights, Akers had been confined to the habitable volume of the shuttle when he was not outside on a spacewalk. After the launch of STS-79 on 16 September 1996, he had not only *Mir* to explore, but the Spacehab Research Double Module in the payload bay as well. "It was kind of like moving out of a three-man tent with seven people in it into a three-bedroom house," he said. "I mean, there was just all kinds of space. It was neat just exploring over on *Mir*." Working as one of the operators for an IMAX camera on board and as the mission specialist in charge of transferring more than four thousand pounds of supplies from *Atlantis* to *Mir*, Akers had opportunity to float all over the Russian station that reminded him, many times, of being in a cave. "You'd go back through those tunnels, and there weren't any windows," he described. "There were things stored against the wall and strapped down that even Russian cosmonauts didn't know what it was. They didn't have any way to get rid of things, other than kick them overboard, so a lot of things were just stowed."

If the flight of STS-79 is best remembered for anything other than being the fourth mission to dock with *Mir*, it is the fact that it lifted John Blaha to the station and finally brought Shannon Lucid back home. The flight had originally been slated to lift off on 31 July 1996, but was delayed when mission managers opted to swap out *Atlantis*'s SRBS due to a potential problem with the adhesives holding them together. The change was made after the stack was rolled back to the VAB as Hurricane Bertha approached. Once back out on pad 39A for a planned 12 September launch, yet another wonder of nature—Hurricane Fran this time—forced a second rollback. Finally, on 16 September, the flight got under way.

Lucid had been in space for 188 days, setting a NASA record and world record for a woman. If there happened to be a difference in their demeanor—Blaha looking forward to beginning his adventure and Lucid relieved to be ending hers—Akers insisted that he did not notice it.

This was going to be Akers's last chance to bask in such a unique environment. Although he had not told anyone yet, he felt reasonably secure in the fact that he was not going to fly again. "Generally, the more times you flew, the longer it was between flights," he said. "You've got a lot of folks who've never got to fly yet. Every time one of us old guys who had flown would fly again, I felt like we were taking up a seat that some new person could go do, and probably do as good or better job than I could." Akers's final journey to orbit had a much different feel than his others:

That fourth flight was a lot more relaxed in terms of the mission getting accomplished than the first three. In the first one, we had to deploy Ulysses *in this ten-second window. You've got to do this just right. Don't screw it up. On the* Intelsat, *obviously, there was the stress and pressure of doing that right—and then Hubble. The* Mir *mission was a lot more relaxed once we were there. In fact, we got everything moved earlier than we had planned on that mission. Then, just having so much room, it was a memorable flight. I was blessed. Every crew I flew with, we had a great crew and everybody got along. It was the same on that last one. I had nothing but good experiences in space, but I do just remember that one being more relaxed.*

Brothers Tom and Ray Magliozzi, hosts of the National Public Radio auto-repair program *Car Talk*, were stumped. They simply could not figure out the problem that "John, from Houston" was having with his government vehicle. Its engine started out loud for a few minutes before getting quieter, and then it would quit altogether. Back and forth the three of them went, trying to get the issues straightened out. Finally, it dawned on the Magliozzis that they were, in fact, talking to astronaut John Grunsfeld and that he was not actually in Houston—he was on orbit, "bumping around the insides of the shuttle *Atlantis*" during the flight of STS-81 in January 1997.

Almost every flight plan managed to include time for crew members to grease the public-relations wheel, and this one was no different, even while docked with *Mir*. If Grunsfeld was comfortable with such duties, that was fine. But on his second spaceflight, he was taking to a mission that did not include a lot of astronomy. Studying the cosmos was his life's work, but on STS-81, he and physicist Jeff Wisoff were responsible for the Biorack suite of biological experiments out of the Spacehab Research Double Module located in *Atlantis*'s payload bay. The relative lack of astronomical research did not matter to Grunsfeld, because, quite simply, he was back in his element. "My first mission had been sixteen days in space, and on the last day, I didn't want to come home," Grunsfeld said without the faintest trace of insincerity in his voice. "I just wanted to stay in space. I felt like I had found my true home. I felt more comfortable in space than I *ever* do on the ground."

Grunsfeld was intrigued by the opportunity to visit Russia during training for STS-81, and then after launch, it was again as if he was headed to a foreign country in the form of the station itself. "Here, we were not just going and orbiting the earth, we were actually going somewhere," Grunsfeld continued. "It was really kind of like visiting a Russian cabin in the woods. When we arrived, it wasn't the emptiness of space and a beautiful earth. There was a place we were going to and people we were going to see and visit. It was quite exciting." Once docked with *Mir*, Grunsfeld and Wisoff filmed an hour-long tour of the station.

If he was surprised at all by the conditions, Grunsfeld figured it was the result of the cultural differences between Russians and Americans:

If you live somewhere like Chicago or New York City, that's a multicultural city. When you go visit somebody from a very different cultural background—say you go visit a friend whose parents are of Russian origin or Chinese origin—the first thing you're going to notice is it smells different. Why does it smell different? Well, because the fabrics are different. The furniture is different. The foods that they cook are different. It doesn't mean it smells bad. It smells different. Well, you go into the Mir *space station, and it smells different than the shuttle. The shuttle has kind of a sterile feel, and eventually, it has a human smell. One of the modules smelled like a gym. That's because that's where they hang up the gym clothes after they work out. They had foam-covered walls for padding, which collected moisture, which grew mold. But it wasn't like it was the run-down, ramshackle space station that appears in the movie* Armageddon. *It was nothing like that.*

Atlantis delivered Jerry M. Linenger to the station, where he replaced Blaha on the long-duration crew. Blaha, who had commanded the flights of STS-43 and STS-58 with Lucid as a mission specialist, would later express disappointment with the handover he received from her when they swapped places on *Mir*. "I'll never quite understand why she did not give me all the information she had learned during her long stay on the *Mir*," Blaha wondered. "She gave me a handover, but it was not a complete handover. It's not what I would do for my brother or sister if I were living in a cabin on a mountaintop. My goal would be to help them start with the knowledge I had when I was finishing."

Blaha began writing training memos while on the station, e-mailing them back to the ground every few weeks or so to help those who would come after him. And when Linenger arrived, the man whose place he was taking was still dissatisfied that not enough time had been allotted for an adequate handover. In his autobiography *Off the Planet*, Linenger explained that Blaha held other, more serious, frustrations:

John told me that living on Mir *was an endurance test. There was nowhere to go; you were trapped. And no matter how bad things became, no matter how uncooperative the Russians were or how poor the ground support was, there was nothing that you could do about it. He insisted that it was important for me to accept my circumstances early. Whether I liked it or not, I was stuck onboard for the duration. Since the only person you can truly depend upon is yourself, when an uncomfortable situation arises, do something about it, overcome it. Do not let problems linger, because they never go away by themselves.*

In truth, nothing could fully have prepared Linenger for what lay ahead. A substantial fire in the station's *Kvant-1* module on 23 February 1997 caused smoke to spread throughout much of the station. Scrambling to strap on a respirator mask, Linenger was sure that he was about to die. "I never thought that my life would, in such an unexpected moment, just abruptly end," Linenger wrote. "Well, I thought to myself, I guess this is how it happens. At some point, we all die. I just never expected it to be now." The fire was extinguished, but it took some thirty-six hours to clear the air.

Linenger did not return home for another three months.

Keeping souvenirs of spaceflights had been a long-held tradition ever since the earliest Mercury missions. Charlie Precourt was no different during his command of STS-84 in May 1997, the sixth flight to dock with *Mir*. One was a memento given to him by a former foe, and the other was a reminder of the accident that could very well have cost Linenger his life:

*I brought Jerry home, so I got there right after the fire had occurred and they had cleaned it up completely. There was no remaining evidence of a fire on board, but Vasily Tsibliev, the [*Mir*] commander at the time, was nice enough to give me a tour and show me where everything transpired. You know the little pin they pull out of a fire extinguisher to make it work? He had about sixteen or eighteen of them on a little chain that he had made for souvenirs, and he gave me one. And he gave me one of his EVA gloves that he had worn outside on one of their many EVAs to repair the leaks in the system and whatnot. That was really a very, very memorable moment. I have his EVA glove in my living room, and it's a souvenir that I'll cherish forever.*

It was not the only fond memory from the flight. Never before had international cooperation been as important in the space program, especially after the fire, and so a meal was planned featuring foods from the native countries of each of the explorers on board *Mir*—Precourt, pilot Eileen

Collins, Linenger, and Mike Foale (who held dual citizenship with his native Great Britain) representing the United States of America; Tsibliev, Sasha Lazutkin, and Elena Kondakova (who flew to *Mir* on *Atlantis* along with the rest of the STS-84 crew) from Russia; Jean-François Clervoy from France; Carlos I. Noriega, born in Peru and a member of the U.S. Marine Corps; and Edward T. Lu, born to a Chinese family and raised in the United States. Over the course of two or three hours, *Mir*'s base block became an impromptu dining hall. Clervoy, asked by Precourt to organize the big event, went so far as to come up with a very nice menu written in both English and Russian. Flown for every crew member was a copy of the food listing, each of which was stamped post-office style on the station and autographed. The menu included a delicious smorgasbord of international delicacies:

Appetizers—kosher dill pickles (Russia); shrimp cocktail (America); cottage cheese with nuts (Russia); block of duck foie gras (France); smoked beef and pork jerky (America and Russia); and dried cured beef and pork (China).

Main course—Pete's barbeque beef (America); chicken fajita with guacamole and picante sauce (Tex-Mex); cassoulet from Gimont (France); duck confit in piperade sauce (France); bacon with lentils (France); tortellinis in foie gras sauce (France); and duck in cep mushroom sauce (France). Appetizers and the main course were both served with Tex-Mex tortillas and black bread from Russia.

Dessert—ice cream, assorted flavors (America); chocolate Space Shuttles (America); pineapple and strawberry cookies (China); and Red Delicious apples (America).

And last but not least, drinks—apple cider (America); chicha corn drink (Peru); hot chocolate (America); kona coffee (Hawaii); and herb tea (France).

"We had a super party," Clervoy remembered. "I asked everybody in the crew to take food from their original country with them, so we could have a good dinner in *Mir*. We had a party for three hours one evening. No one called from the ground. That was probably the best three hours. You can have fun with colleagues during spaceflight."

The kind gift exchanges and dinner party notwithstanding, conditions were not always perfect while *Atlantis* was docked with the station. A Thermal Conditioning System failed, leaving humidity at or very near 100 percent. Some on board the station developed headaches as a result of the high humidity, but their bodies somehow managed to adapt. Blobs of water accumulated in some corners, and Clervoy would later recall that the combination smelled something like an old wine cave:

There was a strong smell of moisture, of mildew. It's funny, because during the five days we were docked, the hatches between Mir *and the shuttle were open continuously, all the time, twenty-four hours a day. From the cockpit of the shuttle, you could tell, "Oh, somebody is coming from* Mir.*" As one person was floating inside the shuttle from* Mir, *that person was pushing a mass of gas and cabin air in front of him or her. That would propagate through the connection of modules to the cockpit, and we would smell the mildew. It was quite funny. It was not the best condition, but we knew it before the launch. We knew that* Mir *was not in the best condition, but it was ready to welcome us.*

Linenger, who received a much-needed haircut from the stylish Clervoy, was about to return home after a harrowing four-month stay on orbit. He was replaced on the *Mir 23* crew by Foale, who would experience an emergency every bit as serious as Linenger's fire, if not more so. As Tsibliev guided a *Progress* resupply capsule to a docking on 25 June 1997, it collided with the station's *Spektr* module, damaging a radiator and a solar array. That was bad enough, but the station was also knocked into a spin and depressurized. The hatch to *Spektr* had to be sealed, rendering it useless, before the rest of the modules could be repressurized.

NASA had on many occasions stood on the precipice of calamity only to bring its astronauts back home safe and sound, ready to move on to the next challenge. The effort had always been worth the risk, but after Linenger and Foale nearly died, what was the advantage in continuing to work with the Russians? The fire and collision were not minor in nature, and because they took place during successive increments, it was almost impossible not to wonder if they were symptoms of larger problems within the Shuttle-*Mir* program itself. Collins remembered, "There was a ton of concern. Even within the astronaut office, there were a few astronauts that wanted to stop the *Mir* flights. They thought we ought to just say good-bye and not go to *Mir* any more. Then, there were those who said, 'No, we need to keep flying.' It was quite a bit of controversy. This was '97. The fire was in the February time frame. We flew in May and then the [*Progress* collision] accident happened in June. So from early '97 all the way to the end of *Mir*, there was always controversy."

So why continue working with the Russians? Collins had her personal opinions:

*I am certain you will get different answers from different people. The Shuttle-*Mir *program came out of the Clinton-Gore administration. So one could say that was the reason we stayed in. I also believed there were great advances from the program. We built trust with the Russians. We worked on problems together. We helped them. They helped us. We built trust on many levels. We learned technically. Astronauts and cosmonauts worked well together; we became friends. Engineers and flight controllers from both countries had to learn to work together. I was one who believed the program should continue. True, I had some self-interest, as I flew on* STS*-84, between the two accidents. Some people say the astronauts always want to fly, and others need to protect us from our own risk-taking! Hogwash. Of course we are not going to fly if we think we might get killed. Yes, the USA had already invested in plans for* ISS*. So in all, all of those factors are reasons. Another president may have pulled us out. But it was what it was, and it turned out to be a highly successful program. Yes, even the accidents taught us much. I think we stayed in because we felt the gain was worth the risk.*

The matter spilled over into the realm of Washington politics. During hearings in September 1997, F. James Sensenbrenner, chairman of the House of Representatives' science committee, argued against sending astronauts to do long-duration stays on the station.

Dave Wolf was supposed to replace Foale the very next month, but not if Sensenbrenner could help it. He was fully against spending months on *Mir* "being an assistant Mr. Fixit." David Weldon was a Republican member of Congress from Florida, and his district included the Cape. Initially a supporter of the working relationship with the Russians, the accidents gave him pause to reconsider. "I became concerned after the fire and decompression that the amount of knowledge we had gained was sufficient and the risks no longer justified a continued U.S. presence," Weldon charged. Ultimately, the science committee recommended against sending Wolf, but left the decision—and the astronaut's fate—up to NASA administrator Dan Goldin.

Wolf was going to fly.

Almost immediately after returning from STS-66, Scott Parazynski jumped into a backup role for Jerry Linenger's long-duration stay on board *Mir*. At six feet three inches tall—if he did not slouch, he joked—he had always been slightly outside the height limit to beat a hasty retreat in the Russian Soyuz craft had some sort of emergency arisen.

The capsule was cramped even in the best of circumstances, but for Parazynski, it would have been nothing short of misery to somehow fold his lanky frame inside. He considered himself a test case—there had always been a chance that he would be deemed too tall to fly inside a Soyuz. A 14 October 1995 NASA press release confirmed that concern. He had been bumped. "After discussing all our options and reviewing the available data with the Russians, it is clear that they do not have the latitude or sufficient modification capability on the Soyuz to allow Scott to return to Earth in the vehicle with a level of risk we would be comfortable with," Frank Culbertson said in the announcement. "Our Russian colleagues share our disappointment in this situation since Scott has achieved such a high level of performance and respect in the Russian system."

For Parazynski, much had already been invested in the Shuttle-*Mir* venture. He spent months learning the Russian language and had moved to Star City himself. However, he knew that the reasoning behind the decision was sound. "Sure enough, after careful analysis, they concluded that if I'd used the Soyuz in an emergency landing, I might have broken my neck and/or my knees," he said. Parazynski was out, case closed.

After that, it was back to Houston to get back in line for another flight, which came in the form of STS-86, the seventh shuttle mission to dock with *Mir*, which also included commander Jim Wetherbee; pilot Michael J. Bloomfield; mission specialists Jean-Loup J. M. Chrétien, Wendy B. Lawrence, and cosmonaut Vladimir Titov. The flight also took Dave Wolf, who replaced Lawrence on the *Mir 24* long-duration stay, to the station and brought Mike Foale back. The assignment did not mean a four-month station stay for Parazynski, but there was consolation—he and Titov were on tap to perform the first joint U.S.-Russian EVA.

Parazynski had already traveled the world many times over, and he found little or no Cold War animosity awaiting him in Russia. His Russian counterparts were warm, he said, and forthcoming with whatever information he might need. Less than a decade after the fall of the Soviet Union, what did get Parazynski's attention were the conditions in which he was living and training. "There was a lot of decay," he remembered. "It wasn't nearly as advanced or well-maintained as I imagined it would be when I got there."

Another surprise was the distinct difference in the Russian and American approaches to technology. The Russian style was much like the trusty pickup truck with a quarter of a million miles on the speedometer, in stark contrast to NASA's sparkling Corvette, right off the showroom floor. Despite the perceived lack of sophistication, Parazynski insisted he never felt even a moment's pause in heading to *Mir*:

I realized then, and I still believe, that the Russian program has been very evolutionary, as opposed to our revolutionary programs. The shuttle was a completely revolutionary approach to flying in space—a winged vehicle, reusable and so many things had to be invented from scratch. But the vehicles that are flying today from Kazakhstan are really a gradual evolution from the vehicle that Yuri Gagarin flew in April of 1961. It's really a very simple vehicle. If you look at the welds on a Russian spacecraft, it's built to be bombproof. The American approach to design is to understand the physics and the engineering limitations and material properties. With a computer-aided design, build it to a safety factor of whatever it needs to be and then do a modest amount of testing. The Russian approach is build it like a brick outhouse and then try to blow it up. If it doesn't blow up, it's probably safe enough to fly.

Atlantis left pad 39A late on the evening of 25 September 1997, an event that took place only after Dan Goldin gave his approval earlier in the day following extensive safety reviews in the wake of the problems that had plagued the Russian-built station. Three days later, Wetherbee docked with *Mir*. Parazynski found the conditions to be—what's the best word to describe it?—homey, much like visiting his great-grandmother's abode in upstate New York. Hindsight being twenty/twenty after being bumped from having a shot at a longer stay, he could not help but to admit, "I was glad that I was only going to be there for about five days, as opposed to five months."

His long-awaited first spacewalk, "the ultimate astronaut experience," he called it, was very nearly over before it began on 1 October 1997. As he tried to test his safety tether, it refused to retract. He pulled out a few feet of cable, expecting it to reel back in, much like the seat belt on any standard Earth-bound vehicle. It would not, no matter how hard he tried. Parazynski gave the device a good, old-fashioned shade tree mechanic's whack. He pulled a little more line out and then tried to push it back in, hoping that would free the jam. Nothing worked, with some fifteen feet or so of unsecured tether floating free.

The root cause of the issue had taken place well before the flight. During servicing back on the ground, a lubricant had been applied that gummed up the entire spring-loaded mechanism in the chill of space. Parazynski improvised by calling upon his experience in the great outdoors. "I had two waist tethers," Parazynski began. "They're about three to four feet in length, I suppose. You put them on a handrail, and that tethers you to the spacecraft. Then, you take your other safety tether and put it on another handrail and move around kind of like an ice climber going up a mountain."

The rest of the EVA proceeded unimpeded, with the international spacewalkers installing a 121-pound solar array cap that a future crew would use to seal a leak in the damaged *Spektr* module. They retrieved four Environmental Effects Experiment packages located on the outside of the station, and then did an engineering test flight of the Simplified Aid for EVA Rescue (SAFER) jet pack. In all, Parazynski's debut EVA lasted five hours. The problem with the tether would not be his last spacewalking drama.

Dave Wolf was not supposed to have made the trip to *Mir* quite so quickly.

As Wendy Lawrence's backup, Wolf was headed to the station himself on board STS-89 in January 1998. That plan would have allowed for four months of additional training, this time as the prime crew member. The pace would not have been a leisurely one by any stretch of the imagination, but Wolf could at the very least have had a bit of breathing room and time to make a planned trip to St. Petersburg with his girlfriend and family. The two major accidents on board *Mir* changed all that in dramatic—and lightning fast—fashion.

Less than two months before STS-86 was to fly Lawrence to *Mir* for a four-and-a-half-month stay, NASA announced on 30 July 1997 that she had been replaced by Wolf. The reason? According to the press release, NASA and Russian space officials had jointly agreed "that it would be mutually beneficial to have all three crew members on the *Mir* qualified for spacewalks in the event additional assistance is needed from the U.S. astronaut." The problem for Lawrence was one of practicality—at about five feet three inches tall, she was too short to fit into the Russian Orlan spacesuit. It was the exact opposite of the issue that kept Scott Parazynski from making a long-duration stay. Lawrence and Wolf were informed of the decision in Russia by fellow astronaut Frank Culbertson, who was serving as manager of the Shuttle-*Mir* program at the time.

The move was not necessarily a shocker for Lawrence, who had begun Russian language training in June 1995. A few months later, in September, she was informed by a flight surgeon that the Russians had released a memo stating that the country's minimum height requirement had been changed from 160 centimeters to 164—a little over five feet four inches. Still, Lawrence was eventually granted a waiver from Valery Ryumin, the Russian Phase I program manager, stating that she could begin training for a flight to *Mir* as long as she was never considered for a spacewalk in the Orlan suit. But when the decision was made to have the American crew member readied for a potential EVA, she was dropped from the long-duration mission. Lawrence received a suitable consolation prize from Bob Cabana, chief of the astronaut office. Not only would she still fly STS-86, she would get a second flight out of the deal on STS-91:

*That's why a lot of reporters just couldn't understand why I was not devastated. I knew that I was getting two space flights out of it. Right after landing, Charlie Precourt [who would serve as commander of STS-91, the final Shuttle-*Mir* docking] and two of the crew members from 91 were there serving as family escorts, Dom Gorie and Janet Kavandi. I literally hadn't been back on Earth more than about four or five hours, and we were already having a crew meeting. I looked at Charlie and said, "Hey, just give me a couple weeks to wrap up the debriefs and I'm ready to go. I mean, that wasn't a bad deal for me. Dave Wolf got a spacewalk out of it, and I got to fly twice and got to fly on 91 as the flight engineer, which is a job that I really enjoy doing, so I think we were both satisfied with how everything turned out.*

Wolf's next few weeks were about to hit warp speed as the launch of STS-86 loomed. "Wendy was walking out of the cottage when I walked in, and she had a funny look on her face," he remembered. "I knew something was wrong. I went in and [Culbertson] said, 'Are you ready to go?'" Wolf was, in fact, not ready at the moment to conduct a Russian spacewalk. He had been fully trained on American EVA procedures while preparing for a contingency spacewalk on STS-58, but the spacewalk methodology for a *Mir*-based EVA in a Russian suit and with Russian life-support systems was quite a different beast. Culbertson needed to know by the next morning if Wolf felt he was up for the challenge. While it was the tallest of orders, Wolf leapt at the opportunity to tackle the challenge. Years later, Wolf remembered the highly compressed preparations for the flight:

We met with all the management, trainers, EVA instructors and put together a program that would run me through the EVA programs seven days a week, all day, in about two and a half weeks. It was pretty brutal. My body was covered with scabs from being in the suit, wearing abrasions. I would require lots and lots of diving, vacuum chamber and laboratory work, as well as course work. We decided we could do it, and we did it. They would do round-robin instructors on me. We completed it just in the nick of time, and I flew back to the U.S. pretty much straight into quarantine. I was not thrilled that I didn't get a first-class seat back, because I was awfully tired.

All the while, Wolf remained involved in what was taking place back in Washington. Goldin called to make sure he was comfortable with flying, and he was. He knew there was a risk, but there was *always* going to be a risk with any spaceflight.

As the hatch between *Atlantis* and *Mir* was closed and the orbiter began to back away after undocking, Wolf could very plainly see STS-86 commander Jim Wetherbee and pilot Mike Bloomfield looking through their overhead windows, and felt as if he were a kid being dropped off on the first day of summer camp. His parents were driving away in the family station wagon, not to return for more than four long months that held who knew what. There had been serious and potentially deadly problems on the station, and they had spawned many calls in Congress, in the media, and even among more than a few in the astronaut office to end the venture then and there. Wolf was no dummy, though. He was not about to undertake a suicide mission, although, incredibly, he fully accepted "a good chance of losing the vehicle while I was there, but I didn't expect to die doing it." End the program with the Russians now, and that was that. Game over. "It's easy to be good partners when things are going well," Wolf said. "This is the time that testifies to the true strength of the partnership, and it's *exactly* the time to continue on."

The faith that Wolf felt in the program could very well have been fatal. No less than three times, *Mir* experienced total power loss with Wolf on board. On 13 November 1997, one outage caused the station's motion control system computer to shut down, and even after scavenging fresh batteries from another module to the base block, they were not enough to fully stabilize the structure. For the next couple of days, battery chargers would be temporarily turned on whenever *Mir* happened to drift into a position suitable to collect the sun's power-giving solar energy, or if the docked Soyuz module thrusters could somehow manage to tip it into the correct orientation. A couple of weeks later, the power went out again. "You don't know quiet until you've been in a spaceship with no power," Wolf quipped. Keeping the structure up and running proved to be tricky at times, Wolf remembered. "To recover the vehicle required twenty-four-hour attention to the power systems and manipulating the power relays, as well as Soyuz thruster management, to capture enough sunlight to bring the batteries back online," Wolf said. "This was all without essential space-to-ground communications. One had to be critically careful to not deplete the resources of the Soyuz lifeboat, because I was providing the only limited thruster and communication capability. I was trusted as a full part of that team."

So aware was Wolf of the problems *Mir* could and sometimes did encounter, he made sure to keep handy a survival pack containing such things as cable cutters and so forth and to always maintain a clear path between his bunk and the Soyuz escape craft. Then came 14 January 1998.

As a child, Wolf had watched in awe as Ed White made America's first spacewalk, and he had dreamed of making one himself ever since. Alongside *Mir 24* commander Anatoly Y. Solovyev, the most accomplished spacewalker in the history of human spaceflight, Wolf saw that his dream was quickly turning into a nightmare. After inspecting the radiators of the station's *Kvant-2* module during a nearly four-hour EVA, Wolf and Solovyev tried to make their way back into an external airlock that would not completely seal and repressurize for the ingress. The tiny captured residual air pressure was too high for their suits' cooling units to function, but far too low to get out of the spacesuits altogether. As they worked to resolve the situation, trapped outside, their suit resources became depleted. Their suit temperatures skyrocketed and poisonous carbon dioxide began to rapidly build up, and even after scrapping, without umbilical life support, into a backup airlock, there was little relief. The shape of the airlock and the stiffness of their still-pressurized suits made it impossible to connect additional cooling units for themselves. Because their visors were now heavily fogged, Solovyev and Wolf were forced to figure out a way to connect each other's cooling units by feel alone through the heavy gloves.

Years later, Wolf's account of the incident was chilling. "We were trapped at vacuum for some twelve hours, and we only had communications intermittently over ground sites and even that ceased as we phased off to the west," he began. "So we didn't have a lot of help from the ground, and the suit completely ran out of its resources. We had to ditch into a secondary volume, which required disconnecting umbilicals. That gave us just literally minutes to live at that point to get in there, maybe five to eight minutes, maybe ten. No one knew. An odd peace came from knowing how I was going to die."

Wolf credited the high stress and thermal conditions that he endured during the intense few weeks leading into the mission with saving his life. He had been placed in a Soyuz capsule in the middle of the Black Sea. He had trained in bulky spacesuits in Star City facilities that were quite often not well air conditioned. There were high-heat stresses in confined spaces. If he could go through something like that and not panic, anything else would be, if not a piece of cake, then survivable at the very least. The umbilicals were cumbersome, hard to move, inflexible, much like the suits that were at once keeping them alive and choking them to death. They each managed to wrestle the hoses into place on the other's suit. "Anatoly Solovyev was the most experienced spacewalker in the world and I was the most junior spacewalker in the world, and we saved each other's lives," Wolf said. "To this day, we look at each other with a certain look. All we have to do is smile and we can tell the whole story." Wolf later concluded:

When I look back at that situation, I shudder and I can't believe almost that I didn't panic. I was literally facing the end in a matter of minutes. I look back and I don't understand that, but I know that at the time, both of us were just working through it step by step, attempting to resolve the problem in a methodical manner. That's what astronaut training gives you the ability to do, somehow. I kind of make an analogy to a car wreck, where you're not nervous during the wreck, but afterwards, you're shuddering. I didn't think that much of it even on orbit until I came down and thought back on it. It truly just makes me shudder to this day.

Wolf's ticket home was to come in the form of STS-89, which docked with *Mir* on 24 January 1998 at an altitude of some 214 nautical miles. With *Endeavour* came the Molecular Biology of 3D Tissue Engineering Space Bioreactor (BIO3D)—an instrument for which Wolf had seventeen different patents now used heavily on Earth. Wolf would call the effort his life's work, figuring out how to use controlled gravitational conditions to enable the construction of human tissue outside the body. So deeply involved in such regenerative medical research was Wolf, he would admit feeling just a little bit of sorrow when he became an astronaut. When the reactor came up, it was already loaded with growing breast cancer tissue. Almost immediately, however, the machine stopped working. "We had to take the instrument apart and repair some valves and do some rather detailed surgery on the machine," Wolf said. "So I kind of once again felt this full circle of satisfaction, to make a contribution on orbit in which the foundation had been laid many years before."

Already a recipient of the NASA Engineering Achievement Award of the Year, Wolf and two others were inducted into the Space Technology Hall of Fame in 2011 for their twenty-five years of work developing this breakthrough medical technology that had come to be known as Intrifuge CellXpansion and the basis of multiple commercial companies.

STS-89 commander Terrence W. Wilcutt had four rookies on his crew, and at least one of them, James F. Reilly, expected him to be loud and proud, a stereotypical gung-ho, do-it-my-way-or-hit-the-highway leatherneck marine.

That was not the case at all, at least not once Reilly got to know his new boss. Wilcutt made sure that the tasks were assigned, and then he got out of the way to let veterans Bonnie Dunbar and Andy Thomas, as well as freshman fliers Reilly, Joe Frank Edwards III, Michael P. Anderson, and cosmonaut Salizhan S. Sharipov, figure out their solutions. Reilly appreciated that kind of faith in the abilities of a largely untested crew:

Here are four people who have never gone to space, and there really is such a thing as space stupid. Fortunately, it never happened on any of my crews, at least not that I saw. What Terry would do, he would assign you the task and then pretty much just let you go. He would ask periodically how things were going, did you need any help, that kind of thing. He became very much a hands-off day-to-day manager. That really stood out in my mind. In fact, I try to use Terry's management philosophy to this day—to give people as much leeway as you possibly can, with as little input from me as possible. They know how to do their jobs better than I do. I'd just as soon let them figure out the best way to do it, and then we as a team will get together and decide that the course of action is going the right direction. If not, then the team will decide how we're going to change it.

Edwards was a 1994 ASCAN classmate of Reilly's, and the two of them remained goods friends nearly two decades later. Anderson, who died in the STS-107 accident, was, according to Reilly, "one of the greatest guys I've ever had the pleasure of serving with." The missions were always important, but it was the experience of working alongside friends and colleagues that Reilly liked most about spaceflight.

Mir had certainly gone through its fair share of issues, but the station had gone through evaluations on what had happened, the responses, and how the system as a whole recovered. "Between the fire and the collision, we had a pretty well-tested machine," Reilly said. "It was very robust, to take a hit like it did from the *Progress*. It was incredible the vehicle stayed together. So we were pretty comfortable overall by the time we went there."

Although the personalities of the astronauts and cosmonauts turned out to be surprisingly very similar, the bigger differences were in their mission control centers and how they related to crews. "Mission control in Houston was very much a partner with the crew on orbit," Reilly said. "We trained together. We socialized together. We spent a lot of time together. We knew each other really well. It was a team effort. We had a lot of conversations back and forth if there was any question about what we were doing." It was not quite the same with TSUP—the Russian acronym for its flight control center, pronounced "soup"—and its cosmonauts. "The Russians were a little bit more like, 'You're going to do what we tell you to do,'" Reilly recalled. "As a result, sometimes there was a disconnect that would occur between folks on the ground in Russia and the folks on orbit. You couldn't really dispute a particular task when you were dealing with the Russians. It was certainly a more dynamic environment in that respect."

Thomas was about to get an extended up-close-and-personal encounter with the Russian style of flight management. He had been serving as Dave Wolf's backup when Wolf was bumped into a prime slot, and that, in turn, moved Thomas into a long-duration stay of his own. Rather than grumble about his original backup assignment, the native Aussie took to the role with great interest, because, after all, he was getting to spend a year in Russia training at Star City. Also, the experience could not hurt when it came time to start handing out flight assignments to the International Space Station. "I figured anyone who had a leg up on getting involved with the Russian culture and the Russian technology and the Russian spacecraft systems would be well positioned," Thomas said. "I thought it would actually open a lot of doors, even though there was not going to be a flight for me at the end of it." On top of it all was the intellectual challenge:

Learning how to operate a spacecraft was one thing. Learning how to operate a Russian spacecraft in Russian, this was just starting to get really bizarre. It was enormously appealing as an intellectual challenge. It was interesting. A lot of people, my contemporaries in the [astronaut] office at that time, shied away from it. They did not want to do it. It was too difficult. They didn't want to travel. They didn't want to spend time in Russia. People did not want to learn Russian. We still have people thinking that way. But that was going to be the new game in town, so I thought it was a very positive step. Of course, it turned out to be even more positive than I thought.

Thomas was strapped into the mid-deck seat farthest from *Endeavour*'s hatch during the ascent of STS-89 on 22 January 1998—the first flight to *Mir* on an orbiter other than *Atlantis*. He would not return to terra firma for 141 days. Of his four spaceflights—which included a science mission, an EVA to help construct the ISS, and the first flight following the *Columbia* accident—Thomas considered his stay on *Mir* the most personally rewarding experience of his astronaut career. It was unusual to be working alongside comrades from a foreign land and speaking almost exclusively in a foreign tongue, but it was also a decided change of pace from a shuttle flight. "Another aspect of flying a long-duration flight versus a shuttle flight is that it's a much more serene kind of experience," Thomas said. "It's much more peaceful. When you fly a shuttle flight, the minute those engines shut down, you unstrap, you go to work, and you're going to work until the wheels stop on the runway. It's just go, go, go. But when you're on a *Mir* flight, there's time to savor the experience and really enjoy the uniqueness of the environment you're in—the zero gravity, the unique view."

It is not that Thomas neglected his work duties on board the station in order to press his nose up against the window and stare at the magnificence of the landscapes flashing by far underneath him. That length of time—nearly five full months—is a long time to spend isolated in any one place, and Thomas did not hesitate to admit that he had not known what to expect when the adventure began. "You don't have a lot of options at your disposal," he said. "I thought about it beforehand and said, 'Well, I need some books. I need some movies.' I needed some recreation more than anything else." Along with the books and movies, Thomas brought with him a few art supplies for sketches and drawings. With him, too, was a calendar upon which he planned to mark off days as they passed. He never actually put the calendar up, because it dawned on him that he might seem a little too eager to leave the station. That was not the case, not at all, although he did feel a certain detachment from the life he knew as an Earthling:

What you want to do is disconnect yourself from the day-to-day tribulations back on Earth. Forget about the car payment, the rent, the mortgage, getting the cat to the vet or whatever you've got to do. Someone else does that, so you just put them out of your mind. All you have to do, and should do, is just focus on the environment you're in. Figure out how you live in terms of getting your day-to-day needs met—where's the food? Where's the galley? Where's the potty? Where are your clothes? Get a good balance of work and recreation. Once you do that—provided you're able to do that, some people cannot—then time starts to just flow. People that have wintered in Antarctica have reported the same thing. You just get into this routine where time just flows pleasantly. It becomes your life. Once you make that transition, the experience is then very appealing. If you don't make that transition, it's very hard. We had some U.S. people on Mir *who had a very hard time because they couldn't make that transition.*

Time eventually flew by so quickly for Thomas that he encountered at least a couple of surprises as his time on the station wound down. Once, he got a call from the ground asking him if he knew what day it happened to be. Was it Thursday? Friday? Monday, maybe? That was not it. Then what? he wondered. It was Thomas's one hundredth day on orbit, and he had no idea the milestone had been approaching, much less that it had already arrived—his calendar was tucked away somewhere, out of sight and mind. Also, there had always been a sense that his last month on the station would pass by slowly as he looked forward to going home. Not so. "I thought the month of May would go really slowly. I thought I'd be counting down," Thomas declared. "But I couldn't believe suddenly May was there, come and gone. It was just in the blink of the eye that the shuttle was there, and I was going home."

Charlie Precourt brought *Discovery* into a smooth capture with the Russian station on the afternoon of 4 June 1998. Thomas's taxi ride home had arrived, in the form of STS-91, the eighth and final shuttle flight to dock with *Mir*. The first phase of a permanent working relationship between former enemies was over. A joint release by the two space agencies summed things up: "The Phase 1 Program endured through a fire, a collision, several power outages, and other significant contingencies and last-minute adjustments. U.S. and Russian space programs bridged cultural, linguistic, and technical differences and created a joint process for analysis, mission safety assessment, and certification of flight readiness. This collaboration resulted in a joint program spanning more than four years that capitalized on a combined four decades of spacefaring expertise both in Earth orbital and inter-cosmos exploration to build the foundation for an International Space Station."

There had been problems aplenty during the Shuttle-*Mir* program, and each and every one had forced the two nations to work together in ways that just years before had seemed absurd. The time had come for an ever greater challenge. The International Space Station was about to be born.

5. A Home on Orbit

As intriguing a project as the International Space Station became, it was an outright miracle that the structure ever made it off the drawing board.

Just one vote—216 to 215—defeated a June 1993 bid in the U.S. House of Representatives to choke off $12.7 billion in funding for a proposed permanent space station, then called *Freedom*. Once the Americans and Russians started working together on *Mir*, the major near-catastrophes of a fire and collision prompted calls in Congress for NASA to stop flying its astronauts there. A major component of the ISS—the Russian *Zvezda* service module—was delayed. That served in some corners as still more proof that the agreement was in serious trouble right from the outset. In the rush to make the ISS happen, would corners be cut on safety? As the issues were sorted out, shuttle missions scheduled to construct the station were shoved back in the flight rotation. Overruns skyrocketed costs for the agency into the billions of dollars. There were problems on both sides, to be sure.

"It was on *very* shaky ground," said Jerry Ross, who served as the lead spacewalker on STS-88, the first ISS construction shuttle flight. "You never know how tenuous a program can be. It's at the whims of Congress, the administration, and how well we progress with our development and control our costs. There were lots of different issues you had to address."

Much was riding on this, not the least of which was the shuttle's very legacy. As the program neared its completion, many pointed to the ISS as one of the program's most important accomplishments. No other machine could have carried such large segments to orbit, as well as the number of crew members needed to piece them all together. After the *Challenger* accident, commercial users of the shuttle were written off and told to go find their own rides into orbit. The military left and bought its own expendable rockets to do the work once done by orbiter crews. The Spacelab modules that fit so snugly in the cargo bay were canceled in order to concentrate resources on developing equipment and procedures for station-based experiments.

Ross laughed when asked where the shuttle would have been without a station to help build. "I don't know where we would have gone," he said. "Without a reason to go fly in space, to carry up the components of the station, we would have been a shuttle without anything to take into orbit. It would have been a very interesting time to see how that would have evolved."

As chief of the astronaut office's EVA and robotics office, Ross was in charge of developing the means by which the ISS would be constructed. After almost ten years of requests, a larger water tank was built to accommodate full-scale mock-ups of station components. He devised a plan in which at least three runs were conducted in the new pool to evaluate virtually every aspect of station assembly and maintenance, with two potential spacewalkers working each trio of simulations.

From there, a consensus was reached on whether or not the parts and/or tasks were adequately designed. If not, they decided what kinds of changes needed to be made. A broad cross-section of crew members was selected for the tests, in terms of physical size, strength, and prior EVA experience. Rookies took part in the runs, and so did veterans. Big ones, small ones, it did not matter. The goal was for any of them to be able to pull off a construction mission.

From that growing cadre of EVA-capable crew members, Ross advised Bob Cabana, then the chief of the astronaut office, on assignments for the first seven shuttle flights to the station. On at least two of his three most recent spaceflights—possibly STS-27, although a rumored spacewalk on the Department of Defense flight has never been confirmed, and certainly STS-55 and STS-74—Ross went without conducting his beloved work outside. And after returning from STS-74, a Shuttle-*Mir* mission, Cabana asked him to get back in line for another flight to the Russian station. "This is the only time I ever turned down a flight," Ross admitted. "I said, 'Bob, I've been working my entire astronaut career to help develop the team and the expertise that's going to be required to go assemble the station. I think this is the wrong time for me not to be concentrating on doing that.'" Cabana did not hold the refusal against his fellow veteran astronaut and in fact named Ross to his STS-88 crew.

After launching on board *Endeavour* on 4 December 1988, STS-88 raced to catch up with the *Zarya* functional cargo block, which shot out of Kazakhstan the previous month on an unmanned Russian Proton rocket. The shuttle carried with it Node 1, called *Unity*, which Nancy Currie maneuvered into place on *Endeavour*'s docking port before arriving at the most improbable construction site mankind had ever known. Then, on 6 December, Currie again used the shuttle's robotic arm, this time to capture *Zarya* and mate it to *Unity*. Ross and Jim Newman conducted three spacewalks to attach cables, connectors, and hand rails to the structure.

Four days later, Cabana and Russian cosmonaut Sergei Krikalev floated into the brand spanking new station at 2:54 p.m., with the rest of the STS-88 crew in tow. The work in getting the nodes up and running had been, "actually, not that hard," Ross said. "But it did take the entire crew most of the time we had available on orbit to do that. After we'd done the EVAS, then we had to do quite a bit of work inside in terms of removing structural bolts to give us access to areas. We had to install the electronics for the early communications system." The crew knew before launch there was a battery charger that was not working properly, and that it had to be changed out. Once in space, some other discrepancies were discovered in the functional cargo block, but Krikalev was able to identify and correct them with little or no problem.

After completing his third and final spacewalk, Ross had officially tallied forty-four hours, nine minutes in the void. It gave him a temporary lead in EVA time, but with a mind-boggling amount of work left to do on station construction, the mark would not last forever.

First, the bad news.

Zvezda's delay meant that Brian Duffy's fourth and final shuttle flight—and his second in command—was delayed and that the crew of STS-92 would wind up training together for some two and a half years before finally making it to orbit. No shuttle astronaut was ever satisfied with any kind of slip. There was too much to do, the tasks at hand too important to sit around on their thumbs, waiting. The frustrated declaration to light this particular candle and get on with it had long been uttered many, many times by countless spacefarers.

Now, the good news, the silver lining to STS-92's cloud.

The holdups had been disappointing, but what they did do was forge a bond between the six men and one woman who finally did strap into *Discovery* for liftoff on 11 October 2000. Along with Duffy were Pamela A. Melroy, making her first spaceflight and, in so doing, becoming the third woman to pilot a shuttle flight after Eileen Collins and Susan Kilrain; and mission specialists Leroy Chiao, Bill McArthur, Jeff Wisoff, Mike Lopez-Alegria, and Koichi Wakata. "I have best friends for life now, where we've shared these incredible common experiences," Duffy remarked. "The human experience, I think, is the part that I walked away with more than the technical experience." Such deep relationships existed not only between the crew members but with their training teams as well. They worked together and socialized together *a lot*—dinners, movies, traveling, bowling, maybe sharing a beer on Friday nights. "It was a crew that was extremely close," McArthur said.

Delays or no delays, the long training flow turned out to be a good thing, Duffy remembered. "It didn't break our hearts, because we were having such a good time together as a crew," he remarked. "We were really a tight group. We still have a great time when we get together, and it's been ten years. We were happy to continue to hone our skills because we were having so much fun together. We got better and better as a team. I had some of the younger folks coming up to me in the astronaut gym saying, 'Wow, I wish I was on your crew,' because we were having so much fun."

Each of the STS-92 crew members had unique functions to play, and not just in tasks they would perform during the flight. Their personalities complemented each other, McArthur marveled.

Duffy was the laid-back commander. "It was hard to get him excited," McArthur said. During their time together on STS-92, Melroy came to respect her commander as "a really, really amazing leader" in the sense that while things were going perfectly and seemed almost effortless, it was never about Brian Duffy. The credit *always* went to the team. Duffy never came across as a heavy-handed dictator. The rest of the crew was left feeling as if the decisions had been made together. "Oh, my gosh, what he achieved by doing that and the way people valued and liked working with him was just phenomenal," Melroy continued. "The biggest thing I learned from Duffy is that you don't have to have a heavy hand to move mountains. You can do it by just understanding people, organizational dynamics, and just key-ing things up in the right way so that the right thing happens without you having to ever say a word. That is *very* powerful. That's something I still strive for, even now."

Chiao and "Mike L.A." could quote movie lines nonstop, keeping things loose when the going got tough or monotonous.

Wisoff was Mister Detail. "He was able to manage detail to a degree I've never seen before," McArthur said. "Pretty soon, we all realized we could try to be just like Jeff and get really frustrated because we couldn't pay attention to detail the way he did, or we could just be really grateful that was a special skill that he had."

Wakata had the nicest way imaginable of telling his crew members that they had screwed up. "We know that Japanese society places a very high premium on courtesy," McArthur said. "It was really fun working with Koichi, because if you were supporting him on a task and you didn't do it well, he would ask you to perform the task differently in the most polite way possible."

McArthur was everybody's backup, and as flight engineer, he sat directly behind the center console on the flight deck. That allowed the army veteran to keep a close eye on the two air force pilots just in front of him. "I could watch each of them, and if I thought they were about to kill us, I would just tell them to stop doing whatever they were doing," McArthur said with a laugh. He used a long telescoping pointer to manipulate various circuit breakers and switches just out of reach, and he liked to tell people that he could also use it to whack Duffy or Melroy on the knuckles if they were about to make a mistake. Not only was McArthur mission specialist two, he was also the number two spacewalker and backup robotic arm operator. The crew enjoyed the *Austin Powers* series of films, and so McArthur picked up the nickname of Robert Wagner's character—Number Two.

Those who met Melroy as a child usually learned two things very quickly. First, they knew her name, and second, they knew that she very much wanted to be an astronaut one day. Melroy applied to the office three times, although she knew deep down that her first time out of the box was pretty much a long shot, because she had not yet graduated from the air force's test pilot school at Edwards. Still, she was "rapturous," if for no other reason than because her references had been checked. Melroy was at long last accepted as part of Astronaut Group Fifteen, "The Flying Escargot," in 1994.

Eileen Collins blazed a path for Melroy, in that she helped members of

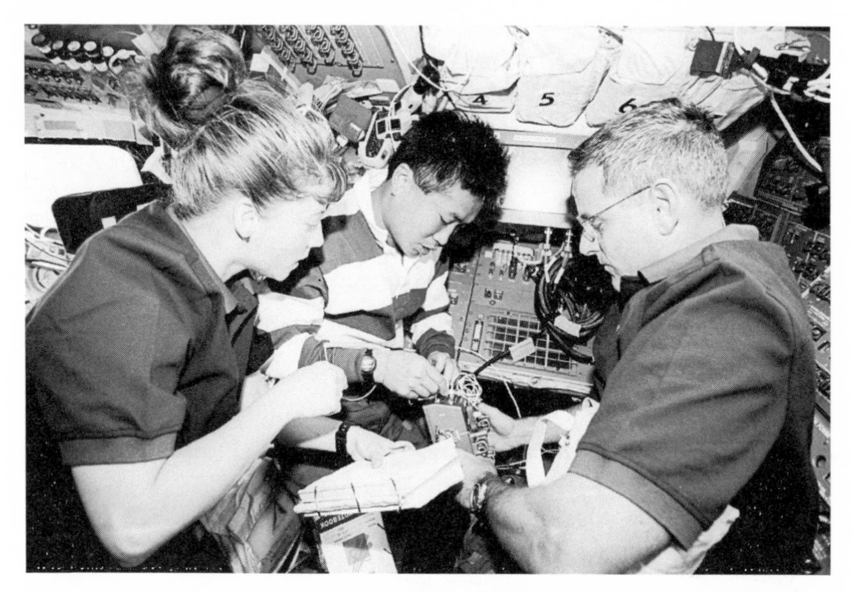

17. Pam Melroy, Koichi Wakata, and "Number Two" Bill McArthur confer during the flight of STS-92. Courtesy NASA.

the media get silly questions about being a female shuttle pilot mostly out of their systems. As a result, Melroy did not face many uncomfortable queries about the subject, although there were always going to be a few who wondered about such things as how a woman's body functioned in space as opposed to a man's and if it was weird to be a lone female among a bunch of decidedly type A personality alpha males. Would there be enough privacy? Melroy did not mind those queries so much, because they were real issues that most everyone encountered at one time or another.

When someone once told Melroy about the kinds of things NASA's very first female astronauts were asked, that is where she would have drawn the line, and then some. "There were really, really, really stupid questions about things like their boyfriends and the size of the flight suit that they were wearing and what would NASA do if they gained weight," Melroy began. "It was this really funny undertone of physical appearance being so central to their value to the agency." She then stopped, imagining herself caught in such a situation and laughed, "Ooooo, boy. Man, I'm glad I wasn't there then. I'd probably smack somebody upside the head if they asked me a question like that, I'd be so offended."

The stranger predicament for Melroy was the fact that as an astronaut, in many people's eyes, she was an example for other women to follow:

I have tried to get used to the whole idea of being a role model. To me, there is nothing else in the world that I could be other than a woman and an astronaut. This is what I was born to be. It's hard for me to remember what it was like when it wasn't a given that women could be astronauts. To me, the stranger things are the people who just seem so amazed that I would even be interested in doing this. That's a harder thing for me, because it's just unfathomable why people would want to know why I would want to do something like this.

Man or woman, it did not matter. STS-92 had work to do, but before the crew could get down to the bulk of it, its KU band system failed on the flight's second day. That meant that no live video could be downlinked, and worse, no radar for the rendezvous. Overcoming the problem became fairly complicated, Duffy explained:

As far as the radar failure goes, we actually have procedures in the event that should occur and we practice them regularly in the simulator. We had a laser system in the payload bay that we could use when we were within range and within the cone of the reflectors onboard the ISS. It was very useful late in the rendezvous, but pretty useless the rest of the time. Other than that, we also had a handheld laser in the cabin that Leroy used to get range and range rate information when we were close enough (again, late in the profile). The rest of the time we relied on some information supplied from Houston and the procedures to get us close. Once we came into the daylight, it was all seat-of-the-pants flying on my part. It ended up being more stressful and difficult than I expected because the conditions we experienced in flight were well outside any we'd practiced, so it was a challenge to save the rendezvous and thus the mission. Fortunately, all worked out and we continued to complete a very successful mission.

Once docked with the station, plans called for the installation of the Z1 truss and a third Pressurized Mating Adapter that would be used as a docking port on subsequent shuttle flights. The truss served as the foundation for what would become the ISS's permanent backbone. McArthur had trained for a contingency spacewalk during the lead-up to his last spaceflight, STS-74, but none was ever required. He got his chance on STS-92, and during the first of his two spacewalks alongside Chiao, McArthur experienced his welcome-to-EVA moment. "Koichi is maneuvering me on the robotic arm, and as he's doing that, I swing into a position in which I couldn't see anything except black space. I couldn't see the shuttle or the space station," McArthur remembered. "I couldn't see any of the earth. It gave me that little bit of a sense of floating out in a void. My comment was, 'My toes are curling right up.' That was my reaction, to lift my toes up, to try to be more firmly attached to the foot loops on the end of the arm."

Although this was the fifth flight to the station, it was by no means whatsoever becoming routine. "Oh, my gosh, no," Melroy exclaimed when just such an idea was floated. "Oh, my heavens, no." Over the course of her career, Melroy flew three times to the ISS, and each time she found profound differences. The station was obviously bigger each time, but that was not all. The way they worked both with the Russians and with mission control evolved over time, and so did life while on the station. If ever there was a work in progress, this was it. "I would characterize that mission as a highly developmental test," Melroy said of STS-92.

There were plenty of firsts early in the station era: the first time a common berthing mechanism had been used, the first time a certain procedure had been run, the first time software had been upgraded, and so on. A rookie, Melroy simply did not fully grasp at first the true complexity of what she and the rest of the STS-92 crew were attempting to do. "I was just like, 'Oh, well, they trained me to do this, so I'm just going to go do it,' not fully appreciating that the training was pulled out of someone's rear end," she said. "They were just guessing what the best way to do it was going to be."

Duffy, meanwhile, was enjoying himself as much as possible on what he knew would be his last spaceflight. It was his choice, and he was fine with it. "I loved what I was doing, but I was aware that from the day you start, at some point, there's going to be a last day, too," he figured. "I didn't know what I would be doing in the future, but I'd done way more than I ever dreamed of doing at NASA. It was probably time for me to get out of the way and let some of the other folks that were ready to step in do it."

Duffy relished his moments on orbit, especially after no less than three landing wave-offs—two of them coming on 22 and 23 October 2000 for stiff crosswinds at the Shuttle Landing Facility at KSC, and another due to rain showers within thirty miles of the runway at Edwards on 23 October. On each of those days, time was spent swapping back and forth between preparing for reentry and staying on orbit. The rest of the time was pretty much theirs to do with as they pleased—and McArthur went scrounging for food:

On day eleven, we're going through all the food that was left over from the first ten days. This is the food that you really weren't interested in eating the first time, so you now get a second chance and you eat what looks good that *time. Now, the next wave-off day, you're looking through the food that you've rejected twice. Leroy and I, we found some of the worst hot dogs in the world—we've discovered the hot dogs that no one was willing to eat. And we had some flour tortillas left over from early in the mission, and they were a little bit moldy on the edges. So we just broke off the edges and threw a little mustard on the hot dogs and enjoyed lunch that way.*

At long last, the crew came home to Edwards on 24 October. "The last half of the entry was in daylight, and we were down low and still going *really* fast," Duffy said. "That was really fun to watch. We went over Los Angeles at about 100,000 feet, still going Mach 5. I heard that we boomed LA with our sonic booms pretty good." For a couple of air force veterans at the controls, coming into Edwards was the very best of both worlds—a homecoming in every sense of the word. Rolling out on final approach, the scenery looked the same from the shuttle's cockpit as it had from the air force aircraft that Duffy and Melroy had piloted into Edwards many times.

To Melroy, the experience of her rookie spaceflight in general was "magic," and to come in at Edwards made it that much more special:

I'm smiling, because I was absolutely three feet off the ground for six months to a year after I landed. You just couldn't hurt me. You couldn't make me sad. You couldn't piss me off. I had such a great time on that flight. We had a ton of fun. The wave-offs helped a lot, because we actually were able to get some sleep. We came in very well rested, which was a wonderful treat you don't always get. Edwards was a place where I had done the best flying I had ever done in my life, the most fun in airplanes that you could possibly have. It's a very, very special place to me. To come in on this gorgeous day over Los Angeles, seeing the lake bed from a hundred miles out was amazing. It just felt like the most amazing homecoming.

Their landing pattern brought huge mines in nearby Boron, California, into sight, and Melroy made the call as she had so many times before. "We're abeam the mines," she told Duffy, meaning that they were straight out the wingtip.

18. Five-time shuttle commander Jim Wetherbee had strict rules for communications among his crews during countdown and ascent. Courtesy NASA.

Only this time, it was a shuttle wingtip.

After landing, Lopez-Alegria took a look around and then said something that spoke volumes about the flight of STS-92 and its crew—and maybe NASA itself, come to think of it. "Mike L.A., I think, put it very, very well," McArthur said. "When we finished, he said, 'You know—it will never be this good again.'"

As a Space Shuttle mission commander, Jim Wetherbee was determined to keep the line of communication open between ground teams and the shuttle cockpit. When the time came to fly STS-102 in March 2001—his fourth command, tying him with Hoot Gibson for most in the history of the shuttle program—conveying information back and forth was so important to Wetherbee that he developed guidelines on the use of humor for his crew to follow on the launch pad.

Up until the standard twenty-minute hold, the veteran skipper allowed a certain amount of joking by those under his command as long as it was not offensive. "Humor is important when you're in a stressful situation, to help you relieve some of the tension," Wetherbee said, though clearly, he was not about to let the crew module become comedy central. Once inside the twenty-minute hold, the rule was tightened to just quick one-liners.

Chatter over the comm loops was increasing, so there was no time for any lengthy tall tales. Finally, coming out of the final scheduled hold at T-minus nine minutes, Wetherbee's rule was still more specific. Nothing whatsoever was allowed to be spoken unless it was absolutely essential to the mission. That meant no jokes, no anecdotes, no one-liners, no nothing, other than the business at hand. "The responsibility I had as a commander, I *will* listen to you, I *will* pay attention, and I *will* act on what you say," Wetherbee explained. "Your responsibility as a crew member, because of that commitment I make to you, you'd *better* be telling me something that's operationally relevant. Going uphill in powered flight, absolutely nothing was spoken unless it had something to do with the mission. So every bit of communication we had was about the energy state, the engines, the propellant, the electrical system, the computers, or the trajectory. *Only* operationally relevant words were spoken on ascent."

Wetherbee was not the only one setting records on STS-102. A Pressured Mating Adapter needed to be moved to make way for the *Leonardo* logistics module that *Discovery* was delivering to the station, and then the base for a robotic arm that was to be delivered by a later flight had to be affixed to the side of the *Destiny* lab. The work fell to James S. Voss and Susan Helms, who along with Yury Usachev had been delivered to the ISS to replace the *Expedition 1* crew that STS-102 would bring back to Earth. Winding up at eight hours and fifty-six minutes, no spacewalk had ever lasted longer.

Voss and Helms were about to embark on a long-duration stay on orbit, while their STS-102 counterpart, Andy Thomas, had already done a nearly five-month stay on *Mir* in 1998. This time, however, he was able to add one more deed to his already impressive résumé. During the flight's second EVA, Thomas and crewmate Paul W. Richards attached what amounted to a toolbox and a spare ammonia coolant pump to the outside of the station. The experience of a spacewalk was much different from his long stay on *Mir*:

It's completely different, completely different. It's different from a long time on a space station, where you have more time to savor the experience. You don't get a lot of time when you're doing an EVA to savor the experience because you're just busy. The timeline is compressed. You have a lot to do, and they keep you going. An EVA is highly orchestrated. You've trained and trained and trained. You do it almost by memory, each step. You know *where the handrails are, because you've trained on a very high-fidelity mock-up. In fact, that's what surprised me about it. It had this sense of, "I've done this before. I've been here before." It was all because the pool training is such high-fidelity training.*

For Thomas, his spacewalk "just seemed natural." He had once dreamed of becoming an astronaut during his childhood in Australia, so he could not have asked for much more than that.

Given the fact that that STS-95 had been so heavily focused on life sciences, it was the one mission that seemed to mesh best with Scott Parazyski's core background as a physician. Yet almost from the time that he first stepped foot on JSC, he had trained on run after run as a spacewalker.

He supported medical operations and payload development to an extent, but his real bread and butter was all things EVA—tool development, procedures, life support, it did not matter. Yet the sum total of his actual EVA experience was the five hours that he and Vladimir Titov spent working outside *Mir* during the flight of STS-86, and he wanted more. Parazynski got it on STS-100, which he flew in the spring of 2001.

On the ninth flight to the International Space Station, Parazynski arrived at a structure that was a bit smaller than what he had found in *Mir*. The greatest difference was power—the ISS was brand-new, light, and bright. It was far less crowded. In *Mir*'s *Kristall* module, the main pathway from the shuttle to the central part of the station, there were places where occupants almost had to suck in their stomachs to squeeze in between stowage of used hardware and trash. That was not a problem on the ISS. "I felt very much at home," Parazynski said. The task at hand was relatively straightforward. *Endeavour* carried the Canadarm2 robotic boom, which Parazynski and Hadfield were to attach to the ISS during a couple of spacewalks. Although Hadfield was a rookie spacewalker, the Canadian wore the red stripes of the lead EVA astronaut on his spacesuit. Like so many others before and after, Hadfield was caught unaware of the incredible view before him:

The opening of the hatch is probably step 750 of the day. And steps one through 749 were all boring and minuscule. Each one was on a checklist and you had to do every one right, so you were very painstaking. But suddenly, you do this one step, and suddenly, you are in a place that you hadn't conceived how beautiful this could be. How stupefying this could be. And by stupefying, I mean it stops your thought. I knew I couldn't keep notes up there and I would forget stuff, so I sort of resolved to myself that I would verbalize and attempt to, as eloquently as I could, express what I was feeling, and not have missed such an amazing experience. And yet when I listen to the transcript of what I said, most of it was just, "Wow!" It was so pathetic. But the experience was overwhelming.

The arm, almost fifty-eight feet in length and weighing almost four thousand pounds, featured seven different joints able to rotate 540 degrees. The first EVA took place on 22 April, the day after Rominger got a hard capture with the ISS. The *Expedition 1* and STS-100 crews had not yet greeted each other, due to a difference in air pressure between the two crafts that had been maintained in order to prepare for the upcoming spacewalk. Hadfield and Parazynski installed a UHF antenna on the station to aid in EVA communications before moving on to the arm, which had to be unfolded. While still secure in its pallet, the two veteran astronauts attached one end to the ISS's *Destiny* module. Before heading back inside, they connected temporary cables to provide power, command, and video and secured fasteners to keep the arm rigid.

These were no easy tasks. Instead, Parazynski remembered them as some of the most difficult he ever performed during an EVA:

I've had two occasions in my spacewalking career where I thought, "This really could go south quickly." One of them was on this flight. It was something that I had to do at the hinge line of this folded-up robotic boom. I was in a foot restraint, and Chris had to release these shoulder joints of the robotic arm. I had to do a clean and jerk, and push this robotic-arm system up about thirty to forty degrees, so that we could get underneath them and install bolts to rigidize this new system. It was really the limit of my physical ability. It was every bit of energy that I had to get these joints started moving. I don't know if it was the very cold temperatures or the overall system, but it was a pretty scary moment. The first time, it didn't budge. The second time, more force didn't budge it. Then, the third time, I gave it everything I had. It did finally move.

Two days later, there was more work to do. "[Parazynski is] almost like a monkey, scampering hand over hand way up high on the station, way back down again," Hadfield said of the second EVA during a pre-flight interview. "Probably the most delicate handwork will be done by Scott during EVA-2 when he is doing all of that wiring which permanently powers the Canadarm2 onto the space station. And that is sort of like changing spark plugs in your car with hockey gloves on your hands. Your hands get stiff working on this type of task." As Parazynski rewired connections to the arm, a circuit on the arm's backup power system failed to respond to commands from *Expedition 2* crew member Susan Helms. Rather than try to whack the arm like he had done with the balky tether mechanism back on STS-86, Parazynski and Hadfield tried another relatively simple fix. They disconnected cables at the base of the arms and then reattached them, much like rebooting a home computer or modem. It worked.

That particular problem out of the way, there was another issue to come that temporarily delayed work on the arm. The hard drive failed on one of three systems-management computers on board the ISS on 24 April, which cut off communications and data transfers between the station's Flight Control Room and the ISS itself. Before the faulty computer was replaced with a backup, communications were routed through *Endeavour*.

The day before undocking, the station's newly installed Canadarm2 grappled its pallet and handed it back over to the shuttle's own Remote Manipulator System for stowage in the payload bay. It was the first-ever exchange between two robotic systems in space.

Scorch.

That is all to which STS-104 pilot Charles O. Hobaugh would admit, that "Scorch" was indeed his nickname. With a smile, he refused to divulge any details whatsoever on the story behind the moniker, allowing only that it was a "closely held family secret." Naturally, the denial left all kinds of theories in its wake. "That's just one of those things I'll never reveal," he laughed. "It's not something you want to be proud of." His crewmates, if they knew anything in the first place, were evidently sworn to secrecy. "That's a deep, dark secret," said Jim Reilly, a sophomore mission specialist on STS-104. "Knowing that he flew Harriers, we figured he probably scorched something with his nozzles. There are all kinds of theories and stories, some of them pretty colorful, some of them almost profane, but we never did get the true story from the horse's mouth, so to speak."

Commander Steven W. Lindsey might not have had a colorful moniker like his pilot, but he had more important things about which to concern himself. He joined the astronaut corps in 1994, the same year as both Reilly and another STS-104 mission specialist, Janet L. Kavandi. The three of them were good friends, but on the tenth shuttle flight to the ISS, one's command over the two others was introduced into the dynamic. "It was really an interesting challenge for him," Reilly confessed. "He was having to figure out how to herd the cats, which is what we were." Lindsey, Hobaugh, Reilly, and Kavandi were joined by one other crew member, lead spacewalker Michael L. Gernhardt. "The personalities we had were considerably stronger as well, so there was a real challenge for him. But he did a great job," Reilly added. "He was not real hands-off, but he was very much a relaxed sort of manager. He would let the team discuss a problem, and then if the team was on the right path and was heading toward a solution, he'd generally not say anything at all. He'd let the crew continue down that path. If we'd start to diverge a little bit or if there was something we were missing, then he would make sure that he'd step in and give us that information until we got ourselves guided onto the right path."

By the time they got to orbit, Gernhardt and Reilly had trained seven hours or so for every sixty minutes they spent outside on the three EVAS they performed on STS-104. They got along "extremely well," according to Reilly, and formed such a working relationship that they could just listen to each other and know exactly where they were in relation to one another. Asked to describe what it was like working with Gernhardt, Reilly, without hesitation, deadpanned, "Mike is an *interesting* guy," adding, "He is one of those really, really sharp guys, so when he's really engaged, a lot of times, he is very different in how he approaches things." Different in what respect? "Just being focused," Reilly said. "There are internally focused and externally focused people. Mike was very much an internally focused person in how he approached working and training. It was a difference in really how we accomplished our tasks. You just kind of have to adapt to each person's methods and personality. I think we did a good job. We actually got way ahead on our EVAS, so it apparently worked."

The first two times out, Gernhardt and Reilly installed the ISS's new *Quest* airlock. They spent the last spacewalk of the flight fitting it with a nitrogen supply tank. Susan Helms and Jim Voss, in the midst of an extended stay

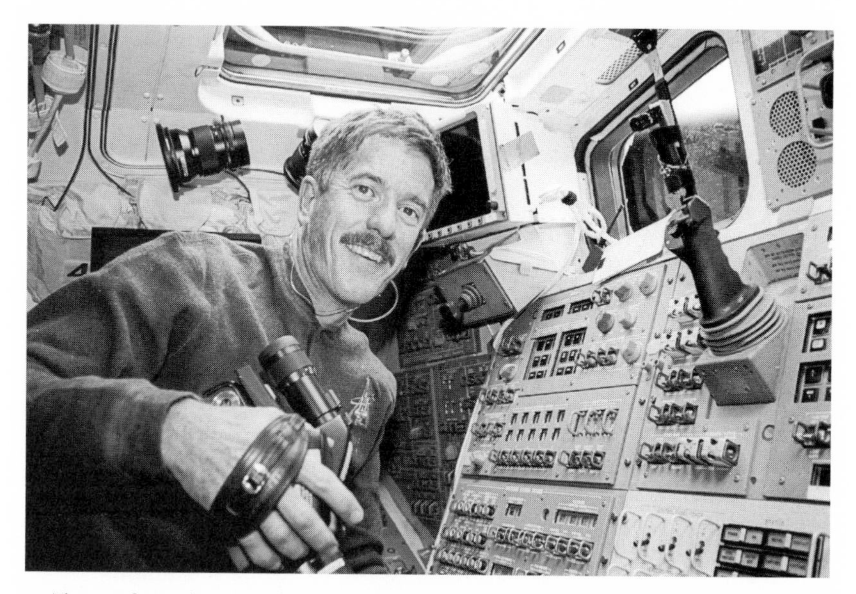

19. The view from orbit was currency to Jim Reilly, so if he could catch a chance to take a peek out of *Atlantis*'s windows during the flight of STS-104, he was going to do just that. Courtesy NASA.

on orbit, worked the robotics to move the airlock out of *Atlantis*'s payload bay and then to its appointed position on the station. The four of them knew each other very well, but they had not actually trained together for the EVA in nearly three months. "It was an interesting experience, in that when it did come time to do the task, it was almost like we'd never been apart," Reilly said. "It was just a joy to watch that whole system operate, in terms of our performance. I was intrigued by that, fascinated by the dynamics of how the team worked and how we gelled. *All* the shuttle and station crews worked pretty much the same way."

Installing *Quest* was the flight's major technical achievement, but the flight also had an emotional impact on Reilly. The up that he knew on the ground was now sometimes down, and his Earth-bound down was now up at times, and it all depended on how his head and feet happened to be pointed at any given moment. It was easy to get used to microgravity, floating everywhere there was float on the station. Back home, it took a little getting used to being once again sucked to the ground. On the first spacewalk of his career, the geologist-turned-astronaut was hanging off the bottom of the station, looking up at planet Earth. There, he had a view of the bluish-green light show of the Aurora Australis being played out near New Zealand, unobstructed by nothing other than the visor of his helmet.

Another meaningful moment took place on his last spacewalk of STS-104, when he and Gernhardt clambered up the station's solar-array truss to check out a mechanism that allowed the wing to swing back and forth in the direction of the sun. He halfway expected to feel some sort of falling sensation as he worked his way along the truss, but it never happened. Instead, he looked out and there was *nothing* in front of him other than the earth. The horizon dipped away, leaving nothing but the deep blackness of space. "You just have to sit there for about ten seconds, just stare at it and take it all in," Reilly said. "It hits you at an emotional level. It's not something you sit there and say, 'Oh, that's kind of cool.' No. It hits you, and you go, '*Man, that's impressive.*'"

Daniel C. Burbank, serving as CapCom during the EVA, asked Reilly to turn a bit to the left so the camera on his helmet might capture the scene. "I was able to just kind of pivot with just my wrist that was holding on to the truss, so that my cameras were pointing in the direction I had just been looking," Reilly continued. "So the folks in mission control got a chance to see what I was seeing. That was pretty cool, to know that they were able to see that."

These were just two of the many "gee whiz" moments Reilly enjoyed in space, when he could see the earth in mind-numbing details he could never have imagined. High-resolution photographs from on orbit are impressive, but they simply cannot compare to the view that actually being there afforded Reilly and shuttle astronauts like him. The sheer enormity and clarity of the panorama before him was currency, and if a spare ten seconds could be had to look around, he was going to take them. Those were memories in the making.

Reilly would wish that he had been a poet, so that he might somehow have been able to better describe the things he had seen from on high. Take consolation, J.R. No wordsmith—not cummings, not Sandburg, not Frost, not Angelou, not Poe, not Ginsberg—could fully capture such incredible things.

Seven. The perfect number.

John Young flew in space six times—two Gemini missions, two Apollos (including three moonwalks during *Apollo 16*), and two on the shuttle. He was a real-deal NASA superstar, and some in the astronaut office had always wondered if some sort of glass ceiling might exist at six flights; it was possible to fly six times and *match* Young's record, and a few had done just that. But actually topple Young from atop the woodpile? That was a good question, and Jerry Ross had the guts to seek out an answer. "I asked the question a couple of times, 'Am I topped out, or is there potentially hope for another flight?'" Ross said. "Fortunately, the answer always came back, 'There is no magic number or limit. If you're interested and keep doing a good job, we'll put you in the hopper for consideration.'" For Ross, that seventh round-trip to space came as a member of the crew of STS-110, which flew in May 2002. No other human had ever launched so many times, and when Franklin Chang Diaz tied Ross's mark just two months later, it made little difference to the veterans. They would just keep flying into their golden years.

Or not.

Ross always tried to treat each mission as if it were his last, but until STS-110, there had always been another one right around the corner. The handwriting was on the wall. Even if he did fly again, the wait was more than likely going to be a lengthy one. Some rather large astronaut-candidate classes had been hired—nearly 100 total in quick succession from 1996, 1998, and 2000 combined—and that left a backlog of eager-beaver rookies standing in line to fly. "It was becoming obvious that there was going to be a generational turnover in the office at some point, and that my chance for too many more flights was going to be pretty limited," Ross admitted.

Atlantis flew Ross to orbit the first time, and she would take him on a final trip that began 8 April 2002. In all, Ross flew *Atlantis* five times and *Columbia* and *Endeavour* once each. Although he found no particular difference in flying one over the other, *Atlantis* was his favorite bird to fly. "It was my first, my last, and my most, so it's going to be my favorite, no matter what you say," he said. In the workhorse's cargo bay on STS-110 was the S0 truss, a forty-three-foot-long segment that formed the first major element of what would become the station's backbone after being mated to the top of the *Destiny* module.

Also lifted to orbit was the Mobile Transporter, essentially a set of railroad tracks that was eventually used to move the station's robotic arm up and down the length of its finished truss. The pieces were installed by Ross and partner Lee M. E. Morin and another team, Steve Smith and Rex J.

Walheim, during the flight's first three EVAs. For the first time, the station's arm was used to ferry the astronauts about the structure. There was no discernible difference for the spacewalkers themselves, although the operators were inside the station and had no windows to look out and check the progress of their work. Instead, they had to rely on television views of the scene outside. "There had to be a little bit closer voice coordination for flying on the station arm, and you had to be a little bit more vocal in helping the operators by giving them more clearance kind of calls," Ross said.

Also, the flight's four spacewalks marked the first time shuttle-based EVA astronauts used the *Quest* airlock exclusively. Ross was not a fan. He recalled traveling to the Rockwell-Downey plant in California during the early stages of the station's design. An engineer asked Ross if he wanted to see the station airlock.

You mean a mock-up of it?

No. *The* station airlock. The actual flight hardware. The piece that is going to the station.

"They already had it built, before they ever asked any of the flight crew members about what the thing should look like," Ross said, then added, "Given our choice, the next time we build an airlock, it won't look the same." To one of the most storied spacewalkers in spaceflight history, the hatch's configuration was awkward and not well thought out. He would know, right? The shuttle airlock was basically a small tunnel. A hatch at one end led to the mid-deck of the orbiter, while the other opened into the cargo bay. There is just enough room for a team of spacewalkers to get situated with good visibility and range of motion to open the hatches to their fullest range of motion. As for the *Quest* airlock, Ross explained, "The problem is the hatch moves in such a way that you're trying to raise it up above your head. You have no visibility and very limited reach. It's just an awkward way to do business. There's no way that you can get your body in any really good orientation to make it work comfortably."

During the flight's fourth and final EVA, Ross and Morin fitted a fourteen-foot beam between the airlock and S0 truss that allowed future spacewalkers easier access to the assembly. Ross tested switches on the truss. The pair installed floodlights on the *Unity* and *Destiny* modules. They also attached a work platform on the station; installed electrical converters and circuit breakers; and attached shock absorbers to the Mobile Transporter's railcar. It was Ross's ninth spacewalk on the books, and at a total of fifty-eight hours, eighteen minutes, only one other person—Russian Anatoly Solovyev—had a greater number of EVAs and more time outside than Ross. At its end, Ross was struck by the impact of the moment. "I can tell you, frankly, coming in at the end of my last spacewalk was difficult," he admitted. "I knew it was most likely over."

Still, although he had accepted the fact that it was at best a long shot, Ross hoped for another flight after STS-110. Then came 1 February 2003. He would never fly in space again.

Mom and Dad were taking the kids on the trip of a lifetime. Only this was no minivan; it was the Space Shuttle *Atlantis* and the destination was the ISS, not Disney World.

"Dad," STS-112 commander Jeff Ashby, and "Mom," pilot Pam Melroy, had joined the astronaut office together back in 1994. For some, that might have created an awkward divide between a pair of gung-ho stick-and-rudder jockeys, but Melroy said that was never the case between her and Ashby. "He took his duties of helping me be prepared to be a commander very seriously, and tried to give me as many opportunities as he could," Melroy said. "He sent me to certain meetings and things like that, just to make sure that I was exposed to the kind of stuff that I would be doing as a commander. It was a really close, wonderful friendship that I treasure."

Her style as a shuttle pilot was very much like Ashby's as a shuttle commander, very much detail oriented and super organized, or, as she put it, "really wanting to get down into the weeds on a lot of stuff." They had similar thought processes. Ashby and Melroy were basically on the same page about most things, such as how various tasks needed to be accomplished, how much insight they needed to have in various training events, and, when it came down to it, what constituted whether or not somebody was doing a good job.

Melroy was in sync with Ashby, and while that was certainly a positive, she was also finding the experience of preparing for her second spaceflight enormously different from her first. The tasks would appear to have been relatively similar at first glance—STS-92 delivered the Z1 truss to the station and STS-112 brought up the Starboard 1 (S1) segment to provide support for its radiators, as well as a cart for the Mobile Transporter that had been installed by the crew of STS-110 six months earlier. Despite that, the missions were different in both style and substance. Melroy described the emotional roller-coaster of her rookie trip into space this way: "On my first flight, I didn't actually know whether I could do it or not. You really don't. You're worried, 'Am I going to make a mistake? Are they ever going to invite me to fly again? Will I mess something up and ruin my career as an astronaut?' It was a wonderful feeling to land and say, 'Hey! I made it all the way through and as far as I know, I didn't screw up.' It's just a sense of knowing that you can do it. It's going to be okay. You do know how to do this."

The second time around, Melroy was fully aware of what it meant to train on the ground and then actually put it into practice while on orbit:

I had a much greater awareness of all the things that could *have gone wrong on my previous flight but did not. I just lacked experience. I just assumed that when an instructor taught me something that they must know* everything. *It's only when you get to space that you realize the limits of what someone who has never been to space can truly teach you, or how much they can truly understand about the space station, the way it operates and the way* you *operate in space. I began to realize that there's a whole lot more resting on the crew than I fully appreciated.*

Of the six people on the STS-112 crew, three had flown before and three had not. The only other experienced astronaut was Dave Wolf, along with freshmen Piers J. Sellars, Sandra H. Magnus, and cosmonaut Fyodor N. Yurchikhin. "Dave's last flight had been to the Russian space station, so he hadn't actually done a shuttle mid-deck op in a while. So I felt very much this burden of responsibility for a lot of things running smoothly was on Jeff Ashby and me. All the things that could go wrong were much more visible, so it was a lot harder for me on that flight because I felt the crushing burden of this responsibility to make sure that I supported the commander. I really remember feeling exhausted a lot during training for that flight."

During the flight's 7 October 2002 launch, a large piece of foam was shed from a bipod ramp on the External Tank that caused a four-inch-wide and three-inch-deep dent near the bottom of the left SRB. With no changes, the decision was made to keep flying.

STS-107 lifted off just three months later. This time, the consequences of a foam-shedding event from very nearly the same location would shake NASA to its very foundation.

As he began training for his first trip to space, John B. Herrington went straight to the source to get the scoop on what it would take to be a good crew member. What do you want me to do? What are my tasks? How can I be a successful supporting member of the team? And who better to ask than his STS-113 commander, Jim Wetherbee? The mission to be was Wetherbee's fifth as a leftseater, breaking him out of a tie with Hoot Gibson for most commands in the history of human spaceflight.

Wetherbee's response to his eager rookie was simple. Herrington had one priority, and that was to know and understand Wetherbee's weaknesses. "He took that to heart," the veteran crew leader marveled. "He recognized that is what's needed on a team, to cover for each other." Astronauts usually were diligent about sharpening their own skill sets, but helping out a crewmate was going the extra mile. To Wetherbee, it was not about coming across as an arrogant know-it-all busybody. Instead, it was simply a matter of being a good teammate. Going into detail, he explained, "We had very specific ways of thinking that we would talk about. I believe the best crews will not only know the systems and have detailed knowledge of how they operate, what to do, how to respond to emergencies, how to read procedures, and on and on and on. But we would also talk about other things, like communication. How do we communicate with each other effectively? How do we sense when a crewmate needs something?"

Wetherbee implemented a "two-person rule" in which nothing could be done in the cockpit during flight without backup. They *always* had backup, helping and watching. The rule was explicitly stated and "most of the time followed," Wetherbee said. "It only worked when you are deeply committed to the rule. You have to *want* to have somebody else help you, and you really, truly have to want to help somebody else." Alone on the flight deck and trying to hold to a critical schedule, or even if it was an easy procedure where the likelihood of making a mistake was minimal, it did not matter. The Wetherbee Two-Person Rule was to be followed, always. STS-113 sophomore pilot Paul S. Lockhart was among those on the flight who took the rule to heart, Wetherbee said:

Paul had such an altruistic way of thinking about things, where the only thing that mattered to him was the mission. His personal glory or satisfaction, or the building up of his credibility, ego, or reputation didn't matter to him. It just flat didn't exist. The only thing he cared about was helping the team be successful. Shortly after we got on orbit, he had a relatively simple procedure to do, but he waited until I was available to watch him and guide him through it. It wasn't because he didn't have confidence. It was because he deeply thought that it was the best thing to do. Even though I had flown many times, I returned the favor to him. A few minutes later, I had to reconfigure the coolant system. I knew I could do it. I had done it successfully. It's a pretty easy procedure. But I decided I was going to return the favor, and I waited until he was available to watch me through it.

The flight delivered Ken Bowersox, Nikolai M. Budarin, and Don Pettit to the station, and brought Valery G. Korzun, Peggy Whitson, and Sergei Y. Treshov back home. Also going up was the massive Port 1 (P1) truss, a bookend mate to the segment installed during STS-112. According to Wetherbee, the two-person concept worked so well that the crew of STS-113—which also included mission specialist Mike Lopez-Alegria—made the fewest mistakes of any mission with which he had ever been associated.

The experience of launch was almost always at or very near the top of the list of an astronaut's most-frequently asked questions. STS-113 struck out for the station on 23 November 2002, and Wetherbee had slept peacefully before his wakeup call. As a naval aviator, he had long ago reconciled himself to the dangers of landing a jet fighter on the pitching and rolling deck of an aircraft carrier. There was no greater risk associated with spaceflight as there had been in the navy. "You're either dead, or you're not," Wetherbee concluded. He trained countless hours for this spaceflight, and he was as ready as he was ever going to be to face the mission at hand:

The time you must deal with the fear of the job is not when you first are selected to become an astronaut. There's an exhilaration that lasts for many months. It's not even the first time you get assigned to a mission, nor is it leading up to your first flight. It's really the night before your first launch attempt, when your head hits the pillow and you recognize that you've run out of time to get any smarter and it's too late to quit. The biggest challenge I think an astronaut has is not fear of death. It's fear of making a mistake. I was able to go to sleep peacefully that night, thinking, "If the worst happens and bad stuff starts happening, I'm going to spend my last nanosecond on this planet trying to help save the crew and the vehicle. If I fail, well, then I've died doing the only thing I've ever wanted to do since I was ten years old."

Wetherbee prided himself on his ability to stay in the moment, focusing on the tasks immediately before him. The future, for all intents and purposes, did not exist. When he woke up in the hours prior to STS-113's liftoff, he hopped into the shower and then went to a weather briefing. No worries about disaster. The crew headed to the suit-up room, and during the walk out to the astrovan, he smiled and waved because that was expected of him. There was concern, but it was based solely on getting to the van without tripping and, yes, whether or not the zipper on his suit was zipped. Wetherbee made sure not to bang his head getting out of the van at pad 39A, thinking only in the present, the next few seconds. Once and only once did the commander allow himself to get out of a purely operational mode:

Just before getting into the world's slowest elevator to ride up to the 195-foot level, Dan Brandenstein [the commander of STS-32*, Wetherbee's rookie flight] would walk around the side and look up at the vehicle, which was bathed in the twenty-million candle-power Xenon lamps. He would just pause and think to himself about what he was about to do. I picked up on that, and I did that on all five of my commands. It was the one time I would come out of my technical concentration mode. I would step around, look at the vehicle, and I would think, "Okay,* Endeavour*, it's you and me now. I'll take care of you, if you take care of me." It's the one kind of emotional time where I would pause and reflect. This is a living vehicle, and together, we're going to be successful.*

Cramming himself into the elevator with six crewmates and various support personnel, Wetherbee's all-business focus quickly returned. Nothing outside of a ten-minute window mattered, and he cared only about strapping in correctly. His connections were exactly the way he wanted them. Comm was fine. There were a thousand switches and circuit breakers to visually verify, not once, not twice, but three times. Who had time to worry about the things of life and death? The first time, a checklist might be used. The second time, the switches' various patterns had been memorized, and if something looked out of place, it was corrected. The third time, the process got down to the specifics of how each system worked. Is this the configuration that will route the correct coolant to the proper pipes? "You're focused in the moment," Wetherbee explained. "You're not thinking about dying or exploding or making a mistake and killing yourself." Then, two minutes before launch, his focus tightened even more. "You've gone so deep into this concentration mode of being in the present, between now and the next ten seconds is all that matters," continued Wetherbee, who was now on a roll, very much in the moment. "As much as I love my wife and my family, they do not exist in my mind as I sit there on the launch pad getting ready to launch."

Wetherbee willed himself to sense everything, to see everything, to know exactly if there happened to be a fluctuation in the hydraulic system. He had trained himself to know if there were the tiniest of changes in the dials, displays, and digital numbers spread all around him. If the crew said anything—no joking, remember?—he tried to respond instantly. Then came T-minus zero:

When they ignite the Solid Rocket Boosters, it's the most powerful feeling you could ever imagine, but nothing has changed in your mind. Nothing is different between one second earlier when you are on the earth and one second later, when you have seven million pounds of explosive propellant underneath you. You're focused on what you're doing right then, right now, staying in the present. You're not worrying about dying. You're looking at every engine instrument. You're sensing how the vehicle is moving. You know if the hydraulics are overheating or overcooling. You know if the computers are working properly. You're double checking and cross checking. As you're going uphill, if you lose one engine, you do something. If you lose two, you do something else. It changes, depending on your energy state. You always have to anticipate what's coming up. The combinations explode in your mind, because there's so much to think about. But you're able to handle it all, because you're staying in the present and you're trained. I can remember feeling every rumble and vibration going uphill, but it doesn't register as an emotional thing. It registers as a technical thing.

Finally, maybe five minutes after main engine cutoff on each of his six flights, Wetherbee was able to come down off the ledge and allow himself a quick "My God. What a ride!" Such mental mastery was Wetherbee's most rewarding experience during his six spaceflights.

The cooperation that the commander worked so diligently to build extended not just to the crew on orbit, but to ground teams as well. Brandenstein, the only spaceflight commander Wetherbee had ever known, taught him another lesson. Before each of his flights as a commander, Wetherbee and his crews sat down for at least one face-to-face debriefing with those who would be working in mission control. They looked each other in the eyes and shook hands, as Wetherbee got across one most important point.

The job of an astronaut is easy, he began. An astronaut climbed on top of a rocket and stood a very real chance of living and dying as a consequence of his or her own decisions and mistakes. That much was easy, he figured. Working a console in mission control was much different. "What the flight controllers do in mission control is make decisions where *other* people live or die," Wetherbee stated bluntly. "They have to live with the consequences of those decisions." His intent was not to add pressure to a job where split-second demands were already off the charts. "My point was, 'All we're asking you for is for you to give it your best shot. If you make a decision based on your training and experience and it goes wrong and we don't make it, try not to let it affect you. Don't stress over it. Don't worry obsessively about it. Make the call and then move on with your life if it goes bad.' I honestly don't know if that helped or not, but I really believed that their job was harder."

For the fifth and final time in his astronaut career, Wetherbee guided the shuttle to a landing on 7 December 2002. Forty days later, the shuttle *Columbia* and the seven-member crew of STS-107 headed toward the heavens.

6. "The Debris Was Talking to Me"

The badly damaged piece of debris was just one of some eighty-four thousand recovered following the loss of the crew of STS-107 and the Space Shuttle *Columbia* on 1 February 2003. In their own way, each piece of twisted metal, bit of tile, and scrap of fabric and scorched paper that rained down upon Texas and Louisiana that morning was a wrenching reminder of the sad chapter that had just been added to the annals of human spaceflight. This particular item provided no forensic clue as to the cause of the tragedy, yet it was perhaps the most powerful testament to the fate of the team that had flown the mission, which included commander Richard D. Husband, pilot William C. McCool, payload commander Mike Anderson, mission specialists David M. Brown, Kalpana Chawla, and Laurel B. Clark, and payload specialist Ilan Ramon, the first Israeli astronaut.

This was a videotape, picked up by chance on the side of a country road near Palestine, Texas. On it were thirteen minutes of footage showing Husband, McCool, Chawla, and Clark on the flight deck as they prepared to come home. At no point was there any indication that something was critically amiss, although flashes of the hot gases of reentry were already beginning to flash outside the cockpit windows. All four were seated and attired in their helmets and bright orange pressure suits, each with an orange chemical light tube on their right sleeve already activated. They chatted away without a care in the world. As crewmates Anderson, Brown, and Ramon settled in for the ride down on the mid-deck, unseen and unheard on the tape, the remaining quartet concerned themselves with checklists, gloves, the camera, and, eventually, the view outside. There have been many haunting images throughout NASA's history—anything pertaining to *Apollo 1* or *Challenger* had always been cause for a difficult pause—and those final calm moments on the flight deck of *Columbia* were no different.

Seconds before the fire broke out that claimed his and his crew's life, *Apollo 1* commander Gus Grissom had wondered in great frustration about how America could possibly make it to the moon if bugs in the communications system could not be ironed out. Every bit as profoundly ironic were the conversations between Husband, McCool, Chawla, and Clark as they discussed the super-hot gases beginning to envelop the orbiter. At about 8:45:54 a.m., the exchanges began.

"It's really neat," said McCool, the pilot who was one of four members of the crew making a first trip into space. "It's a bright orange-yellow, out over the nose—all around the nose."

"Wait 'til you see the swirl patterns out your, you know, like left or right windows," his commander told him.

McCool's reply was a simple but awed "Wow!" Seconds later, Husband added, "Looks like a blast furnace." As the first telltale signs of gravity began to register once more—McCool let go of a card and it fell rather than floated—this was a crew whose emotions could not exactly be described as giddy, because they were far too professional for any such nonsense. Yet they were clearly pumped, happy with what they had accomplished over the last sixteen days. At 8:46:51 a.m., McCool turned his attention back to what he was seeing out his windows.

"This is amazing," McCool continued. "It's really getting fairly bright out there."

"Yep," Husband replied. "Yeah, you definitely don't want to be outside now."

Chawla chimed in, "What? Like we did before?" There had been no EVAs on the science research mission. Could she possibly have been referring to some training incident during a sim back on the ground in which the crew had not landed safely? We will never know. Although her quip drew a round of laughter from her colleagues, Husband was still all business. By this time, Clark, seated directly behind McCool, had the recording camera. As the two of them lightheartedly discussed being able to see each other, McCool told Clark, "Now I can see your camera." Husband stopped the conversation in its tracks with a quick and rather stern "*Okay* . . ." Sufficiently rebuked, Clark's response was immediate. "Stop playing," she said, and that was that. Within moments, the recording showed signs of digital distortion and ended at approximately 8:48:01 a.m., as *Columbia* was over the Pacific Ocean southwest of Hawaii.

Four minutes later came the first sign of distress, when a sensor in the left Main Landing Gear wheel well showed a temperature spike. Husband's last discernible transmission was a broken "Roger . . ." at 8:59:32 a.m.

Forty-six seconds after that, the shuttle was seen disintegrating in the skies over west Texas.

Like so many children of the 1960s, Rick Husband could not really remember a time when he did not want to be an astronaut. He was all of four years old when he first expressed the sentiment, and at the time, the only astronauts were the famed Mercury Seven—America's original spacefarers. He was a step ahead, even then. He watched launches on the television of his parents' home in Amarillo, Texas. He learned all he could about his astronaut heroes and the rockets they flew. Even better, there were plastic model kits on the store shelves. Years later, just before the flight of STS-107, he would remember and relish building a replica of a Gemini spacecraft. From Mercury to Gemini and on through Apollo, he followed the program every step of the way:

It was just so incredibly adventurous and exciting to me that I just thought, "There is no doubt in my mind that that's what I want to do when I grow up." And at the same time, I was very interested in airplanes and flying. And, you know, I'd be out in my backyard playing. Any time I heard any kind of an airplane, it's like stop what you're doing and take a look and see where's that airplane? What kind is it? Where is it going? How high is it? How fast is it going? It's the kind of thing that has just been such a part of my life in what I wanted to do when I grew up.

While Husband was a student and air force ROTC cadet at Texas Tech University in the late 1970s, NASA was in the process of hiring its first group of shuttle astronauts. He fired off a letter to the agency asking about the office's requirements, and when Husband got a packet of information back, he took it to heart and laid out a plan of action in order to one day become an astronaut pilot.

Join the air force and go through pilot training—check.

Graduate from test pilot school—check.

Get master's degrees in engineering—check.

Apply to the astronaut program—not so fast there.

Three times, Husband tried and failed to get into the program. The second time, the astronaut corps was on a hiring freeze after the *Challenger* mishap. When Husband again applied in May 1991, he fibbed when asked on a questionnaire if he had ever worn hard contacts. To some, the lie was nothing about which to worry. To Husband, a man of great character if ever there was one at NASA, guilt weighed on him. He got into his first testing and interview round, but, again, fell short. In March 1992 he tried one more time. He prayed. He wrote extensively in his journal. Most important, for Husband's own conscience, he told the truth this time about once having briefly worn hard contacts. He passed his eye exam, and in December 1994, out of some 3,000 applications and 120 interviews, Husband received word from NASA that his astronaut dream had finally come true. Husband was promoted to lieutenant colonel, and the good news did not end there.

His wife, Evelyn, also discovered that she was pregnant with their second child after years of miscarriages and struggling to conceive. During an interview prior to the flight of STS-107, Husband said, "I think apart from NASA, the most enjoyable part of my life has been my time with my family." He pointed out that his marriage and the birth of their two children were the most exciting and memorable events of his life, "just the awesome experience of seeing a baby come into the world and just being so overwhelmed with God's goodness in blessing us with two wonderful children."

Joining Husband in The Flying Escargot 1995 class of astronauts were Anderson and Chawla. Of the group's ten pilot astronauts, Husband was the last to fly. When he was named in 1998 as pilot of STS-96—the first flight to dock with the ISS—his classmates gave Husband a party. Because the nod also meant that he would no longer have to park so far out at JSC—he could now pull right up to Building Four South with the cool kids—the cake was in the shape of a parking space.

Kent Rominger commanded Husband's first spaceflight, and was chief of the astronaut office when his former crewmate perished. "Rommel" recalled his time with Husband at a memorial service on 4 February 2003 at JSC:

Rick was a terrific human being and a great leader. He was my pilot on his first flight. I grew to really appreciate all of his talents, gifts and laugh at all of his Amarillo sayings. His favorite saying, and I can hear him saying it right now, is, "You know, I feel more now like I did than when I first got here." And when we were training, it took me six months to learn how to say that. As a matter of fact, Rick e-mailed me six days ago from orbit and that was in the subject line, that saying. He was a naturally gifted pilot. I was envious, even though I was his more experienced commander. . . . He molded seven individuals from different parts of the world with diverse backgrounds, various religious beliefs, into an incredibly tight knit and productive family.

About fifteen months after returning from his first flight, Husband was named as commander of the STS-107, the Space Shuttle program's final mission devoted exclusively to scientific research. It would be far from NASA's sexiest mission. There was no Hubble Space Telescope to be rescued or serviced, no John Glenn to return triumphantly to space, no space station with which to dock. Yet despite several slips that ultimately pushed the flight's launch to 16 January 2003, the crew set out to train with a rich vitality. As with every mission, the commander set the tone. Because Husband was in the midst of his first command, rookie McCool came to feel a great sense of synergy with his flight's leader:

I'm trying to figure out the ins and outs and what my roles are and getting very good at them as the pilot. Rick is in the same situation, but on the commander's side of the house. So, it's really made for, I think, a close relationship and bond between he and I. As we sit side by side during ascent and entry sims, we're both figuring out our roles and really developing together. So, in that development, we kind of see where each other needs help or needs some prodding or where we're strong. We just have, I think because of this relationship, developed a very close harmony—almost to the point where I can feel things on Rick's side without having him say anything to me. And he can do the same with me. And I find that frequently, he'll be reaching across to flip a switch because I'm busy over here, but he just senses that it's time to do it. Or I'll flip up a display for him because he's busy. But we've got that harmony now, and it's really because we've developed and grown in our roles together.

Husband's charismatic demeanor touched others at NASA. Fellow astronaut Jeff Ashby spoke during another tribute on 28 February 2003 at First Baptist Church in Lufkin, Texas, the epicenter of the massive debris search. Ashby began his remarks by saying that it was Husband's memory that gave him the strength to stand before the large gathering, and then paid his fallen comrade one of the highest compliments one pilot could give another when he told the congregation that the STS-107 commander "was a very natural stick-and-throttle guy." Ashby continued:

Rick was a tremendous leader with a very "aww, shucks" demeanor and an Amarillo slang. Sometimes, we kidded him that we could understand Kalpana better than we could understand Rick's English. Rick had an enviable faith in God, one that I have absorbed somewhat over the last week. I'm told that he was very close to his church. He had an old Camaro that he drove and we kidded him about it. But he loved that car. I'm told that he donated it to his church sometime back, and a couple months later, went back and asked the pastor if he could buy it back from him. He missed it too much.

If there was a temptation to completely idealize the lives of the members of the STS-107 crew in the wake of their untimely deaths, the truth in at least Willie McCool's case was a bit different after the divorce of his parents. "We had to grow up young and early," Kirstie Chadwick told *Runners World* writer Steve Friedman a few years after her older brother's death. Friedman wondered in the piece if McCool somehow became the sum total of the traits of those so instrumental in his childhood: his father's intelligence; the perfectionism of his adoptive father, a marine naval aviator who built models and flew radio-controlled planes with McCool; and his mother's drive.

McCool looked out for his sister following their parents' divorce, became an Eagle Scout, and was later accepted into the United States Naval Academy where he became captain of the cross-country team his senior year. At Annapolis, McCool was considered a passable member of the track team—good, but not necessarily great. That did not keep him from trying. Al Cantello, who coached navy runners for a half a century, had never seen an athlete quite like McCool. "My earliest, vivid memory of him?" Cantello asked Friedman rhetorically. "I was screaming at him, something about a crappy workout. Now, most kids, they'd listen, then say, 'Yes, but . . . yes, but . . .' But not Willie. He just looked straight at me, with those big, steely eyes, taking it all in."

At the time of his graduation, McCool had turned in the twenty-sixth fastest time in the history of the school's five-mile cross-country course. A commemorative plaque in a school trophy case listed only the top twenty-five.

The same determination that he exhibited as a cross-country runner was also present in the classroom—McCool graduated second in his class of 1983.

He married a former flame, Lani, in 1986 and accepted her two children from a previous marriage as his own. They had another son a year later. The very next day, he left for a six-month deployment on board the USS *Coral Sea*. One of the twenty-four types of aircraft that McCool eventually flew was the EA-6B Prowler, and while at the navy's test pilot school in Patuxent River, Maryland, he somehow managed to yank the craft out of a so-called death spin. To this day, McCool's actions are a case study for aspiring test pilots at the famed "Pax River" institution in what to do when their lives are so far out on the line.

After joining NASA in 1996—along with Brown and Clark—he felt right at home:

The military and NASA are a lot alike when you talk about working together as a team. In naval aviation, you work as a squadron. You work with an air wing. I flew the EA-6B Prowler. We had a crew of four, and we advocated crew coordination and working together as a crew. All those lessons that I learned in my aviation career and my navy career about working together as a team just seemed to naturally apply and work well with NASA. NASA does the same exact thing. We operate as a crew in the same way as we did back in my navy days in the EA-6B Prowler. The astronaut office, the folks here at JSC, operate in the same fashion that we had learned to operate as a team within the squadron and within the air wing. So I think they dovetail quite well.

McCool continued running, coaxing his and Lani's son Cameron to go along with him. The father told the son stories to pass the time, and to take the youngster's mind off the pain of long-distance running. Sometimes, the tales were about books McCool had read. Others, he told his son about his own life—his time as a plebe, the ice cream–eating contests he endured as a senior midshipman. Cameron eventually gave his dad a birthday coupon for fifteen "complaint-free runs, to be used whenever you want." McCool later told a NASA interviewer that he truly appreciated time with his family in the great outdoors. "My most enjoyable experiences are going out with my wife and my boys back country backpacking in the Olympic Mountains, or the canyon lands in Utah and just enjoying life without outside distractions, enjoying each other and enjoying the environment," he said.

"We love to do that frequently, whenever we can. Unfortunately, I don't get enough of it here, recently, with all the training. But those memories prevail. And they're something that I look forward to doing in the future when we get done with this mission."

McCool was never late, for anything. Instructors could remember only once him arriving past when he was due—he had been trying to kill a cockroach for Lani. In November 2007 a memorial to McCool was unveiled at the 3.1-mile mark on the naval academy's five-mile cross-country course. At the pace McCool set, the tribute is sixteen minutes from the finish line.

Serving as the flight engineer, Kalpana Chawla was seated directly behind the main console on *Columbia*'s flight deck during the fateful ascent and reentry phases. She was born in Karnal, Haryana, India, in 1961, the first year human beings were strapped into rockets and shot into space. Although the town in which Chawla lived was small, it was home to a flying club, and it was there that the first seeds of a career in aviation were sown. She saw the tiny Hindustan HUL-26 Pushpak two-seater airplanes taking off, and as she and her brother chased them as far as they could on their bicycles, wondered where they might be headed. The interest grew further still when she got to ride in the Pushpak and a glider that the flying club also maintained. She saw a plane flown by pioneer Indian aviator Jehangir Ratanji Dadabhoy (J.R.D.) Tata hanging in an aerodrome, and it captivated her imagination. Much like her STS-107 commander, she laid out her plans in high school. Beginning in the eighth grade, she concentrated on courses designed for a career in aerospace engineering—physics, chemistry, mathematics.

She was and always would be fascinated and inspired by explorers. She read several books written by participants in Sir Ernest Henry Shackleton's famous crossing of the Antarctic continent in the early twentieth century. She studied another transcontinental voyage that took place more than a hundred years before Shackleton's journey, this one of what would become the United States by Meriwether Lewis and William Clark. She marveled at Patty Wagstaff's three U.S. National Aerobatic Championships and the work done by author Peter Matthiessen. Yet could she have seen herself as an explorer of the heavens as she was growing up in India? Not a chance:

The astronaut business is really, really far-fetched for me to say, "Oh, at that time I even had an inkling of it." Aircraft design was really the thing I wanted to pursue. If people asked me what I wanted to do, I remember in the first year I would say, "I want to be a flight engineer." But, I am quite sure at that time, I didn't really have a good idea of what a flight engineer did. Flight engineers do not do aircraft design, which was an area I wanted to pursue and did pursue in my career. It's sort of a nice coincident that that's what I am doing on this flight.

She eventually received her pilot's license as well, taught by her husband, instructor Jean-Pierre Harrison.

Of the seven STS-107 crew members, only three—Husband, Chawla, and Anderson—had previous spaceflight experience. The petite Chawla flew STS-87 on board *Columbia*, serving as the prime robotic-arm operator as she deployed the *Spartan-201* solar physics spacecraft to study the outer layers of the sun's atmosphere on 21 November 1997. Two and a half minutes after the release of the satellite, it became apparent that its Attitude Control System was not functioning correctly due to a missed input into a computer overseeing the task. When Chawla attempted to regrapple *Spartan-201*, she did not get a firm capture signal. The problems did not end there. When Chawla backed the arm away again, it inadvertently gave the experiment-packed craft a rotational spin of about two degrees per second. Commander Kevin R. Kregel tried to give Chawla another shot at capturing *Spartan-201* by firing *Columbia*'s thrusters to keep pace with its additional spin, but the thrusters instead imparted still more rotation and increased the chances of a collision between the satellite and the orbiter. The satellite entered a shutdown mode, and two days later, Winston E. Scott and Takao Doi did a seven-hour, forty-three-minute EVA to capture the misbehaving platform by hand.

A subsequent investigation determined that a number of factors helped cause the incident. Chawla acknowledged what happened and moved on from the incident. "Something like that, you don't ever forget," Chawla said following her first spaceflight. "So even though they came up with a lot of recommendations which can make you feel okay, you never feel okay. You always feel eternally guilty. I think you can live ten lifetimes after that and still feel the burden of something like that." Although she was almost certainly privately concerned about her future with the agency, she did not give it a public face, continuing, "I think for me my attitude is always not to think about stuff like that. Just keep on training to do your best, and whatever happens, happens."

Named to the crew of STS-107, Chawla's keen sense of humor sometimes disappeared during training sessions. "She was the studious one of the bunch," Ashby said at the Lufkin service. "She would focus very, very intently on the training. Her training team told me that she would tell the rest of the crew, 'Cut the comedy. Let's get serious,' to which they would all just laugh."

Although Laurel Clark did not become an astronaut by accident, it seemed to be something very close to it. Growing up, her eyes were not on the stars but on the environment, ecosystems, and animals as she studied zoology at the University of Wisconsin–Madison. Struggling to make ends meet as a first-year medical student at the same school, she was quickly using up her savings at the same time as she was also looking at some huge student loans that were growing almost daily. She received a recruiting packet from the navy in the mail offering to pay her tuition in exchange for a service commitment, and when she signed up, Clark intended to put in her three years with the navy and then be done with it. Instead, she spent more than fifteen years in the military.

After graduation from medical school and naval training, she began life as an undersea medical officer in 1989. "The submarines that we serviced were out of Holy Loch, Scotland," Clark recalled. "While submarine crews, like astronauts, are selected from a pool of people who turn out to be very healthy, in the end, they select people. If you have medical problems, then you're not allowed to continue in the submarine service because you're out at sea for long periods of time. Even still, things happen. People get appendicitis and can get infections."

While in Scotland, she dove with the elite Naval Special Warfare Unit 2 SEALS. Picturing Clark in the midst of a submarine rescue is right out of a Hollywood script, the heroic doctor rushing to the aid of a just-as-heroic sailor in desperate need. She almost certainly would have balked at the suggestion, but every case was different for Clark, if for no other reason than the vast assortment of variables once she arrived on the scene. If an evacuee was not able to walk, the only way off a submarine most of the time was through a narrow hatch. Move them the wrong way just the slightest bit, and the situation could get that much worse. Once off the sub, they still had to get to another ship and then to land. This was often done in another country, with a different medical system. It made for a perfect, action-packed script.

Clark was next assigned as a flight surgeon for a marine AV-8B Harrier squadron based in Yuma, Arizona, and in the same role with an advanced training squadron in Pensacola, Florida. She at last had one other iron in the fire:

I met someone who actually knew about the mission specialist program and they said, "You know, somebody with your background would be a good candidate for the astronaut program." I really didn't think that that was really true at the time. But while I was in my flight surgery tour, I decided that it certainly didn't hurt to ask the question, so I filled out the many, many pages of paperwork to become an astronaut. I wasn't accepted right away. I was interviewed the first time and not accepted. They always interview many more people than they need. Then, the second time I was interviewed, I was accepted. To sum up, it was circuitous. There were several different factors—luck, wherewithal and just sticking with it.

Accepted into the astronaut corps in 1996, Clark went through training with her usual gusto. She had been here before, whether it was at her first "real" job at McDonald's or during five years spent as a lifeguard. There had been plenty of submarine rescues. She dove with navy SEALS. And now she was an astronaut. This was a woman who packed a lot of life into her nearly forty-two years:

I think it's important that you do the best you can and take responsibility for your actions. Be precise, be caring, communicate well what you need and listen to what other people need. I also tell people to find out what you like and do what it is you love to do, because the realities are that you may or may not be able to do exactly what you want to do. Not everyone's in luck to be an astronaut. Certainly, not everyone who wants to be an astronaut gets to be an astronaut. But if you do what it is you love to do—whether it's flying in airplanes or cell-science research or being a physicist, you'll do a really good job at it, because you love it. I myself never thought about being an astronaut until I was in my thirties. It was something that came about much later. I feel fortunate every day that I've been chosen to do what I'm going to do. I think that sometimes life takes you in very unexpected ways.

Clark was also remembered as the queen of paraphernalia for the flight of STS-107. She had different pastel-colored shirts with the mission's shuttle-shaped blue logo, to go with matching jewelry. Always smiling, if Clark knew someone was in the building, she would never call or e-mail. She would go and meet face-to-face. Yet for all the professional successes Clark enjoyed, it was in being a mom that she said she found her greatest joy. Married to Dr. Jonathan Clark, a NASA flight surgeon, they had son Iain, who was eight when his mother flew on board STS-107. Ashby remembered during his remarks at the church, "No matter how hectic her day, Laurel was also known for always leaving in time to pick up her son from school." She told an interviewer before the flight, "Motherhood's been incredible. I tell my son all the time that my most important job is being his mother."

Dave Brown had much in common with Clark. He studied biology in college, she was a zoology major. Both were physicians in the navy. Both had incredibly interesting résumés—Brown was a naval doctor *and* aviator, and he had been a member of the varsity gymnastics team at the College of William and Mary in Williamsburg, Virginia. He had also worked as an acrobat, unicyclist, and stilt walker for Circus Kingdom. They were members of the same astronaut class. Both were spaceflight rookies when they joined the crew of STS-107. Brown was seated behind McCool during the launch, while Clark was on the mid-deck, center seat. They swapped for reentry. And while Clark had filmed her crewmates on the flight deck as *Columbia* started its last plunge into the earth's atmosphere, Brown had documented some 150 hours of the crew's training for a planned movie of the experience.

Brown's omnipresent camera was familiar to virtually everyone in and around NASA's human spaceflight community. Rominger remembered that "Doc [Brown's nickname] loved cameras and always had a camera with him. Riding out to the pad, I've never seen anybody as intense at making sure he filmed every bit of what was going on with his crew as Doc. Usually when he was filming folks, he would tell them, 'Just act like a little brown squirrel.'" His remarks were very closely echoed by Ashby during the Lufkin service.

The footage became *Astronaut Diaries: Remembering the* Columbia *Shuttle Crew*, airing on 14 May 2005 on the Science Channel. The documentary would have been a fascinating glimpse into the making of a shuttle crew, regardless of its fate. Knowing what happened to them made it one of the most powerful media efforts in the history of the shuttle program. There had never been such an intimate portrayal of a spaceflight crew preparing for an upcoming mission than Brown's film work. The draw was not just that they were astronauts. They were astronauts *and* real people doing the things that real people do. Husband, Anderson, and Brown were shown during the presentation getting lost trying to find the historic Anne Frank House in Amsterdam, and once they got there, they were not exactly sure how to pay for parking. The crew made a spaghetti supper together. Clark laughed and said during a crew hiking expedition that a crewmate's attempt to make brownies looked like "bear scat." Viewers rode the astrovan with the crew to the launch pad. Just before climbing aboard *Columbia* on the morning of launch, Chawla helped tie Clark's hair into a ponytail. "Making Dave's movie was closure for me personally," said Doug Brown, the astronaut's brother. "People don't know that astronauts are people too. I hope people can relate, and that this program can relate that it's still difficult and dangerous work."

There was another similarity between Clark and Brown. Like Clark, Brown never envisioned himself standing alongside giants like the Mercury Seven and the moon-going heroes of Apollo. He would always remember going for a flight in a close family friend's small airplane, and from the precise moment the wheels of that craft left the runway, Brown loved to fly. But to actually become an astronaut? They were a cut above anything he could imagine himself doing at the time:

I absolutely couldn't identify with the people who were astronauts. I thought they were movie stars, and I just thought I was kind of a normal kid. So I couldn't see how a normal kid could ever get to be one of these people that I just couldn't identify with. While I would've said, "Hey, this is like the coolest thing you could possibly do," it really wasn't something that I ever thought that I would end up doing. It was really kind of much later in life, after I'd been in medical school and I'd gone on to become a navy pilot, that I really thought, "Well, maybe I would have some skills and background that NASA might be interested in." I went ahead and applied. I think growing up, I really underestimated myself. I was really a bit wrong about things that I could do. I'm glad I figured out kind of later in life that if I wanted to pursue that, I could.

When Brown was selected for flight training in 1988, it had been a full decade since a flight surgeon had accomplished the same goal. Two years later, he was a full-fledged naval aviator ranked first in his class. That was

Brown, in a nutshell. He spent four years at the Naval Strike Warfare Center in Fallon, Nevada, and he would call his time there some of the most enjoyable of his career. He was flying two different kinds of high-performance jets and riding his bicycle to work. This was not just a quick trip down the block either. "I lived in kind of a rural area," Brown said. "I'd ride my bicycle to work thirteen miles each way past all these ranches and cows and alfalfa fields. I actually rode my bicycle about 2,500 miles that year. For a guy who likes to fly airplanes and be outside and do interesting stuff and be around challenging people, that was pretty neat, that four years I spent in the navy in Nevada."

After joining NASA in 1996, Brown was taken aback by a piece of advice he received from none other than John Glenn. There had been no time during Glenn's legendary first spaceflight in 1962 for sightseeing or reflection, so when he told Brown nearly four decades later, "When you get up there, you need to make sure you look out the window," Brown took it to heart:

As I thought about it, that's really what all the astronauts say. When you look out the window, you're not looking at the stars or the moon. You're looking at the Earth, and when you look at the Earth, invariably people say that they think about people. They invariably say they think about the people that they kind of know and care about. So, I think the best advice I've gotten is to make sure you really appreciate how wonderful an experience that it is because we're going to be so, so busy. So take some moments to kind of look out the window and really think and appreciate how special this experience of going into space is going to be.

As a young child, Mike Anderson had a wide variety of interests. He loved the sciences and music, writing, and reading literature. Growing older, however, his focus was narrowed to basically two areas—science and aviation. His father was in the air force, and living on various bases, Anderson was always around airplanes. He saw them taking off, landing, and flying over his house every day. The noise might have bothered some more sensitive kids, but not him. To Anderson, it was all "just a fascinating thing to me as a kid." His course was set:

My interest in aviation and my interest in science were, I guess, two of the things I really latched on to, and two things that I just couldn't shake as I grew older. So one day, just sitting down and thinking about, "How can I combine my two strong interests, my interest in science and my interest in aviation?" At that time, we were going to the moon and doing some really fantastic things with the space program. To me, that was just the best combination of the two. Here you have these men that are scientists and engineers, and they're also flying these wonderful airplanes and these great spaceships, and they're going places. To me, that just seemed like the perfect mix and the perfect job. So very early on, I just thought being an astronaut would be a fantastic thing to do. Of course, you don't know how to go about something like that. You just sort of pursue your interests, and you pray about it and hopefully one day all things will kind of fall into place and you'll have a chance to make those dreams come true. Fortunately for me, it did happen that way.

For Anderson, college life at the University of Washington was the best of both worlds. He majored in physics—which he determined to be the broadest field of scientific study—and astronomy. Even better, he was in the ROTC program, so the air force was providing a scholarship to help pay for it all. After four years in the air force working to improve the branch's communications systems, he was accepted to flight school in 1986. He would end up with more than three thousand hours in various incarnations of the KC-135 cargo transport and T-38 jet aircraft, and after picking up a master's degree in physics from Creighton University in Nebraska, he felt ready to take on NASA. He was accepted into the astronaut office in 1995.

Relatively speaking, Anderson did not have to wait very long for his first flight assignment. STS-89 flew in January 1998, the second-to-last mission to dock with the *Mir* space station. Serving as the mission's flight engineer during both launch and landing, Anderson had perhaps the very best view in the house—not that he was particularly keen about it. "I'm probably different than most astronauts," Anderson admitted. "I really don't enjoy launches. I think a launch is a terrible way to get into space. But right now, it's the best way to get to space we have—actually, the only way. So I'll take that ride." Later, he elaborated:

When you launch in a rocket, you're not really flying that rocket. You're just sort of hanging on. I really shouldn't say that I don't like launches. I guess I should say, "I understand the serious nature behind a rocket launch." I mean, you're really taking an explosion and you're trying to control it. You're trying to harness that energy in a way that will propel you into space. We're very successful in doing that, but there are a million things that can go wrong. When you really sit down and you study the Space Shuttle and you really get to know its systems, you realize that this is a very complex vehicle and even though we've gone to great pains to make it as safe as we can, there's always the potential for something going wrong. We try not to think about those things. We train and try to prepare for the things that may go wrong to do the best we can, but there's always that unknown. I guess it's the unknown that I don't like. The benefits for what we can do on orbit, the science that we do and the benefits we gain from exploring space are well worth the risk. I don't like launches, but it's worth the effort. It really is.

Andy Thomas took the ride up on board *Endeavour* to take over for *Mir 24* crew member Dave Wolf, who returned to Earth with Anderson and the rest of the STS-89 crew. "He used to joke a lot about taking me up to *Mir* and leaving me there," Thomas remembered. Of Anderson's personality, Thomas continued:

He was an understated person. You didn't really have an appreciation of his skills because he didn't have a need to advertise them or to have a lot of bravado. It's only when you worked with him—and I flew with him quite a lot in the aircraft—that you started to get an appreciation for his background and his talents. He was looking forward very much to subsequent flights after Columbia. *He used to talk about it. He was hoping very much that he could do a flight like we're doing now, a flight to the International Space Station. That was something he felt very strongly about and wanted to do. It was very sad, his loss.*

The introverted side of Anderson's personality was one that Rominger and Ashby also took heart. "He was the quiet type, unless you asked him about his family or his Porsche," Rominger said at the JSC service. "And perhaps because he was quiet, we all loved to see him laugh. And when he laughed, we laughed with him even harder. He knew just when to drop a great punch line." Again, Ashby repeated the memory at the Lufkin tribute almost word for word. Such character made Anderson a good choice as payload commander for STS-107, which had a flight plan filled to the brim, and then some, with research and experiments. He said in a pre-flight interview:

Who do you train for which payload? How do you take advantage of each crew member's strengths, to assign them to the payload that's most appropriate for them? How do you choreograph the operation on orbit? We have just this limited amount of time up there, and we want to make sure that we get the best use of that time. My job, basically, was to try to pull this mission together, try to make sure that the crew was well trained, try to make sure that payloads were well integrated into the Space Shuttle, that all the requirements were taken care of and that we were going to get the best science we could get out of this sixteen-day flight. It's been a very busy job. I really try to think of my role as one of helping the rest of the crew do their job, trying to make their jobs easier. If I do my job well, then their job should be just that much easier. We should have a better time on orbit and have a much more successful mission.

To the very depths of his being, Ilan Ramon felt that his position on the STS-107 crew made him a representative of the Israeli people to the entire world. That is why, despite the fact that he did not ordinarily follow a strictly kosher diet, he did so during the flight. That is why he carried with him a drawing of the earth as seen from the moon that had been made by Petr Ginz, who lost his life at the age of sixteen in an Auschwitz gas chamber. It is also why he had with him in space a tiny Torah scroll that had been given to Yehoyahin Yosef as a boy in the notorious Bergen-Belsen concentration camp. Yosef was taught from the scroll for his Bar Mitzvah by the rabbi of Amsterdam, and after surviving the nightmare, he fought for Israel's independence. He later became a professor of planetary physics who helped manage his country's Mediterranean Dust Experiment (MEIDEX) that flew on board *Columbia*.

Ramon, one of five rated jet-fighter jockeys on the STS-107 crew, was the lone combat veteran. He was one of eight Israeli F-16A pilots who bombed and heavily damaged a nuclear reactor eighteen miles southwest of Baghdad, Iraq, on 7 June 1981. He would always feel a kinship with the people of Israel:

Personally, I think it's very, very peculiar to be the first Israeli up in space, especially because of my background. But my background is kind of a symbol of a lot of other Israelis' background. My mother is a Holocaust survivor. She was in Auschwitz. My father fought for the independence of Israel not so long ago. I was born in Israel and I'm kind of proof for them, and for the whole Israeli people, that whatever we fought for and we've been going through in the last century—or maybe in the last two thousand years—is becoming true. I was talking to a lot of Holocaust survivors. When you talk to these people who are pretty old today, and you tell them that you're going to be in space as an Israeli astronaut, they look at you as a dream that they could never have dreamed of. So it's very exciting for me to be able to fulfill their dream that they wouldn't dare to dream. It is very exciting, very exciting.

People of the Jewish faith had flown many times on board the shuttle, and one—Judy Resnik—died in the *Challenger* accident. Still, an Israeli citizen had never received a spot on a crew until Prime Minister Shimon Peres met with Bill Clinton in December 1995. "Not only is the world global, space is certainly global, open to all nations and all peoples," Peres said. "So, on one of my visits with President Clinton, I asked him if this were possible. He said he'd check into it, and a few hours later, I received a positive reply."

After thousands of applicants were screened, Ramon, wife Rona, and their four children reported to Houston in July 1998 for his training. In many ways for the Ramon family, the move was not an easy one. Not only were they half a world away from home, they were also among a people whose language they had not spoken. "For them, it was a big adventure," Ramon said. "Leaving the state of Israel to live in the U.S. is by itself a very special experiment, which is not so easy. Some of them did not know a word in English and we just stuck them in public school and they had to struggle their way through." As much as his position on the STS-107 meant to Israel, Ramon came to feel that it maybe meant even more to Jewish communities in America:

I meet a lot of Jewish organizations and people here in the U.S., and some time, I feel maybe it's even more important and they appreciate maybe much more than the Israelis this kind of event. For them, it is a big event. It's a big issue. . . . I didn't think it over too deeply, but I think there are several reasons. One of them is probably that there is so much going on in Israel that they don't have too much time to appreciate this. And the second is that in Israel—and I learned it only from being here—you don't appreciate so much your freedom and your ability to do whatever you want to do because you are in your state within Israel. When you go out of Israel, you suddenly feel how much the state of Israel is important for the Jewish communities.

One cannot begin to comprehend what the families and friends of the STS-107 crew went through in the days, weeks, months, and years after the catastrophe. But life went on, and the pain began to ease in many ways.

Something like this could never happen again, could it? For Rona Ramon, another overwhelming heartbreak awaited. Her eldest son with Ilan, Assaf, followed in his father's footsteps and became an F-16 pilot in the Israeli air force. He was named outstanding cadet and valedictorian of his class, graduating on 25 June 2009. Less than three months later, the twenty-year-old Assaf died on 13 September when his jet crashed during a training exercise. He was buried the next day alongside his father. "Don't tell me to be strong," Rona Ramon said. "I have already been strong." When she spoke at the funeral, her words were a study in agony. "I'm so angry, I'm trapped," she continued. "My Assaf, this is my grave. This is my spot. You should have buried me here as an elderly nice lady with a million grandchildren. They promised me they would look after you. They promised to look after Dad. I know that you're looking after him now."

STS-107 was originally baselined by mission planners for 11 May 2000, yet by the time the crew was finalized with a 1 December 2000 press release, the flight had already been delayed eight times for one reason or another. The slips were not over, not by a long shot. Cracks were discovered in the flow lines of *Discovery*'s main propulsion system, so NASA ordered checks of the other orbiters as well. A mission to service the Hubble Space Telescope got priority, as did missions to the ISS. "Missions are always reviewed," said Wayne Hale, whose first official day as launch integration manager of the shuttle program was, incredibly, 1 February 2003. "There's always a consideration of, 'Do we need to fly this mission? Do we need to do different things on the mission? When do we need to fly it?'" He continued:

Most of the flights in that time period involved construction of the International Space Station, and we had a real schedule constraint on when you could fly to the space station to get the parts up there. It's got to be choreographed with the other vehicles that visit from other countries, particularly the Russian vehicles. It's all got to be negotiated. Space station construction really kind of drove the whole schedule. The STS-107 mission, which was a pure research mission, had nothing to do with the International Space Station. It wasn't flying in an orbit that was close to the International Space Station. It really didn't have a time constraint, but it had to be flown. It didn't have a place it had to fall in the sequence, so it got bumped around quite a bit.

By the time Husband and his crewmates finally made it into space, they had faced no fewer than eighteen postponements. The disappointment of facing so many delays could very well have destroyed the crew's morale and work ethic—if they were not going to fly any time soon, then why train so hard? That was not the way this bunch operated. Clark and Brown at one time or another both compared the process to a marathon. "The finish line is only about a mile away, and suddenly they tell you, 'Oh, no . . . we moved the finish line. It's actually ten miles away,'" Clark quipped. "You have to kind of regroup and re-pace yourself so you can finish the whole marathon." Other crews became closely knit units during training, but *Columbia*'s final crew seemed all but a family. They were young and relatively inexperienced, with only three previous flights between them. They joked that Jerry Ross, who flew for a record seventh time on board STS-110, had more than twice the number of missions himself as they did combined.

Some of the most telling images that Brown captured with his trusty video camera took place during a ten-day hiking expedition to the top of Wind River Peak in Wyoming in August 2001 that was part of the National Outdoor Leadership School. These were seven very highly skilled individuals, yes, but they were also a team that had a mission to perform together on this hike and eventually in space.

The danger was much like that facing a professional sports team with a payroll in the hundreds of millions of dollars. Separately, the players were superstars, but could they play as a cohesive unit and ultimately win it all? There were no cell phones and no immediate emergency services to fall back on in the event someone got hurt. There were no restaurants along the way, forcing them to prepare their own grub even if it did turn out to look like bear poop. With sixty-pound packs on their backs, they slogged through the Wyoming wilderness for miles on end each day. They told stories, got altitude sickness, fished, and swam. Brown struggled to make pizza and the rest of the crew gave him a hard time about his efforts, but they ate every morsel. They helped each other across streams and played cribbage, with McCool losing most of the time, to his great consternation. They took turns leading the foray and played a sort of hide-and-seek game with their instructors. Clark studied everything, up to and including a pile of moose manure she found along the way.

On the day they were scheduled to summit the 13,192-foot Wind River Peak, Anderson grew wary. "The shuttle was his livelihood, and he didn't want to risk any sort of injury," said John Kanengieter, NOLS's professional training director. "The boulders are unstable, so they are a hazard. All of them had a lot to lose if something happened on that climb. Mike was looking at it and asking, 'Why take that risk? I have a wife and children to support.'" The group considered splitting up, but, in the end, that was not an option. They were, after all, the crew of STS-107 and they were one. They rose at 4:45 a.m. and made it the last 2,500 feet to the mountain's peak six hours later. Husband would later say of the trip:

We got to see some incredible scenery. We got to learn a lot about how each of us, as individuals, deal with the kind of situations that they put us into. It's a physical challenge with the backpacks and the walking up and down. It's also a challenge learning how to keep track of all of your equipment personally and then learning to work together, pulling together and learning more about each other. When you come back, you have just about gotten to the point where you know each other very, very well. You know each other's strengths and weaknesses, so you can maximize that during the rest of your training flow.

Packed into the flight plan were more than eighty experiments, with much of the work scheduled to take place in the Spacehab Research Double Module that would be located in *Columbia*'s payload bay. Another structure spanning across the payload bay contained the Freestar package of tests, and attached to the top of Spacehab were even more instruments. Even with more than two weeks to work, most, if not all, of the crew experienced nagging concerns over being able to get to it all. Two teams—Husband, Chawla, Clark, and Ramon formed the red, while McCool, Anderson, and Brown were the blue—would work around the clock. "I think that the biggest concern I have about the long-duration mission is not so much the duration, but the density of the whole mission," McCool admitted. "It's very tightly choreographed, and there's so many payloads. Each one, in terms of time, has interdependencies on the other one being done. If there are any hiccups or delays, it's just going to ripple through the timeline."

The work was to be tied into three basic scientific disciplines, one of which was earth science—the study of anything having to do with our home planet. Included in this field was Freestar's Shuttle Ozone Limb Sounding Experiment-02 (SOLSE-02). SOLSE-02 featured a new technique to study

20. Godspeed the crew of STS-107, clockwise from bottom center: commander Rick Husband, Kalpana Chawla, Dave Brown, pilot Willie McCool, Mike Anderson, Israeli Ilan Ramon, and Laurel Clark. Courtesy NASA.

the vertical distribution of the ozone layer and, once proven, would be implemented on future weather satellites. Israel's MEIDEX experiment studied various properties of atmospheric desert dust particles over North Africa, the Mediterranean, and the Atlantic Saharan regions. To understand the influence of the sun on the earth's climate, the Solar Constant Experiment (SOLCON-3) measured the amount of incoming solar electromagnetic radiation.

Next was physical science—in a nutshell, the study of nonliving materials. Crystal growth was researched. The crew looked at the impact that the liquefaction of sands in coastal areas can have on buildings and structures, especially during earthquakes. There was a combustion module on board to study different varieties of flames in microgravity, soot production, and fire-suppressing techniques.

Finally, life science is just that, science having to do with living organisms. There were to be several studies conducted on cancer research and astronaut health. Clark, Brown, Anderson, and Ramon served as test subjects during the flight, while Husband, McCool, and Chawla—as commander, pilot, and flight engineer—were spared in order to save them for the landing. Remarked Anderson:

They sat us down and explained to us what our role would be, not only as the scientist doing the experiments, but also as subjects of the experiments. They outlined, in great detail, what would be expected of us in that role. When you're talking about getting on a Space Shuttle mission, you'll do anything, so, of course, you sign up for anything that they ask you to do. But when reality actually hits and you actually start becoming that subject and they actually start poking and prodding you a little bit, then you start to realize what you've actually signed up for.

The integration of the complicated and intricate STS-107 flight plan was truly an international effort. Participants in the Space Technology and Research Students (STARS) program from schools in Australia, China, Israel, Japan, Liechtenstein, and the United States contributed tests involving spiders, silkworms, inorganic crystals, fish, bees, and ants. Said Chawla:

It really is incredible to see that there are all these countries that are participating in this research. Basically, they have one goal, which is to better understand these processes and then be able to use the benefits that come out of them. What's really interesting in a scientific community is when you go to one place and you know about some of the rifts some of these people might be having. But in this room, these six scientists from six different countries are together. They are trying to do something which is totally mind-boggling. To sit with them and talk to them and understand their fears and concerns on if their assumptions are wrong, but if everything that they've done is right and some big benefit can come out of it, it's just tremendously satisfying to have been there and be a part of that process and to help them carry out their experiments in space.

In all, an estimated 30 percent of the total data expected was in some way recovered following the loss of the crew, the orbiter, and its cargo of experiments. None of the human life sciences information was saved, while the STARS initiative had about 70 percent of its objectives in hand due to in-flight downlinks of video, photos, and humidity and temperature readings. Three of the instruments on the Freestar payload achieved 100 percent success.

After spending more than two years together as a crew, the men and women who would fly STS-107 met on the night before launch to go over last-minute details. Rominger remembered how Husband chose to close the session, by quoting the sixth through ninth verses of the first chapter of the book of Joshua:

Be strong and courageous because you will lead these people to inherit the land I swore to their forefathers to give them. Be strong and very courageous. Be careful to obey all the laws my servant Moses gave you. Do not turn from it to the right or to the left; that you may be successful wherever you go. Do not let this book of the law depart from your mouth. Meditate on it day and night so that you may be careful to do everything written in it. Then you will be prosperous and successful. Have I not commanded you? Be strong and courageous. Do not be terrified, do not be discouraged, for the Lord, your God, will be with you wherever you go. (New International Version)

Husband led the group in a final prayer before leaving the suit room the next morning. Finally, the crew of STS-107 reached pad 39A at 7:42 a.m. EST on 16 January 2003. This was no dress rehearsal. This was the real deal. They took an elevator to the 195-foot orbiter access level, and ten minutes after arriving at the pad, Husband was the first to board *Columbia*. Next in were Ramon, McCool, Anderson, Brown, and then Clark. At 8:43 a.m., Chawla became the last of the STS-107 crew to take her seat.

Doug Hurley joined the astronaut corps in 2000, but when it came time for STS-107 to fly, he was still waiting for a mission of his own. He bided his time as a Cape Crusader, one of the astronauts responsible for getting the cockpit ready and strapping their brethren into their seats prior to launch and then helping them back out of the spacecraft after landing. He was the group's lead for STS-107, and he took to the job with relish. Tradition called for the commander or pilot to take with him or her the name patch of the lead Cape Crusader, but Hurley thought that Husband had somehow forgotten. He had not. Husband grabbed the insignia from Hurley's bunny

21. Partially visible just above Rick Husband's raised elbow is the name patch from lead Cape Crusader Doug Hurley's uniform. Tradition called for crews to take such mementoes with them on flights, and Husband did not forget. Courtesy NASA.

suit and stuck it to a piece of Velcro on his side of the flight deck. Just before Hurley left the flight deck, McCool looked at him and remarked, "I can't wait for you to be sitting here like I am."

The hatch swung shut behind Hurley, never again to be opened by human hands. After experiencing so many delays getting to this point, to some, it would have been almost appropriate had there been some kind of weather concern. There was not. Instead, it was a beautiful morning, sixty-five degrees and nearly cloudless. Following normal planned holds at the T-minus twenty- and T-minus nine-minute marks of the countdown, *Columbia* took off at 10:39 a.m.

About fifty-seven seconds into its flight and at an altitude of 32,000 feet, the orbiter shot through a wind shear that buffeted its left side. The nose of the vehicle was pushed to the right, which in turn exerted aerodynamic loads between it and the External Tank that were larger than normal, but within design limits. The push packed such a punch that an unexpected oscillation was caused by liquid oxygen sloshing around inside the tank. That kind of thing had happened on other flights, but not so much and not for so long. Two other subsequent wind shears were detected, but both were less substantial. As *Columbia* pushed through the wind gusts, its flight control system kicked in to automatically stabilize the ascent. Those were the kinds of anomalies for which NASA had always planned.

The agency had also seen many debris strikes that ultimately had no impact on the mission. *That*, in and of itself, was a very large part of the problem that was about to take place. At least three pieces of foam from the External Tank's left bipod—one larger and two smaller ones—came loose 81.7 seconds into the flight. The largest piece, twenty-one to twenty-seven inches long and twelve to eighteen inches wide, hammered the underside of *Columbia*'s left wing somewhere very near RCC panels five through nine two-tenths of a second later. The impact tore a gaping hole in the wing that went undetected, and in the aftermath of the accident, a question from the darkest days of the post-*Challenger* era reemerged. How could such a calamitous problem go unnoticed? There had been a strike of some sort, that much was known. The Intercenter Photo Working Group received high-resolution film of the launch on 2 February, and the foam strike was clearly evident. Surely, someone somewhere would need better information, so Robert W. Page, chairman of the photo group, walked just down the hallway at KSC to talk with Wayne Hale, who was serving at the time as the de facto launch integration manager before officially taking over a couple of weeks later. "He came into the office and said, 'We've had a debris strike on the orbiter. Let me show you the pictures,'" Hale remembered.

Houston needed to know, so Hale called shuttle program manager Ron W. Dittemore and Linda Ham, who served as chair of the STS-107 Mission Management Team (MMT). Hale continued:

I think Bob was still in the room. I got on the phone and said [to Dittemore and Ham], "Hey, the review team at the Intercenter Photo Working Group has seen this. There's a debris strike. Pictures are not very good." The best particular camera was not focused properly, and we were working off another camera that really wasn't optimum to see what we wanted to see. So I said, "You need to be aware of it." They said, "Good. We got it. We'll turn the team onto looking at it and working on it." I said, "Okay, I'll see you back in Houston," because I was flying back that afternoon. That was where I left it. Bob and I talked, and he said what a lot of other people said, "I wish we had more than that. Those pictures aren't very good. Is there any way we can get some more views of what happened?" I said, "I don't know. We'll think about it."

According to a report issued 26 August 2003 by the *Columbia* Accident Investigation Board (CAIB), the exchange between Page and Hale was the first of three "discrete" requests for on-orbit imagery of the wing by a NASA engineer or manager. After confirming the strike, the photo group e-mailed a report of its findings and digitized video clips to various NASA and contractor employees, and the response was almost instant. Calls and e-mails flew between JSC and KSC, United Space Alliance, the consolidated contractor jointly owned by Boeing and Lockheed Martin to operate and process the shuttle, and Boeing, the company that built the orbiters in the first place. Engineers weighed in, debating the seriousness of the issue. A Debris Assessment Team (DAT) was formed, but a formal meeting was not held until the sixth day of the flight. A Boeing mathematical modeling tool dubbed Crater calculated the impact's damage, yet its results were discounted. On 19 February, Rodney Rocha, a chief structural engineer at NASA who served as cochair of the DAT, e-mailed a JSC manager to see if the crew of STS-107 had been asked to visually check the wing for damage. No one ever got back to him, and the CAIB would later judge this to be the first of no less than eight missed opportunities to diagnose the problem.

Two days later and shortly after the end of the first DAT meeting, Rocha sent another insistent e-mail to JSC engineers Paul E. Shack, David A. Hamilton, and Glenn J. Miller, with three other people copied on the note. It read, in part: "The meeting participants all agreed we will always have big uncertainties . . . until we get definitive, better, clearer photos of the wing and body underside. Without better images it will be very difficult to even bound the problem and initialize thermal, trajectory, and structural analyses. Their answers may have a wide spread ranging from acceptable to not-acceptable to horrible, and no way to reduce uncertainty. Thus, giving MOD [Mission Operations Directorate] options for entry will be very difficult. Can we petition (beg) for outside agency assistance?"

Rocha concluded his demand by outlining a few options that he determined to be available. He put forth the idea of an on-orbit repair as well as the limiting of high cross-range de-orbit entries. Rocha added lastly that right- or left-hand turns during the Heading Alignment Circle landing phase could have been constrained if there was found to be structural damage of the RCC panels that affected flight control. "In my e-mail request, I said, 'Let's beg for outside agency assistance,'" Rocha added. "'We're highly uncertain about this problem. There's too many possibilities here. Some of them are very bad.' We needed an image. That's all we needed." This was the third and final imagery request to be rejected by the Mission Management Team by 22 January, the seventh day of the flight. "The order to halt those requests was then interpreted by the Debris Assessment Team as a direct and final denial of their request for imagery," the CAIB report concluded. Rocha was "devastated":

How could you possibly say no? The Debris Assessment Team was specifically charged to make a recommendation that it's safe or not safe, and deliver an answer in three days. But without this photo, it's like someone saying, "I want you to tell me how bad that car accident is that you just heard out the window. And you say, "Well, I'll go look out the window," and someone says, "No, you may not look out the window. You do your analysis first, and you tell me if you need to call an ambulance first." How can you possibly get out of that kind of uncertainty? It's impossible. That's where we were, stuck.

Yet another opportunity fell by the wayside as Hale began making inroads on the morning of 22 January with a Department of Defense representative at KSC in response to Page's earlier request. It did not take long for the ball to get rolling. Within the hour, the U.S. Strategic Command (USSTRATCOM) at Cheyenne Mountain Air Force Station in Colorado was already figuring out what it would need to get the images. Hale's call, however, had been made without Ham's knowledge or permission. When she got wind of the DoD request, she began asking around about who had made it. She polled MMT members on whether there was a requirement for a request. When there was none to be found, the DoD official from KSC contacted USSTRATCOM to let them know that, according to the CAIB report, "NASA had identified its own in-house resources and no longer needed the military's help." A personal note obtained by the CAIB was particularly noteworthy. Ham, the note stated, said that imagery was "no longer being pursued since even if we saw something, we couldn't do anything about it. The program didn't want to spend the resources."

Years after the accident, Hale refused to cast an accusatory finger at Ham or anyone else, for that did not seem to be his style. Married at the time to astronaut Kenneth T. Ham, she bore much of the heat after the accident, fairly so or not. There had been many debris strikes on many flights that landed safely, but STS-107 did not. Had the issue on that flight been processed as all the other ones had been? "That is not a short-answer question," Hale began. "That's a complicated question. Part of it is complicated by the fact that the conventional wisdom inside the shuttle program—at least in the management area—was that the shuttle is mature. We understand how it operates. There are no big surprises out there. Therefore, we don't get to devote as much resources to maintaining and operating it." He continued:

Over the past eight or nine years prior to 2003, the shuttle budget had been reduced almost by half. Think about that a second—still flying the same number of flights, but doing it for half the money. So in this mindset, there were reduced numbers of people to deal with anomalies. We had anomalies, dozens of them, every flight. Most of them are minor. A tape recorder didn't work. Some minor piece of instrumentation failed of no consequence. We had enough people that it would be given some level of evaluation, but it wasn't like the early days. So that was the mindset, and in that environment, I think the debris strike got the amount of attention that you would expect for any anomaly in those days. Clearly, hindsight being what it is, it wasn't enough.

Still, the fight was not over. A second meeting of the Debris Assessment Team was held, and afterward, Rocha pounded out another e-mail that read:

In my humble technical opinion, this is the wrong (and bordering on irresponsible) answer from the SSP [Space Shuttle program] and orbiter not to request additional imaging help from any outside source. I must emphasize (again) that severe enough damage (3 or 4 multiple tiles knocked out down to the densification layer) combined with the heading and resulting damage to the underlying structure at the most critical location (viz., MLG [Main Landing Gear] Door/wheels/tires/hydraulics or the X1191 spar cap) could present potentially grave hazards. The engineering team will admit it might not achieve definitive high confidence answers without additional images, but, without action to request help to clarify the damage visually, we will guarantee it will not. Can we talk to Frank Benz [director of engineering at JSC] before Friday's MMT? Remember the NASA safety posters everywhere around stating, "If it's not safe, say so"? Yes, it's that serious.

The e-mail was written but not sent. "My finger hovered over the send key," Rohca admitted. "Like any large organization, you have protocol. You have managers, and you have other people that you should inform." Rocha continued to agonize over the matter, yet he seemed to feel utterly helpless to do anything about it. "There was no management behind me," he said. "So I felt pretty alone and kind of crazy. I'm starting to be outnumbered by the people who think it might be okay or the others that are just kind of giving up on it." He continued, "One of my regrets is that I didn't break the door down. Like, why not just go to the very top and be very demanding and insisting on what we need? Hundreds, if not thousands, of people would've put their brains together and started thinking pretty smartly about how to respond to this. Now, would that have worked? I guess I'll never know."

In fact, Rocha was not alone. Much of the concern seemed to be concentrated on the Main Landing Gear wheel well and what might happen if the foam from the bipod ramp had struck it. Another prime example of the frustration felt in some corners of the NASA engineering community came in an e-mail from Robert H. Daugherty, an engineer at Langley Research Center in Virginia, to Carlisle C. Campbell, a JSC engineer who was responsible for the shuttle's landing gears, tires, and brakes. Daugherty wrote very simply on 28 January 2003, the seventeenth anniversary of the *Challenger* accident, "Any more activity today on the tile damage or are people just relegated to crossing their fingers and hoping for the best?" Those who made the big decisions at NASA dropped their gloves just as fate was about to deliver a stunning sucker punch.

Throughout American history, many phrases uttered either in prepared speeches or on the spur of the moment have come to all but define a particular event.

Out of the Civil War came "Four score and seven years ago . . ." The bombing of Pearl Harbor on 7 December 1941 was "a date which will live in infamy." Almost exactly twenty years later, President Kennedy told us that he believed "this nation should commit itself, before this decade is out, to landing a man on the moon and returning him safely to Earth." Then, famed CBS anchor Walter Cronkite made one of the most famous news broadcasts ever when he announced that the young leader had died from an assassin's bullet on 22 November 1963. *Challenger* was "obviously a major malfunction." Todd Beamer told fellow United Flight 93 passengers, "Let's roll," just before they stormed a hijacked cockpit on 11 September 2001.

For better or worse, Charlie Hobaugh became the voice of the *Columbia* accident while serving as reentry CapCom. Over and over he called, trying to somehow reestablish contact with an orbiter that was no more. There was no panic in his voice, no urgency.

Columbia, Houston. UHF comm check.

Columbia, Houston. UHF comm check.

Columbia, Houston. UHF comm check.

Columbia, Houston. UHF comm check.

The same phrase over and over, but no reply ever came.

Hobaugh had the kind of connection to the STS-107 crew that many in the astronaut office did, members of the same ASCAN class and so forth. He had known McCool in particular for years—they had been members of the same swim team all the way back in a Minnesota junior high school, and he remained in contact with the STS-107 pilot's family well after the mishap. Hobaugh did not seek out his role in the nightmare but accepted it as part of his duty:

I did not go and talk to too many media or press afterward. I stepped out of that whole arena, in respect to the families and the crew members on board 107. We have close ties to all of them, and so I didn't feel it was necessary to go and talk about any of the feelings or anything, because it's really about them and making sure they're taken care of. I think out of Columbia, *we learned a lot. It kind of reset yourself back to reality, to just realize that low-Earth orbit is not an easy environment to operate in. The Space Shuttle is not a proven vehicle that you can just continually cycle rapidly up. It just gave everybody newfound respect that I think we all had, but it just reaffirmed the importance of taking everything seriously. It was obviously a hard time, with a lot of time for reflection and making sure we're committed to the right things.*

Just a few feet from Hobaugh's workstation stood Milt Heflin, the head of the flight director office, in a closed-off viewing room behind the control room. He began working for NASA less than eight months before the fire that claimed the lives of the *Apollo 1* crew, and he had worked the trenches in mission control for the flight just before *Challenger* was lost. Heflin was working for the agency when those tragedies took place, but had no direct connection to them. That was not the case for STS-107.

Along came the *Columbia* accident, and this one happened on his watch.

In years to come, the friendly Oklahoman reflected back on his childhood in Tornado Alley, and the many times had he seen pictures of straw stuck in telephone poles after a vicious storm. "What that reminds me—and I have a degree in physics—is that something that doesn't weigh very much but that's traveling at a significant speed can hurt you," Heflin said.

Only one other person was with Heflin, Ronald C. Epps, the division chief at mission operations for flight dynamics. Years after the fact, it was hard for Heflin to recall when it first began to sink in that something might have been going wrong with *Columbia*, but when John Shannon, working that day as mission operations director, grabbed a thick white notebook of contingency procedures and walked very quickly from the control center, Heflin knew deep down that the situation was dire. Seconds later, Shannon entered the viewing room and confirmed the worst.

We lost them.

At some point in the blur that was the next couple of hours, Heflin found himself in JSC's Teague Auditorium for the first of what became many press briefings in the coming weeks and months. Dittemore was seated to his right, while public-affairs representative Kyle Herring moderated for a room of reporters and JSC employees. Over the next hour and a half, the looks on Heflin's face alternated between utter sadness and sheer determination. Nowhere was that more apparent than in his opening remarks. "This is a bad day," he began. "I'm glad that I work and live in a country where when we have a bad day, we go fix it." Trying hard to keep a stiff upper lip, Heflin continued with the details that were beginning to trickle in. He would later regret beginning that way. "The first thing out of my mouth, I should have said something about the crew and their families," Heflin concluded.

It did not take long for him to begin replaying in his mind what he had seen and heard during the flight, the things he had missed in the backrooms, all the sensor warnings on the left side of the orbiter during *Columbia*'s tragic reentry. There *were* people with concerns about the debris strike, but during the Mission Management Team meetings, it had not been discussed much. He was putting two and two together. The foam hit. That was the culprit. Still, it was hard to accept. "We've had orbiters return with *lots* of damage," Heflin said. "There were a few out there that had a different viewpoint, but a lot of us thought, 'My God, this is *just* foam. It's been a turnaround issue. It has never been emphasized as an issue of safety.' That was so ingrained in my way of thinking. Maybe it was a bit of denial. Are we sure? Surely, foam couldn't have done that, right?"

It could, and it did.

When Heflin got back to mission control, he was met by Jackie Reese, the director of JSC's employee assistance program. She wanted to go into the space station's Flight Control Room and help workers there begin to sort through their emotions. A steely eyed missile man, Heflin said no. The people in the control room were tough. They could take a hit and keep on going. Reese did not stop. "She was like a pit bull on my pant leg," Heflin admitted. "I finally relented. I said, 'Okay, Jackie. Okay. I understand. I'm sorry. Let me go talk to the flight director and suggest to him that you come in with your group and talk a little bit about this.'" The move was a good one.

"It was one of the best things I ever did," he continued. "I'm sorry that I got in her way to begin with. Over the days and weeks afterward, she and her small team spent so much time here on site dealing with these kinds of things with people. I had more than one person come up to me afterward and tell me how much they appreciated the fact she took the time to do that."

The skies over the shuttle's landing strip at KSC broke very foggy on the morning of 1 February 2003, and there was some concern, initially at least, that it would not clear off in time for *Columbia*'s planned touchdown at 9:16 a.m. In viewing stands up and down the runway, there hung in the air a keen sense of celebration, of expectation. There had never been a problem during reentry, going back to the very earliest days of America's human spaceflight program. Yes, it was risky, but only in theory. The astronauts of STS-107 were coming home, and that was that. Soon, it would be another crew's turn and life would go on as it always had.

One cell phone began to ring, and then a couple more. Up and down the runway, word spread like wildfire that something was amiss. Doug Hurley was back at KSC with one of his best friends and 2000 ASCAN classmate Robert L. Behnken, ready to be the first ones on the shuttle to help the crew back out. Jerry Ross was there, too. So was Pam Melroy. And longtime NASA employee Tom Overton. And Wayne Hale, who had never seen a shuttle land in all his years of service at NASA. For all of them, the next few hours were to be some of the most highly charged and emotional of their lives.

9:16 a.m. came and went, the countdown clock now counting up, past the scheduled time of landing. Every tick was like a kick in the gut. "We sat there and sat there. Time elapsed—and no vehicle," Hurley remembered. "So now you're thinking, 'Holy smokes, this is *really* bad.'" Ross said a very quick prayer to himself, spoke to some of the family escorts, grabbed a couple of flight doctors, and called crew quarters to have them shut off any televisions and get security in place.

Bob Cabana, the director of Kennedy Space Center, gathered the bewildered families in a crew-quarters conference room to somehow deliver the unspeakable news. "I was in there when Bob told the families that the vehicle broke apart, that he didn't think it was survivable and that they hadn't heard anything from them since just after 9 o'clock," Hurley said as he recalled those terrible minutes. "Of course, it was an obvious reaction from the families, which is just haunting to this day what they must have endured." As heart-wrenching as that meeting must have been, Hurley and Behnken were then tasked with going to the hotels of the *Columbia* families and packing up their personal belongings. "That was," Hurley began, and then paused. "That was excruciating to do that. You were packing notes to parents that the kids had written. I still, to this day, can't imagine what that must have been like for those families to go through." Like so many thousands of others, that day changed something inside Hurley. "It's one of those things I honestly don't like talking a bunch about, but it is what it is," Hurley concluded. "It has had a sustained and lasting effect on me personally. I don't know how to explain what exactly it changed, but there was my life before 1 February 2003 and then there was my life after. I think there probably are a fair amount of people here who would be able to say that. Maybe it was youth, my loss of that. Or idealism. Something. It really did have an effect."

The call was made that none of the astronauts at KSC could fly their T-38s back to Houston that night because, to be quite honest, they were probably not in any shape to do so. In shock, they loaded into Shuttle Training Aircraft and NASA Gulfstream jets for the two-, two-and-a-half-hour flight back to Texas. Hurley would vividly remember landing in Houston and seeing the American flags already at half-mast. Exhausted in every way possible—mentally, emotionally, and physically—Hurley and Behnken sat down for dinner that night at a Houston-area Pizza Hut, but Hurley would never return. The pizza was good, of course, but there remained in that red-roofed restaurant too many memories of that agonizing day.

After ushering the families onto awaiting jets, Ross was to fly to Shreveport, Louisiana, to assist in the search for debris. Emotionally drained, he did not make it to the airport before the weight of everything that had happened came crashing down around him. "I had to finally just pull over and stop there along the NASA causeway and cry for a while," Ross admitted. "I had to let it out. Until then, I just had to control it because I had things that I had to get done." He had good friends on *Challenger* and *Columbia*. Dave Brown and Mike Anderson had worked for Ross, and he had helped get them assigned to the crew of STS-107. He had flown to KSC with the crew, stayed in crew quarters with them before launch, and rode the astrovan halfway to the pad on 16 January. Collecting himself, Ross made it to Shreveport well after midnight on the day after the accident.

As he had done hundreds of times before, Tom Overton was on VIP escort duty that day. In addition to his own son, Michael, an officer in the navy, with Overton were the widow of a sailor killed during the 12 October 2000 terrorist attack on the USS *Cole* and their teenaged son. Once *Columbia* landed, Overton's son and the sailor's son were headed onto the landing strip itself, where they were to receive an up-close and personal view of the ship just returned from space. It would have been an awesome, once-in-a-lifetime experience, but something was glaringly wrong. The twin sonic booms that would have announced *Columbia*'s presence were noticeably absent. Overton overheard KSC coworkers talking in hushed tones on their cell phones, and countdown clocks began to count up once they passed the orbiter's expected arrival time. A veteran who had been with NASA since before the *Apollo 1* fire, Overton was in a daze. This could not be happening again. Told to return the USS *Cole* family to their motel, Overton managed to shepherd them back to an awaiting car.

I was in the back seat with the young boy, and the mother was in the front seat with the driver. I'm trying to keep a stiff upper lip, but it's tearing me up. I'm crying. I liked so much Rick Husband, a fine Christian man, one of the most well-liked astronauts. I blamed God, I blamed everybody. It's hard to explain what goes through your mind when something like that happens. But here I've got these guests, and I'm choking up. I'm really upset. We got probably halfway back to the hotel, and this young man patted me on the leg and he said, "Sir, I know just how you feel."

It was too much, Overton thought at first. For the third time in his career, NASA was facing the prospect of having to come back from the loss of a crew. The risks inherent in spaceflight were great, but he could not leave like that. He went to Texas to help search for debris, one of more than twenty-five thousand people who searched a total of seven hundred thousand acres on foot. All said, an area about the size of the state of Connecticut was scoured on land and water, as well as from the air.

Approximately 84,900 pounds of debris was recovered, or 38 percent of the orbiter's dry weight. Debris was found as far west as Littleton, Texas, and as far east as Fort Polk, Louisiana. Overton's group, composed mostly of firefighters from Montana, spread out six feet apart and walked field after field, picking up parts. "It almost helped you get through," Overton said. "It felt like you were somehow contributing to the solution. That was a help for me."

Overton eventually retired from NASA in early 2011, but continued his duties as a VIP escort well after the last flight of the shuttle program.

Many of the men and women who worked in the three orbiter processing facilities that stood in the shadows of the VAB at Kennedy spent considerably more time on board the Space Shuttle than in-flight astronauts. United Space Alliance (USA) technicians known as spacecraft operators had more seat time, flipped more switches, and monitored more displays than any mission commander. The shuttle was the pride and joy of those in the OPF, and then some. How many other people in the world could say they worked on the most complicated machine ever constructed by human hands?

When STS-107 flew, R. Lester Hanks Jr. had spent thousands of hours on board *Columbia* working as first a quality control inspector and then as a crew module electrical technician for USA. The spacecraft was his baby, but so were the other orbiters, and they were not his alone. "We took great pride in the vehicles," Hanks said. "You paid very, very close attention to detail when you performed your task. I can't recall anybody that ever took shortcuts. They made sure they did their best job, and if it wasn't right, they did it again until it was right." Hanks had spent four years as the crew chief on an F-16 fighter jet while serving in the air force, a job that progressed into the shuttle forward shop. "It was something I never would have imagined I'd have the opportunity to do," Hanks said. "I was actually shocked when I got the call to work on the Space Shuttle. I was like, 'Wow, this can't be happening.' I really thought I'd wind up with Delta Airlines or United, because at the same time, I was just applying everywhere. I was very fortunate. It was probably the greatest ride anybody could ever experience. The people, the job satisfaction—you can't put it into words. Every day was awesome."

Hanks was supposed to work on the night of 1 February 2003, to help begin *Columbia*'s turnaround for its next flight. It was a task he had performed following many flights, but this time, there was no vehicle to process. Living and working so close to the Cape for so long, he, wife La Donna, and young daughters Brianna and Brooke had long grown accustomed to the twin sonic "boom booms" as the shuttle neared its landing facility. After Hanks and his girls waited in vain in their front yard for the shock waves created by the nose and tail of the shuttle reentering the atmosphere at supersonic speed, he spent the rest of the afternoon and evening watching television news reports, scouring the Internet for information, and calling coworkers.

A hangar at the shuttle's landing facility became the focal point of debris collection and reconstruction efforts, and within weeks, after being initially identified and tagged with a serial number, pieces began arriving at KSC via trucks. Hanks was there when the first piece rolled in, and he stayed on that particular job for most of the next year. Each piece was "sniffed" for toxicity, and then sorted onto shelves designated for where it might have been located on *Columbia* itself. Crew module pieces were placed in a separate room, and those were the toughest for Hanks to handle personally. Not only had that been where he had worked for so long himself, it was also where seven astronauts had lived—and died.

That was a tough job, especially the first couple of months. At any moment, you didn't know what you were going to see when you opened the box or bag. You just had no clue what you were going to see. When you saw something and you recognized it, you knew where it was, it hit you like a ton of bricks. It was

22. Sorting through pieces of *Columbia*'s wreckage could be physically and emotionally draining, but the sad work had to be done. Here, shuttle technicians Les Hanks (right) and colleague Pat Marsh work with a piece of debris in a KSC hangar. Courtesy NASA.

tough to hold back sometimes—you knew that crew was there. You knew that it wasn't pretty, by the looks of what you were seeing. That was tough. It brings it back into perspective just thinking about it now.

Working among the crew module debris was indeed the most personal of tasks, as Pam Melroy was also about to discover. Melroy had already flown twice as the right-seat pilot, and she was looking forward to a command of her own. In the meantime, she was in the process of beginning her own stint as a Cape Crusader. When *Columbia* launched, she was still committed to a public-relations tour for STS-112, which had landed three months earlier. After finishing up her time in the barrel, however, Melroy was standing on the KSC runway, waiting for her friends on STS-107 to return.

Melroy was thrust onto the reconstruction team, where she was lead for the team sorting through debris from the crew module. She hopped a T-38 from Houston to Florida, where she was shown an empty hangar and an empty room set apart for pieces of that part of the destroyed vehicle. Two weeks later, trucks started rolling in with debris from east Texas. "That was an incredibly complicated thing to try to figure out what to do," Melroy admitted. "There were so many times when I just didn't have a clue what we should do with something and how we should handle certain situations. You just kind of work it out." Setting apart a room for pieces of the crew module served a very real purpose. "It is *enormously* stressful to look at the debris that was so closely connected to the crew," she began. "That was just not something that would've been good to expose everyone to all day, every day. That was a really wise decision to segregate that debris and give it some privacy."

Almost daily, questions cropped up. What had happened to the crew? Had their equipment worked as properly as it could have, given the circumstances? James P. Bagian, the former astronaut who was serving as the chief flight surgeon and medical consultant to the CAIB, held meetings and encouraged NASA to do further studies, over and above what the CAIB was doing. Melroy was immediately on board with the proposal:

I felt very strongly at the time that there were a lot of questions. The debris was talking to me. I could tell that there were stories. I couldn't figure out why we'd gotten some things back and not others. It was riveting to me. I mean, it was just fascinating. I knew it so well from spending so much time with that debris, I thought, "We've got to try to understand why we've got so much of the mid-deck floor but nothing of the flight deck floor. We've got to understand that." The investigation was over, but as a little bit of time evolved, some questions started to emerge in various locations about what we were doing in terms of crew survival.

The effort became known as the *Columbia Crew Survival Investigation Report*, and Melroy served as deputy for its engineering analysis. The report was not released to the public until 30 December 2008, nearly six years after the accident. Melroy labored on the team even during training for and following the flight of STS-120, her first command. Three of her ASCAN classmates—Husband, Anderson, and Chawla—were on board *Columbia*, making the separation between her professional responsibilities and her personal feelings that much more difficult. Compartmentalizing had always been an important part of being a good pilot, yet while Melroy readily accepted that she first had a job to do, the fact that she knew these people very well was always right there below the surface. "Being able to segregate your feelings at the moment you need to do it is really important," she said. "You also have to remember that you are a human being and that you can't keep yourself separate from your emotions forever.

"It was really hard. I would not . . . ," she said and then paused, an unmistakable hint of deeply felt emotion in her voice. After a moment, she continued, "I would not want to wish it on anybody, but who else is going to do it, right? Who else was going to do it?" It was a striking instant from an air force colonel, a veteran of more than two hundred flight hours of combat and combat support during Operation Just Cause in Panama and Operations Desert Shield and Desert Storm in Iraq and more than five thousand hours overall. When accidents happened in a flying squadron, superiors *never* wanted others in the same group to conduct the investigation into what took place. The problem was not necessarily that they might have been involved. Instead, they were simply too close to the situation, making it too painful to proceed. Such a situation, Melroy said, was tantamount to doing an autopsy on a close friend. After her work in the fallout of *Columbia*, Melroy knew the feeling very well. "We just didn't have any choice," she concluded. "I'm hopeful that, if we ever have an experience like that again, we'll have a big enough pool of people that we won't have to put somebody in that same situation."

At that, Melroy stopped again, the emotion thick in her voice. Rather than ending the conversation, she pressed on, the way any good soldier or astronaut would do. She was not alone.

Working on the reconstruction team at KSC had a profound impact on Hurley as well. "Every day, I was going in and looking at pieces of what was left of *Columbia*," he explained. "In retrospect, that was not healthy. To see the stuff the way it was and to deal with that day after day, it's taken me a long time to get past not only what the stuff looked like, but the smell and seeing the pieces of the vehicle and pieces of their spacesuits and all that stuff that they recovered. To live that for months took its toll on me personally."

Melroy, Hurley, and every other astronaut in the office had to ask whether or not flying the shuttle was worth the risk. They were not the only ones asking such questions.

Although Wayne Hale had been easing his way into the job for a few weeks, as of 1 February, the title of launch integration manager was his and his alone. He was enjoying the party atmosphere next to the shuttle's three-mile-long runway, catching up with various people he had not seen in a while and astronauts who had moved into management positions within the agency. Hale ran back to his car and shot to the Launch Control Center firing room when the situation began spiraling out of control, where the best information available was coming from media outlets such as CNN. NASA administrator Sean O'Keefe was there, and immediately, a group was set up to start an investigation and recovery action plan. O'Keefe called the White House. "We were just in a state of shock," Hale acknowledged. "It's all I can think of, working through the list of things that we had to do."

By July of 2003, Hale had moved into the role of deputy manager of the Space Shuttle program. It was the kind of role that would have been high profile and even higher pressure under the most ordinary of circumstances, and this was anything but. The e-mails he started writing as a flight director grew to the point where "some topics that just bubbled up inside me and I couldn't hold them in." Once he became deputy program manager, he was encouraged to focus his efforts on a blog through which he became *the* go-to guy at NASA for an insider's perspective. Some thought he was maybe too nice a guy for his own good, especially considering what he was up against, but he came to be known for a keen sense of open honesty and candor. In one incredible e-mail, Hale wrote:

I cannot speak for others but let me set my record straight: I am at fault. If you need a scapegoat, start with me. I had the opportunity and the information and I failed to make use of it. I don't know what an inquest or a court of law would say, but I stand condemned in the court of my own conscience to be guilty of not preventing the Columbia *disaster. We could discuss the particulars: inattention, incompetence, distraction, lack of conviction, lack of understanding, a lack of backbone, laziness. The bottom line is that I failed to understand what I was being told; I failed to stand up and be counted. Therefore look no further: I am guilty of allowing* Columbia *to crash.*

In reality, those were merely symptoms. The root cause turned out to be what Hale called the "*Apollo 13* Myth"—NASA could always solve any problem and bring the crew back safe and sound. "Here I was, and *Apollo 13* was unfolding in front of me," Hale said. "If any one of us had stood up and said, 'Hey, this is not right. We should be more serious about this. We should devote more resources,' I think it might . . . might . . . *might* have changed things. Probably not. But none of us did." One of the things that truly bothered Hale when *Columbia* went down were the press conferences during which various officials talked about virtually everything except the essence of what actually went wrong. "We did a lot of press conferences where we said, 'We didn't understand. Really, we wish we would've done things differently,'" Hale continued. "But nobody was willing to stand up and say, 'You know, we just blew it.' And we did, collectively." He has found the experience of saying such things cathartic, getting them off his chest, out in the open, not hiding anything. Others within NASA management felt incredibly remorseful, and said so in private. Hale, though, was different, calling himself a "blabbermouth." He put his raw emotions out into the open.

"So people think I'm . . . ," Hale said, then reconsidered what he was about to say. Singularly at fault? The fall guy? The scapegoat? Nothing could have been further from the truth. He continued, finally, "I'm really no different than the rest of the NASA managers, so just leave it at that." He concluded, "I don't think there was a lot of finger pointing or trying to determine who was culpable. It was more a sense of 'Oh, my goodness, we've got to figure out what happened.' I think that's the nature of us geeky engineers. We're less concerned about blame than we are about why it turned out that way, what happened, and what we are going to do about it."

All around the agency, there were decisions to be made and fates to accept. Was Hurley cut out to be an astronaut? Did he even *want* to fly anymore? "If someone had asked me in June of 2003, I'm not sure what my answer would've been," he admitted. "Emotionally, at that point, I was just completely out. It was tough." As a naval aviator, Hurley had of course lost friends before in the line of duty—but this was seven people, not just one guy or maybe two, and images of the debris streaking through the sky were replayed on television almost, it seemed at times, on an infinite loop.

Once Hurley finally got a chance to decompress, he was able to put his thoughts into perspective. He saw the work that the agency was doing, and how management was handling it. Although it may sound cliché, what would the STS-107 crew have wanted? "Do you think Willie or Rick or Ilan or KC or Laurel or any of them would have wanted us to stop flying?" Hurley asked, the answer obvious. "Then, you look inward toward yourself and ask, 'Why did I really come here? It's not like I haven't been dealing with these types of risks my entire career.'" The accident reset the agency's mindset. Flying in space is a dangerous business, taking a 250,000-pound orbiter from zero to 17,500 mph and then several days later, taking it from 17,500 mph back to zero. "That's *a lot* of physics that needs to work right," Hurley concluded.

For the third time in its history, NASA was faced with a lengthy and painful delay following the loss of a crew. It would not be an easy path back to orbit.

7. "We Came Home"

Eileen Collins sat on the couch of her Texas home, watching television with her toddler son. STS-107 was scheduled to land in fifteen minutes or so, and this was what Mommy was going to be doing herself the very next month. She had flown the shuttle into space three times before, including once as the first female mission commander in NASA's history, but this would be her first trip since the birth of the youngster. A couple of months past his second birthday, there was only so much he could possibly grasp. As the agonizing moments of radio silence passed, Collins knew all too well that something was not right. She heard radar control in Florida say that they did not have a track on *Columbia*, and she began to flip channels. Like *Challenger* nearly twenty years before, images of the disintegrating orbiter streaking through the sky were on every channel.

Almost immediately, she began trying to diagnose the issue. Surely, an Auxiliary Power Unit had blown up. That had to be it. Collins spent most of the rest of the day in the office, on the phone. A colonel in the air force, Collins had been announced as commander of STS-114 on 17 August 2001, along with pilot James M. Kelly and mission specialists Steve Robinson and Soichi Noguchi of the Japanese Space Agency. The flight, scheduled to launch on 6 March 2003, was set to take *Expedition 7* crew members Yuri I. Malenchenko, Sergei Moschenko, and Ed Lu to the ISS to start their long-duration stays and bring back *Expedition 6*'s Ken Bowersox, Nikolai Budarin, and Don Pettit. Those plans changed almost as soon as *Columbia* went down sixteen minutes from home. Collins knew she would not be flying in space anytime soon . . . if ever:

My first thoughts were for all my friends and their families. You start thinking about your crew. You're mostly just focused on the people. It's just a process you go through. You think about all the people at Kennedy, and how they must feel horrible. And they did, because those shuttles are like family members to them. Before the day was out, of course, when I was calling my crew, I was thinking, "Hey, our mission may not even happen." I knew there was no guarantee that I would ever fly in space again. At the time, I was not even concerned about that . . . it just seemed so unimportant.

As with every other tragedy that had befallen the NASA community, the loss of *Columbia* brought forth an intensely personal kind of grief. For Robinson, the pain went "way beyond a death in the family." Their world stopped, in shock. NASA and the idea of human spaceflight, if only for the briefest of moments, fell away. Mission planning and training did not matter. "We became a family," Robinson continued. "That's all that mattered . . . us, the people." Nine months later, the ferrying of an Expedition crew to and from the ISS was officially dropped from the STS-114 flight plan. In their place went mission specialists Andy Thomas, Charlie Camarda, and Wendy Lawrence.

Thomas was deputy chief of the astronaut office at the time of the accident. 1 February 2003 began simply enough for the Aussie:

It was a Saturday morning. For each launch and landing, the astronaut office operates a small contingency action center. I thought, "Well, I probably should go into the action center since I'm the deputy chief." So I got up, and I have to admit, I didn't shave and didn't shower, because I thought I was only going to be there fifteen minutes, thirty minutes. I went in, and I was still there sixteen hours later. It was a tough day, no doubt about it. It was hard to believe, it really was. The feeling was one of shock, mourning, disbelief. You'd walk the corridors of the astronaut office and you'd find people crying, hugging each other. It was a tough time.

At forty-four strong—thirty-five Americans and nine international mission specialists—the 1996 astronaut selection group was the largest in NASA's history. Camarda, Kelly, and Noguchi were members of "The Sardines," as were STS-107's Willie McCool, Dave Brown, and Laurel Clark. Two years earlier, Robinson had been selected in the same "Flying Escargot" class as Rick Husband, Mike Anderson, and Kalpana Chawla. The new astronauts learned the NASA ropes together. Their families knew each other, and the tragedy impacted them every bit as much as it did the astronauts. Camarda's daughter, Chelsea, babysat for Clark's young son, Iain. Dave Brown took her for a ride in his airplane. Training in Russia when *Columbia* broke up, Camarda received a frantic phone call. "My daughter called me up," Camarda began. "At that time, she was about thirteen years old, and she was just bawling. These were friends of hers. They were part of our family. It was devastating, it really was." Camarda freely declared, completely without reservation, that his shock quickly turned into fury. "More than anything else, I was angry. I was *so* angry," Camarda admitted. "What I had seen just startled me, that we allowed this vehicle to be in orbit and not to evaluate the impact of this foam."

Robinson would one day say that NASA never moved past the *Columbia* mishap. Instead, emotions were channeled toward a safe return to flight and beyond:

We didn't move past it and then go into a different phase. I don't think we are past it. We added those new phases to what we were feeling. The shock and the disbelief stayed there. It sort of hardened our resolve to honor their legacy by continuing the mission, only in a much better way. We had to be honest with our own failings, with the things that we should have seen that we didn't. We had to be honest with the fact that we felt horrible. We had to channel that energy. When you have a disaster happen in your life, the most helpless thing is feeling like you can't do something about it.

It was a full six months or so before Thomas "started to think about what was coming, rather than what had been." The agency and members of the astronaut office came to terms with their emotions, he said. What remained were the intellectual and engineering facts of life that spaceflight is a dangerous business and there would always—*always*—be a deep-seated need for caution. "You mourn the loss, and you separate yourself from that, as you do any emotional loss," Thomas remarked. "At some point, it turns the corner. You *have* to do that. In fact, to not do that is grievously unhealthy."

Rick Hauck, who commanded the first mission after the *Challenger* accident, knew what Eileen Collins was facing as the commander of a Return to Flight mission. He sent her an e-mail, and because he was going to be in Houston, he suggested they have dinner. Collins, all but one of her crew, and some of their spouses met Hauck and his wife for pizza. "There wasn't any technical exchange," Hauck recalled. "It was mostly talking about how NASA was reorganized after the accidents and how responsive NASA was, and how we both had a sense of comfort that things were proceeding in a way that made us confident it was going to be as safe as it could be. We had a nice evening, and just got to know each other a little bit." Hauck and Collins later served together on the NASA Advisory Council. She remembered the same general nature of the discussion. "We didn't talk about *Columbia*'s problems in particular," Collins added. "I had flown as a commander before. I knew the shuttle very, very well. So our talk was more about, 'Hey, this is unique. You're going to be distracted by not just the media, but all the other things that are pulling at you as part of the Return to Flight mission.'"

One piece of advice Hauck gave Collins was the need to get out and visit employees of the agency and its contractors. She and the rest of her crew met workers in the OPF in Florida. They headed to Canoga Park, California, where the shuttle's main engines were refurbished at Rockwell International's Rocketdyne plant. They visited an ATK Thiokol facility in Brigham City, Utah, where the twin SRBs were built and restored. Several stops were made at the Michoud assembly facility in Louisiana, home to the External Tank. It was there that ET-93—the unfortunate tank used on STS-107—had been constructed and where STS-114's own ET-120 was put together. The original four members of the STS-114 crew also ventured to east Texas to join in the search for debris from *Columbia*. As they walked the woods, Noguchi found a piece of tile. "It gave me a chance to connect with what was happening, and it kind of closed the loop in my mind on helping find out what caused the accident and how we could prevent these kinds of things from happening again," Collins said. "God help us if they do happen again. We have learned so much, the risk of spaceflight, in my mind, goes down in the future because of the experience that we've had in the past and what we've learned."

Following *Challenger*, Hauck had been given free rein to attend any meeting he wished as NASA got back up to speed. Collins, however, did not get the same offer. "No, I didn't, because I think at that point in time, that was expected," she said and then added:

I think the culture at NASA had evolved to the point that, because of Challenger, *astronauts got much, much more involved in management-type decisions. It wasn't hard for me to decide, "Should I attend these meetings or not," because I knew I should. It was pretty much expected because of what Rick had done. His crew and the crews after him kind of set the standard. It was something I wanted to do. My problem was deciding which meetings to go to, because there were so many of them.* NASA *management wanted the astronauts at the meetings. They wanted to know as soon as possible what the astronaut office, as a group, wanted on certain very specific technical issues like, "Do you need this piece of hardware to do this spacewalk?" We have to train also, so we don't have time to go to every meeting. It's just not possible. A flown astronaut pilot, Steve Frick, was our crew support officer. He attended the meetings that I just simply could not be at. The astronauts post-*Challenger *got very involved in all the activities around operations and even development of new missions.*

Collins's first command had been an eventful one, to be sure. One attempt to launch STS-93 was called off a split second before the start of main engines, and there was substantial trouble during ascent when it finally got under way a few days later. It had also been the first time a shuttle flew under the command of a woman, and there were a few silly questions over just that fact. For obvious reasons, the silliness for STS-114 was long gone. The queries came fast and furious: Was the crew ready to fly? Why should the shuttle fly again? What kind of risk should the country be willing to take to fly in space? Are the fixes to the External Tank complete? Whether on camera or not, the focus was of a far more serious nature.

It is impossible to know for sure if Collins—or anyone else for that matter—would have retained the command had she not had previous in-flight experience as a commander. Both she and Kelly had flown before in their respective positions, but to Collins, it really made no difference. "Crew members that fly for the first time do a very good job," she started. "I believe that we underestimate their capabilities. For complicated reasons, NASA management would always say, 'We've got to send the most experienced people on a mission.' Well, then you'll never fly the new guy." There was already one freshman spaceflier, Noguchi, on the crew of STS-114. And when three new astronauts were named to the crew, Collins actually requested that they include one rookie, if not two.

After more than thirty years, Charlie Camarda was going to get his chance to fly.

23. STS-114 commander Eileen Collins does a pre-flight check of her NASA T-38. Courtesy NASA.

Camarda graduated from Polytechnic Institute of Brooklyn in 1974 with a bachelor's degree in aerospace engineering. Almost immediately, he went to work at Langley as a research scientist. The year before, between his junior and senior years, he had done an internship at the center that forevermore sealed his love for research and engineering. Camarda received his master's from George Washington University in 1990 and a doctorate in aerospace engineering from Virginia Polytechnic Institute and State University a decade after that. At Langley, much of his work was focused in the area of thermal protection. One innovation, the heat-pipe-cooled sandwich panel, was a "competitor," Camarda says, to the RCC panels that made up the leading edge of the shuttle's wings. Camarda's component consisted of self-contained tubes of sodium. At a forty-degree angle of attack coming through the atmosphere, the alternate panel was designed to transfer heat from the bottom of the wing and then radiate it through the top surface.

Camarda applied to the astronaut corps in 1978, and he was turned down. The dream remained. He had custody of Chelsea, and although he had accomplished many things in many areas, he wanted more than anything else to show her that "in this life, you have to take risks. You have dreams, and you should always pursue your dreams." He wanted to accomplish the goal for himself, but it was also to be a life lesson for his daughter. He applied again in 1996, after more than twenty years with the agency. This time, he made it. Camarda would never forget breaking the news to then-eight-year-old Chelsea:

When the call came, I was in a meeting. I went to the phone, and at that time, the head of the office was Bob Cabana. He kind of kidded around, making like he was telling me bad news. Then, all of a sudden, he said, "You were selected, and we would like to have your response as soon as possible." So I immediately took off of work in Hampton, Virginia, and drove to Virginia Beach. I asked my daughter's teacher if I could take her out of class for a couple of minutes. I took her in the hallway, and I said, "This is it, Chelsea. Daddy's been selected. Are you really willing to leave your mom in Virginia Beach and go to Texas?" She looked me in the eyes, and she said, "It's an adventure!" I raised my daughter not to have any fear, and a little eight-year-old said she was ready to pack up and go to Texas. This was an adventure she wanted to be a part of. We headed off to Texas, not knowing what we had in store for us—whether I would be able to pass all the tests, become an astronaut, and fly. We took a shot.

On 1 May 1996, one week short of his forty-fourth birthday, Camarda was officially announced as a member of The Sardines. Only one classmate, Lee Morin, was older. It would be another nine years before Camarda strapped into the seat farthest from the hatch on *Discovery*'s mid-deck, right next to Lawrence, on 26 July 2005. He had already gained a healthy respect for the dynamic act of putting a Space Shuttle into orbit:

I would've loved to have seen the expressions on the faces of my wife and the kids when they actually saw, heard, and felt the pressure waves on the roof of the Launch Control Center, as they experienced all that energy during launch. It's almost a spiritual event. You can't explain it. I tell you, it's a heck of a lot more exciting watching it than actually being inside of it. It is just awe-inspiring to know that there are people in the tippy-top of that orbiter vehicle when you see and feel and hear all that energy. It shakes the ground. It sets car alarms off. The hair just stands up on your arms. What's interesting is, as you're watching it, tears will just be running down the sides of your face. You don't know if they're tears for joy. You don't know if you're worried. It just happens automatically. It's just awesome.

After having waited for so long to earn his gold astronaut pin, what was it like to be Charlie Camarda during the liftoff of STS-114? "You're gonna have to ask Wendy Lawrence," Camarda said, beginning to chuckle. "She was sitting right next to me, and what Wendy Lawrence says is that I was just laughing and grabbing her hand and shaking her the whole time we were going up."

Camarda would not be laughing for long.

The Merritt Island National Wildlife Refuge, of which KSC is a part, was created in 1963 to preserve the area's more than 1,500 species of plants and animals. For years, tour guides have pointed out potential sightings, including a well-known eagle's nest on the parkway coming into the center. In the minutes before launch, Collins and Kelly discussed a flock of turkey buzzards circling pad 39B, adding that surely, once the main engines and SRBs lit, the large three- to five-pound creatures would get the message and scatter. Most did, but one unfortunate fowl did not and was skewered by the tip of the External Tank, well before *Discovery* itself cleared the tower. Luckily, the buzzard's carcass fell to the ET's backside, away from the orbiter. Some would later make light of the incident, saying that the bird had simply been at the wrong place at the wrong time. The implications, though, could not have been more serious. The piece of foam that brought *Columbia* down two and a half years earlier was maybe half the buzzard's weight. Before the next flight, NASA put in place a bird abatement program. The effort included special radar tracking and quicker roadkill removal around the center, eliminating a food source for the buzzards.

The news was about to get worse. A small chip was lost off a tile from the front landing gear door at some point prior to SRB separation. At 127.1 seconds into the flight and 5.3 seconds after the SRBs were freed, a large piece of foam from the ET's Protuberance Air Load (PAL) ramp broke free. The PAL ramp was in essence a wind deflector designed to protect the tank's cable trays and pressurization lines during ascent. This was no small chip either. At 36.3 inches wide, 110 inches long, and 6.7 inches in diameter, it was one of the largest debris liberations in the history of the shuttle program.

There were no fewer than nine other losses from other parts of the ET, including three from liquid hydrogen Ice-Frost Ramps (IFRs) alone. One released piece measured about seven inches by two inches. Ascent and on-orbit imagery also documented a seven-by-eight-inch divot near the tank's left bipod attachment fitting. While the debris did no real damage to the vehicle, they had a chilling impact of a different sort. How in the devil could this have happened again, after so much study and discussion by countless teams across the globe? Before anything else, the very first recommendation of the CAIB had been to fix the foam-shedding problem. The issue had been addressed in a huge way, yet here was more debris falling off in chunks every bit as large as, if not larger than, the one that cost seven astronauts their lives.

Six hours after liftoff, Collins and the rest of the crew were informed of the liberation events. Individually, some were at first sure it meant the immediate end of the Space Shuttle program. "We were shocked," Robinson said. "Everyone at NASA was shocked. Everyone at NASA said, 'I thought we fixed that. Wow . . . this problem is harder than we thought.' I personally wondered, 'Maybe we were the last shuttle flight, even though we got away with it this time. It didn't hit us. Maybe we don't know how to fix the thing. Maybe it's not fixable. Maybe we're always going to have chunks that fall off.' That occurred to me, and I thought, 'That would be very sad.'"

Their initial concern was not so much for their own safety—video of the event showed very clearly that the debris had missed the structure, and there was to be an unprecedented series of visual and robotic inspections in addition to a docking with the ISS. The crew of STS-114 was as safe as it was possible to be during a spaceflight. The Space Shuttle program itself was not. Thomas wondered:

How could this have happened, after all *of the work that was done on that particular tank and* all *of that engineering analysis? There was* so *much of it. It was unbelievable what they did to that tank and to the whole foam application process. Some of us, we even talked about it and said, "If this plays out, it could turn out that this is the last shuttle mission. Congress might just say, "You weren't able to fix the problem. You're done." That could have happened. It was a low probability, but we did talk about that amongst ourselves a little bit. After talking about it, we thought it was probably not a likely scenario, because the station needed to be built. We still had a lot of work to do. But for me particularly, it was just a sense of frustration. Where did we miss something about that foam application process that let that happen?*

According to Camarda, the crew had received assurances that the chances of foam falling off the PAL ramp were only 1 in 10,000. When it did, Collins was as upset with herself as she was anyone else. On one of their many trips to Michoud, Collins remembered walking around an ET and asking a mid-level manager of the PAL ramp, "What is this? Why is it there? Are we fixing it? We don't we remove it?" She was told that it could not be removed because if it was, aerodynamic loads could potentially break the tank's liquid oxygen (LOX) feed line. "That was all they needed to say," Collins recalled. "I knew if the LOX feed line broke, you'd have an explosion and you'd lose the mission." Satisfied with an answer from people she trusted, she left it at that. When informed of the foam loss, however, Collins never realistically thought it meant the end of the program. Instead, she felt she could have pushed harder for an adequate solution.

According to Camarda, his wife, Melinda, learned of the foam-shedding incident from the news, not NASA. Was foam loss not the same thing that had doomed the crew of *Columbia*? Did the debris strike the shuttle or not? The shuttle fleet was grounded, so it had to be serious. If something was wrong and other orbiters were not able to fly a rescue mission as outlined in the CAIB, what then? "Nobody from NASA ever called her up to tell her," Camarda said. "That was probably the thing that upset her the most. It was kind of poorly done."

Had *Discovery* been damaged by the buzzard or ET foam, it would not have gone unnoticed. The amount of inspection on the vehicle after launch was unprecedented, not just in shuttle history, but since humans first took to the heavens. The launch had been filmed from virtually every conceivable angle—it was a new camera mounted on the ET that caught the debris liberation event. Sensors were added to the shuttle's wings. The robotic arm was used to examine the *Discovery*'s nose cap and wing leading edges. *Expedition 11* commander Sergei Krikalev and flight engineer John L. Phillips started taking images of those sensitive areas with high-powered lenses as soon as the shuttle began approaching the ISS. The photography was then downlinked to some two hundred analysts back on Earth for study.

Then, when the orbiter was six hundred feet below the station, Collins began the first rendezvous pitch maneuver, which flipped the vehicle end over end for still more inspection. Moving at three-fourths of a degree per second, most of the operation was flown on autopilot. The difficult part for Collins was simply getting the shuttle into position for the pitch-around. Six different parameters on the control stick—up and down, right and left,

24. The crew on board the ISS was able to examine *Discovery*'s exterior very closely for signs of damage during its pitch-over just before rendezvous, a safety precaution added in the wake of the STS-107 accident. Courtesy NASA.

and fore and aft, each with a corresponding rate—had to be hit very precisely. While Collins flew *Discovery* into position, she was receiving at the same time information from Kelly and Lawrence, who was using a handheld laser device that measured distance to the station. Once the maneuver was complete, Collins retook control of the vehicle and flew from underneath the station to in front of it and eventually docked.

The moves were tricky, but each one had been practiced many, many times back in the sims at JSC. "Believe me, I made plenty of mistakes in training," Collins admitted. "There was a lot of learning that took place. The engineers in the rendezvous section that helped design this maneuver and helped us learn how to fly it were amazing. They were superior engineers in their creativity, as well as their ability to design something and make it happen." She had intentionally flown herself out of the prescribed positioning during several simulations, just to make sure she could make the necessary corrections. She had also asked simulation supervisors to throw her and her crew as many curveballs as they possibly could. That is where mistakes were supposed to be made—in the sims, as opposed to the actual flights:

I don't think I did this as much on my first command as on my second command, but I asked our instructors to put our crew, intentionally, in a very bad position. If we're handling everything by the book and we're doing a good job, then just make it harder. Force us to make mistakes in training. Not only do we need to learn how to do the mission, we need to learn how to handle mistakes as a crew. I had a process for handling mistakes. It worked so well for me, I still live by it today. I teach it to my kids. Step one, you admit you made a mistake. Step two, you fix it. Step three, you put a plan or procedure into place to prevent it from happening to somebody else. Step four, you learn from it, you remember it, and you move on with confidence. That fourth step, people need work on that. I watch kids playing sports. You make a mistake out there on the field. If you get mad at yourself and can't shake it off, you're not going to be a very good teammate for the rest of the game. I had time on my second command to really think about these. You practice these skills to help us communicate better. Actually, we knew each other so well that many times we didn't even have to talk because we'd been training together so long.

The inspections revealed a number of areas of concern, including damaged insulation near the left cockpit window and a number of protruding pieces of gap filler on the underside of *Discovery*. The gap fillers were designed to cushion tiles during the shaking of the orbiter on ascent. If left in place, the potential was there to create turbulence and a resultant increase in friction and temperature during reentry. That, or they could possibly pooch out the tile, and maybe pop one or more out of place. The insulation near Collins's window was considered a low risk, but managers opted to add a new task to the flight's third and final EVA. After he and Noguchi installed a stowage platform on the outside of the ISS, Robinson would be placed on the end of the arm and extended by Lawrence around *Discovery* to its belly, where he would either gingerly remove the two most concerning gap fillers by hand or saw them off. It would be the first time a spacewalker had worked on the underside of a shuttle, out of the direct view of the rest of the crew.

During their first EVA on 30 July, Robinson and Noguchi worked in the cargo bay on various RCC repair techniques before moving onto the station and installing a base and cabling system for the stowage platform that would be added during their third spacewalk. Finally, they rerouted power to one of the station's four six-hundred-pound gyroscopes. Two days later, they installed another gyroscope. All the while, preparations for Robinson's trek down under continued. By 3 August, everything was in place for the memorable—if rather simple—procedure. "Before Steve went out, I told

25. Out of the view of his crewmates, Steve Robinson made quick work of removing stray pieces of gap filler from *Discovery*'s underside. Courtesy NASA.

him, 'Whatever you do, do not pull a tile out!'" Collins said with a gallows-humor laugh. She never truly considered the gap fillers a safety issue. Surely, shuttles had reentered Earth's atmosphere many times before with such protrusions in place and not known it. This was just the first time an in-flight shuttle had been picked over so thoroughly.

No one had ever trained for gap-filler removal, but Robinson looked forward to the challenge:

This is the great thing about spaceflight. It's very likely you'll end up doing something that nobody's ever done before, which is one of the most appealing parts about the job. So the fact that we were doing something we hadn't trained for didn't faze us too much, because we had trained to do so many things, we had quite a tool bag of skills. We had a fantastic support system of really smart people on the ground, and they were including us in the discussions. So we hadn't trained specifically *for that, but it's sort of like when you learn to fly an airplane. You can land at an airport you've never even seen before, and it's because you've got all the skills you need.*

As he was being ferried into place, Robinson spotted a few dings to tiles here and there, including the one on the Main Landing Gear door. It took less than a minute to remove the first protruding gap filler on the port—or left—side of *Discovery* with nothing more than a gentle tug with his fingers.

Seven minutes later, he was ready to go at the second problem area, which took even less time and effort to free. "It looks like this big patient is cured," Robinson radioed Houston. He remembered several years later:

Here's my chance to see this stuff with my own eyeballs. I was pretty familiar with the underside of the orbiter, but nobody had seen it in space. I was just thrilled. I was like, "I'm getting a chance to inspect my own spacecraft with my own eyes. This is great. So, yes, I took a very good look. I took a lot of pictures, too, in case I'd missed something. From what I could see, which was most of it, it was in great shape. So I had a lot of confidence that we were going to be fine coming home, as long as I got those gap fillers out. Doing that wasn't a big fight at all. They were a little tight, but it was a pretty measured approach. I wasn't going to pull too *hard, and I had tools with me to cut them off if I had to. That wouldn't have been very difficult either.*

Several years later, the memory of the once-in-a-lifetime spacewalk brought a smile to Steve Robinson's face. The repair had worked.

On the morning of 4 August 2005, the last full day *Discovery* was docked with the ISS during this mission, the joint crews paid homage to fallen astronauts and cosmonauts. The tribute, titled "Exploration: The Fire of the Human Spirit," began with Collins. "Those who dare to venture into an unexplored land will have revealed to them things which were never known," she said. "Those who venture out upon the sea will have revealed to them things never heard. But those who venture into the sky upon wings of silence—yes, the ethereal adventurers—theirs is the revelation of things never dreamed. Such are the way of explorers and the surpassing way of the sky."

Kelly then took the microphone:

As we orbit the Earth today, we are able to watch the beauty of the Earth and heavens unfurl before us as we undertake this journey. And we are reminded that it is upon the completion of the journey and the arrival back at the place from whence we came that we can say we truly know ourselves. Sadly, there are those who have been challenged by the adventure of human space exploration but who have not been able to experience that special feeling that comes with returning home. These are the men and women who have come before us, in courage, but who did not complete their journey of exploration. It is to these explorers that we now take a moment to reflect upon, and to whom we now pay tribute.

It was then Robinson's turn:

The spirit of exploration is truly part of what it is to be human. Human history has been a continual struggle from darkness toward light, a search for knowledge and deeper understanding and a search for truth. Ever since our distant ancestors ventured forth into the world, there has been an insatiable curiosity to see what lies beyond the next hill, what lies beyond the next horizon. That is the fire of the human spirit that we all carry. Through that spirit and through realizing its ambitions, the human race has come to find its present place in the world. Previous generations went first on foot, then on horseback. Later came the wooden sailing vessels that opened new continents and new lands. Today we have aircraft and spacecraft. We have shrunk the world in a way that early generations of explorers could never have imagined.

Next was Lawrence:

Likewise, even if the future is equally unimaginable to us, we can be sure that future generations will look upon our endeavors in space as we look upon those early expeditions across the seas. To those generations, the need to explore space will be as self-evident as the need previous generations felt to explore the Earth and the seas. As President Kennedy said of space exploration, "Space is there and we're going to climb it, and the moon and planets are there and new hopes for knowledge and peace are there. And, therefore, as we set sail, we ask God's blessing on the most hazardous and dangerous and greatest adventure on which man has ever embarked. We choose to do these things, not because they are easy, but because they are hard." And, certainly, space exploration is not easy, and there has been a human price that has been paid. As we step out into these new frontiers, we find that it is very unforgiving of our mistakes. The lives lost over thirty years ago with the early steps taken by the crews of Apollo I, Soyuz I, *and* Soyuz II *showed us that. The loss of the crew of* Challenger *reaffirmed the need to be ever vigilant of the risks.*

Camarda was next to share in the moment:

Tragically, two years ago, we came once more to realize that we had let our guard down. We became lost in our own hubris and learned once more the terrible price that must be paid for our failures. In that accident we not only lost seven colleagues, we lost seven friends. Their families never shared in their homecoming. Those seven were driven by the fire of the human spirit within. They believed in space exploration. They knew the risks, but they believed in what they were doing. They showed us that the fire of the human spirit is insatiable. They knew that in order for a great people to do great things, they must not be bridled by timidity.

Expedition 11's Phillips then took his turn:

To the crew of Columbia, *as well as the crews of* Challenger, Apollo 1, *and* Soyuz 1 *and* 11, *and to those who have courageously given so much, we now offer our enduring thanks. From you we will carry the human spirit out into space, and we will continue the explorations you have begun. We will find those new harbors that lie out in the stars and of which you dreamed. We do this not just because we owe it to you, but we do it because we also share your dream of a better world. We share in your dream of coming to understand ourselves and our place in this universe. And as we journey into space you will be in our thoughts and will be deeply missed.*

Noguchi and *Expedition 11* commander Krikalev then translated Phillips's paragraph into their native tongues. Thomas next read a poem composed by Sir Hurbert Wilkins, a veteran of World War I who explored places as far ranging as the outback of his native Australia and the North and South Poles. Finally, Collins returned to conclude the tribute:

And, in closing, for all our lost colleagues, we leave you with this prayer, often spoken for those who have sacrificed themselves for all of us:

They shall not grow old, as we that are left grow old:
Age shall not weary them, nor the years condemn.
At the going down of the sun and in the morning
We will remember them.

At 3:24 a.m. EDT on 5 August, *Discovery* was undocked from the ISS some 223 miles over the Pacific Ocean west of Chile. For most of the next three days, the crew busied itself by stowing equipment and testing the shuttle's flight control system. Built into the flight plan was time to relax a bit, and as it turned out, there was plenty more to come. Weather almost always seemed to be a factor in Florida's summertime heat, and it certainly was for the landing of STS-114 at KSC. Twice, on 7 and 8 August, low clouds and unstable weather conditions forced a wave-off of the crew's return home. And while there was work to be done preparing both days for the de-orbit burn and landing, the extra days gave Robinson the freedom to set up some external speakers on his laptop to create an impromptu deejay station. "Wave-off days are *great*," Robinson concluded, as if he really needed to spell it out. The crew played music, took photographs, told stories, and Robinson became the first person ever to transmit a podcast from space. During the brief episode, he looked back on the flight's unexpected moments:

We've had some surprises. We sure didn't exact that big piece of foam to come off the tank. Fortunately, it missed us. We didn't expect to go outside and get to remove gap fillers from the belly of the orbiter. That was, I would have to say, the most fantastic experience of my life. Just incredible to be way out there on the end of that arm all by myself and see no evidence of humans anywhere. Just me and the space station and the Space Shuttle from a view that neither I nor anybody else has ever seen, and watch the sun come up over the bottom of the Space Shuttle and get to sort of drink in that big view. I'll never forget it, and I'll never be able to describe it adequately, I'm sure. But I feel very fortunate to have been able to get a chance to do that. And also very glad it worked!

Collins practiced approaches to the shuttle runway on a laptop computer. She added:

Most of the time was spent looking out the window, and I would say talking with each other, reviewing the mission, making sure we had any loose ends tied up and relaxing. The last two days of that flight, even though we had only about six hours each day, I had never *relaxed like that in space. I had absolutely nothing that I was behind on, nothing to catch up on. I was able to really think strategically about what we had done. I took some notes. Now, keep in mind, it sounds like we might have a whole twenty-four hours off, but no. By the time you get everything out and you get it ready for landing, like your suits and your seats and all the equipment and your checklists, you've got to put a lot of that stuff away. So there is a little bit of frustration having to do all that work over and over again, like the movie* Groundhog Day. *We kind of laughed about it, "Here we are, living this day over and over again."*

Thomas, too, enjoyed the free time:

When you're docked to the station, it's just go, go, go. Then, when you're getting ready to come home, it's just pack everything up, put everything away, bring out the seats, get the suits out, get everything ready. It's just work, work, work. We'd get up, start working, and somebody would say, "Why are you still working?" I've got things to do. They'd say, "Well, it's time to go to bed." You'd think, "Where did the day go? What happened?" So that last day was very savory. It was nice. We didn't have work. We didn't even pretend to have work. We were just there, and we were just going to enjoy the time. We had time to enjoy being in zero gravity, time to enjoy the view and just make the most of it. I was actually very appreciative of the day. I think the rest of the crew were, too, actually. We just floated around. We would just raid the pantry for whatever treats we could find. We spent a lot of time looking out the window. There was a great view of the Aurora Australis from 114. The aurora was particularly spectacular, so every time we would get down toward the southern hemisphere, we'd be getting the cameras out and looking for that. Then, we were just looking at the view as we were passing over the U.S. We were just enjoying it. It was fun.

Only one STS-114 crew member—Robinson, during the STS-130 ISS assembly mission in February 2010—would ever again fly on board a Space Shuttle. Noguchi spent six months on the ISS as the flight engineer on *Expedition 22* and *23*, having launched on 20 December 2009 on board a Russian Soyuz spacecraft. Thomas had four shuttle flights, a long-duration stay on the ISS, and a spacewalk as a STS-102 mission specialist in March 2001. What more could he ask for? Camarda was a fifty-three-year-old rookie, so he knew going in that his chances to fly again would be few and far between, if ever. Collins also knew it would be her last flight, barring any unforeseen, pie-in-the-sky circumstances. "I never like to say, 'This is the absolute end,'" she began. "So I would think in the back of my mind, 'Weeeeellllll . . . maybe NASA will call me back someday. I might be a tourist in space someday. Maybe when I'm eighty years old, I'll get to be John Glenn and go out into space again.' So I just kind of had little thoughts like that in the back of my mind."

Reality, however, was a different matter. There were some fifty astronauts back in the office who had never flown, and only twenty flights remaining after the addition of STS-135. She would not return for a fifth slice of an ever-shrinking pie. There was one more reason, and it was the most important of all:

It was just very hard on my family. My daughter was nine. My son was four. I had been away so much. My husband was raising them. We had a nanny raising them. We were calling relatives in to watch them. It was getting to be too much. My mother was very sick. In fact, she died three months after we landed. And then my father died three months after that. I just had way *too many personal things in my life that I had been neglecting. It was time for me to move on. I miss it greatly. I made that decision with my head, not with my heart. In my heart, I really miss flying the shuttle. I miss the training. I miss the people. It's been wonderful. It has just been an absolutely amazing, fantastic career. I wouldn't change any of it.*

Yet another round of bad weather forced LeRoy Cain, the reentry flight director, to make a call no one really wanted made. Instead of bringing *Discovery* back from whence she started, in front of friends, family, and hundreds of employees at KSC who called the ship their own, STS-114 would have to land at Edwards Air Force Base all the way out in California. The weather was perfect there, so at 7:06 a.m. EDT, the de-orbit burn began 220 miles above the western Indian Ocean. Collins could not help but think of the friends and colleagues she and the rest of the human spaceflight community had lost:

It wasn't an overwhelming kind of thing. It was just the thought, "Here they *were at this point." My prayers had been with them and their families for the previous two and a half years. I had played through this in my mind, so it was more of just a little reminder on my part. As we came down on the entry, I would call out to my crew in the mid-deck what our altitude was and what our mach number was, so they could be aware. They don't have the instruments down in the mid-deck, so I wanted them to be aware of where we were. As far as our safety, I was very, very confident because we had done such a thorough check of our heat shield. I knew we didn't have any problems.*

Robinson, in particular, was as confident, if not more so, than anyone else on the crew. He had seen firsthand the most critical parts of the orbiter's thermal protection, and he knew them to be intact.

We knew from experience that an orbiter with a good heat shield is going to make it home just fine. I think I would have to say that I probably had the advantage of being more confident than anyone, since I'd been outside and I'd been able to inspect the thing with my own eyeballs. Any pilot, if you can do a pre-flight inspection and look at your bird with your eyes, that's what gives you the confidence to go fly. I had the benefit of that. For sure, as we came through the entry, the 107 guys were definitely on our mind. It was kind of emotional after we landed. Finally we got stopped, and certainly I, and I wouldn't be surprised if everyone else, realized that we had gotten to do what the Columbia *crew didn't get to do. We came home.*

The day after the launch of STS-114, NASA management indefinitely grounded the shuttle fleet to once again rethink its study of the foam-shedding problem. During a press conference that day, shuttle program manager Bill Parsons told gathered reporters, "Until we're ready, we won't go fly again. I don't know when that might be, I'll state that up front. We're just in the beginning of this process of understanding. This is a test flight. This is a flight we had to go off and try to get as much information as we could and see if the changes that we had made to the tank were sufficient. Obviously, we have some more work to do." NASA in general, and the crew of STS-114 in particular, had dodged a very large bullet. The flight before it had not been so fortunate.

To eliminate any sort of foam liberation from taking place was simply not possible, even to the most cautious of engineers. To get the stack off the ground and into space required one of the most forceful events ever created by humans, and once the fuse was lit, parts and pieces were going to shake and rattle. The Space Shuttle, however, was no ordinary bucket of bolts. Not only did it need to get a crew of astronauts into orbit, it also had to bring them back safely. After *Columbia*, the most critical and dangerous possibilities for foam coming loose were thought to have been fixed. They were not. "Just by luck, we didn't do critical damage to *Discovery*, not because we had done a good job of engineering," said Wayne Hale. "That was really a bad day when we saw that. We were lucky, rather than good."

Hale was driving home on the day of the launch when he got a call from a colleague in systems engineering. "Look . . . the video shows a big foam release," the caller said. "We think it's hit the left wing of the orbiter. You need to come see this." Hale immediately did an illegal U-turn in the middle of NASA Road 1 and quickly headed back to JSC. He was not pleased with what he saw. "First of all, I was appalled we hadn't done it right," Hale said. "When I saw the video, it was just devastating, personally, to think we'd spent all this time, done all this work, spent all this money, and we still had a huge foam release that was hazardous." Inspections determined that there was no damage, and if there had been, NASA knew what it could do to possibly save the crew. Still, the fact of the liberations remained. "Our whole theory on foam loss from the shuttle was inaccurate, as it turns out," Hale admitted. "It could very easily have been the last shuttle flight."

Several factors played into determining when the shuttle would return to orbit. Just before Hale was promoted to manager of the shuttle program in September 2005, Hurricane Katrina made landfall on the Louisiana coast near New Orleans on the morning of 29 August. More than 1,800 lives were lost, with total damage in the hundreds of millions of dollars. Located just fifteen miles or so from New Orleans's famed French Quarter, Michoud was hammered by winds of more than 130 mph and waves topping the levee height of nineteen feet. When the mammoth storm finally blew through, the birthplace of the ET had sustained more than $100 million in damages. A crew stayed through the worst of it and operated pumps to remove more than one billion gallons of water out of the facility. Land routes to Michoud were cut off, while electricity and water to the area were nonexistent for a full three weeks. Ninety-four percent of its two thousand employees reported damaged homes, including six hundred that were completely destroyed. On 5 January 2006, the thirty-eight members of the ride-out crew were awarded the agency's Exceptional Bravery Medal. One of the honorees, Stephen Turner, said many of the team members "didn't even know the fate of their own families until days after the storm had passed. The crew worked through many weeks of hard recovery work under very tough conditions. I am very proud to have served with this brave team and of what Michoud means to this community."

Just nine weeks after Katrina, Michoud returned to full operations. Although Hale had first estimated that the natural disaster would delay another shuttle flight until at least September 2006, the estimate, Hale said, "was not very palatable to any of the senior management. We didn't want to go out on that note." Once back into the facility, it was possible to attack the foam issue. Eventually pinpointed was the cause of the most serious STS-114 liberation, and, most likely, the ones from the flight before. The culprit was found to be "a difference in the coefficient of thermal expansion"—which meant, basically, that separate layers of foam on the ET changed in volume during the broad swings in temperature from the fueling of super-cold liquid hydrogen and liquid oxygen and then to the heat of launch. Hale described it in layman's terms:

In essence, there are places on the tank where we applied layers of foam on top of other foam. Because they were different formulations of foam and had different densities and different thermal coefficients of expansion, when the tank chilled down, that would cause the foam to develop cracks. Then, aerodynamics and heating of ascent would cause pieces to come off during the launch phase. If you have a glass of water and you put it in your freezer, the water expands as it freezes and the glass contracts. That's a difference of thermal coefficient of expansion. It will shatter a glass jar if you fill it with water and put it in your freezer. It's similar to that. The tank gets very cold with liquid hydrogen and liquid oxygen in it, and some of these different layers just move against each other. Our goal was to eliminate those foam-on-foam applications as much as possible, and where we couldn't eliminate them, try to go back with foam that was more closely in the same coefficient of thermal expansion.

Another major change was to the PAL ramps, which were removed altogether and replaced with much smaller IFR extensions. The PAL ramp on the liquid oxygen tank that had been uppermost on the ET was 13.7 feet long. Below it was the same kind of deflector on the liquid hydrogen reservoir, and it measured 36.6 feet in length. Both extended about 5 feet into the intertank area—the ribbed ring that connected the two tanks a little more than halfway up the complete structure. The foam removed from the tank weighed a total of 37 pounds. It was the biggest change to the shuttle stack in a quarter of a century.

Remaining, however, were thirty-four Ice-Frost Ramps that still had some in the agency very much concerned. Twelve were located on the liquid oxygen upper portion of the tank, six on the intertank, and sixteen on the lower liquid hydrogen tank. Each liquid oxygen tank IFR was about 1.5 feet long by 1.5 feet wide by 5 inches high and weighed about 12 ounces. Larger ramps on the lower portion of the ET were 2 feet long by 2 feet wide by 1 foot high and weighed about 1.7 pounds. In April 2006 the call was made by shuttle program management to leave the IFRs as they were. According to a NASA release, "The rationale for doing so was based on several factors including the performance of the Ice-Frost Ramps on previous flights. Any design changes would need to be thoroughly tested and certified before modifying the tank. To do otherwise could result in more uncertainty instead of reducing the risk of the tank."

The answer was not good enough for Camarda, who had been named director of engineering at JSC following the flight of STS-114. He wanted the IFRs gone. He made his case during a 17 June 2006 Flight Readiness Review (FRR) for the next flight, STS-121, which was to carry commander Steve Lindsey, pilot Mark Kelly, and mission specialists Michael E. Fossum, Lisa M. Nowak, Stephanie D. Wilson, and Piers J. Sellers, along with *Expedition 13*'s Thomas A. Reiter. Camarda was backing up positions held by NASA chief engineer Christopher Scolese and safety officer Bryan O'Connor. "I got up to the podium and said, 'We should not launch because we should fix the Ice-Frost Ramps. We should eliminate the hazard,'" Camarda recalled. "It was at that time that Mike Griffin stood up and said, 'You know what? I'm the head of the agency and I have to make these tough calls. I think we should launch.'"

Scolese and O'Connor both voted against launch, and while Griffin disagreed, he insisted that it was a professional matter and not personal. Years later, he would say that both were very highly qualified individuals and that the three of them were still good friends. However, he also noted, "I think I had studied the issues behind the handling of foam and the probability of losing another orbiter because of foam, frankly, more carefully than any other single person. Other people had different pieces of it. I think I had studied the whole issue very carefully." His conclusion was a simple one. "I felt that risk of flying was being overstated," Griffin said. For him, it was a buck-stops-here moment. He had heard both sides of the issue, and some sort of decision had to be made. It was what bosses do. He had recommendations from one side, and he had recommendations from another. One way or the other, he was in a position of overruling somebody. "The Flight

Readiness Review team was divided," Griffin said. "Normally, those decisions are unanimous, but they were divided. So the decision came to me. I was the head of NASA—where else was it going to go? So, I decided we would fly." Griffin had also considered Camarda a friend, yet fully admitted that Camarda "was very upset with me for launching." So, too, were several media outlets, including the *New York Times* and *Houston Chronicle*, both of which carried scathing op-eds over his decision to proceed as planned with the flight of STS-121.

Nine days after the 17 June review meeting—and just five days before the scheduled liftoff of STS-121—Camarda sent out an e-mail in which he said he had been fired as the center's engineering director. He was given the news by Michael L. Coats, the director of JSC who had flown three shuttle missions, one as pilot and two as commander. Years later, the dismissal was deeply rankling to Camarda, who was reassigned to the NASA Engineering and Safety Center. Asked if his dissension on the IFR issue during the STS-121 review meeting had anything to do with the surprise move, Camarda said, "They would say no, but I guarantee you that had a lot to do with it." He continued:

Mike Coats pulled me into his office and told me to go work someplace else. Of course, he used some other reason for doing that. It was a terrible thing to do, just days before the next launch. The head of the engineering directorate had all his people ready preparing for the next mission, and you take him out. I begged them to let me wait until the next flight, and he basically looked me in the eyes and said, "You know what? Your second in command can do this job. You go downstairs and work at the NASA Engineering and Safety Center. You're relieved of your job." What was interesting, Mike Griffin asked Chris Scolese to call me up, even after I was reassigned from my job, to make sure I was in mission control during the launch of STS-121 because of this Ice-Frost Ramp issue.

According to Griffin, the decision to reassign Camarda was made by Coats not due to the STS-121 dissension, but rather because of what he termed "other managerial behavior that Mike could not support." Once Coats outlined his reasons, Griffin both accepted and supported the decision to move Camarda regardless of fallout that was almost sure to happen. Griffin continued, "[Coats] said, 'This is going to cause you some problems, because people are going to say he's being punished for speaking out against the launch.' And I said, 'I don't care. If it's your decision as the director of the Johnson Space Center that Charlie needs to be in a different job, then that's what you should *do* as the director of the Johnson Space Center. I'll support your decision, and I don't care what people say.' Mike was right. People did object. You're raising this issue now."

Camarda was obviously not pleased with Coats's decision, but Griffin accepted the fact that such a reaction was to be expected:

Charlie always saw Charlie as being correct, and Mike being incorrect. It's not reasonable to suppose Charlie is going to support his removal from the job. That's not going to happen, so Charlie is always, for the rest of his life, going to believe he was unfairly removed from his job. That's just part of life. There's nothing I can do about any of that. I think Mike Coats was correct. I think Charlie's behavior was incorrect. I am not going to go into what that behavior was, but I would tell you that it had nothing *to do with the decision to fly or not fly on STS-121.*

After all that had happened, why not make discretion the better part of valor and hold off on the launch of STS-121 until everyone was satisfied? "There was a huge debate over whether we'd done our job well enough," Wayne Hale acknowledged. "We had a real fight on our hands to get that flight off the ground, when we'd done everything that we humanly knew to do." What NASA needed, he continued, was information that could only be obtained by proceeding with STS-121, which was also considered a Return to Flight test mission:

The argument that I used, which I think was the one that carried the day with the administrators, is that we had done everything we knew to do. To do anything else, we needed flight data. We needed to understand how the things that we had done would actually perform in the flight environment. There are things you cannot duplicate in ground test. The real issue on that flight was the calculation that said that some foam liberations that could occur could be critical. The probabilities didn't come down to be as remote as we wanted them to be. Part of that was we didn't know how well the fixes would work. We needed to go fly to get some data to improve the accuracy of those statistics. Actually, after we did fly and measured the releases that we did have, our calculations showed that we were within an acceptable level of risk—not completely free of hazard, but within an acceptable level. That's after the fact. Folks were reacting to the analysis that said, "Given the error that we have in our predictive capabilities, we could still be in a catastrophic situation." The fact of the matter was, we didn't know what else to do. We had nothing left in our bag of tricks.

That is not to say, however, that Hale was intentionally downplaying the risk if and when the mission flew:

We went into it, stating up front, that we had done everything we knew to do, but we still had a high degree of risk. As program manager, I put the red flag on that hazard, and I said, "We cannot hide this. We can't have some off-line meeting like they had for the Challenger *launch decision that all the management is not aware of. We need to put this out there. We need this debate in public. We all need to hear all sides of this story at the Flight Readiness Review. We dare not sweep it under the rug, because I could be wrong. The shuttle team could be wrong. We need to hear the dissent to make sure that we have understood it, and know that we could be making the wrong decision."*

The launch of STS-121 took place on America's Fourth of July holiday following a couple of scrubs for bad weather, and during the ascent, at least one piece of debris from an IFR was liberated. The next four flights left the pad with the same design in place, but after that, STS-120's External Tank featured fourteen liquid hydrogen IFRs and four feed-line brackets that had been modified with different foam configurations as a stop-gap measure. ET-128, the first tank to be completed following the flight of STS-114, was used on STS-124 during its late May to mid-June 2008 mission. Its changes ran the gamut and included completely redesigned liquid hydrogen IFRs and liquid oxygen feed-line support brackets.

The launch of STS-121 was not the only source of Camarda's frustration. Infrared thermographic inspections on *Discovery* following STS-114 began in mid-September 2005 and eventually uncovered three major delaminations—areas of chipped coating—and a number of subsurface cracks on RCC panel 8R on the leading edge of the vehicle's right wing. One delamination was found in late July 2006 on the lower portion of the panel, and once it was fully and readily liberated with a dental pick, it was about 6.5 inches long, 0.4 inch wide, and as much as 0.4 inch deep. Just above the apex of panel 8R was another problematic area, this one 6.25 inches long and 0.38 inch wide. With a bit more difficulty yet another region was discovered, upping the ante still more. "This vehicle, if you lose a thumbnail-size piece of this coating in orbit or because something hits it, that wing leading edge burns up," Camarda said. "You get a hole in it, and you basically could lose the vehicle. You don't have to lose much of it."

Camarda discovered the problem through e-mailed briefings by the Leading Edge Structural Subsystem Problem Resolution Team. "I was astounded to see the damage on one of the wing leading edge panels on my vehicle," he said. "No one had ever told me about this, or any of our crew." When Camarda attempted to raise a red flag, he says the situation took a turn for the worse. With little information trickling in, he wrote a 150-page paper on why the delaminations and cracks were a critical and possibly systemic issue. "People were trying to say this only happened on this one panel, and so they continued to fly vehicles without this even being a concern," Camarda continued. "They convinced themselves that this wasn't a problem and continued to fly. Remember, it was a wing leading edge that caused the accident, right?"

I continued to raise this as an issue, to the point where I was vilified. I basically had to go to the Freedom of Information Act to get information, and they still didn't send it. Do you believe that? An astronaut who was on the Return to Flight mission, whose technical expertise was in this exact area, discovers a problem and had to fight an army of people to do the right thing. I continued to doggedly pursue them, to say, "You know what? This is a problem. We need to change out these panels." The NASA Engineering and Safety Center took me off the e-mail list and told me I was not allowed to work on the team that was investigating this problem. I was not only finding the problem, I was helping them solve the problem. I was putting together teams to understand the problem and to figure out what to do. But you want to know something that's crazy? You know how easy it was to solve the problem? All you had to do was change out those panels. Why continue to fly with defective panels when you had spares?

It took at least a year, but similar problems were found on a total of twelve other panels on all three remaining vehicles. All were replaced. "They didn't change them out just to make me feel good," Camarda said. "It was millions of dollars. Each one of those panels cost about half a million dollars, so it was six million dollars plus to re-outfit those vehicles. It really was a problem, or they wouldn't have done that." Camarda landed one more blow before he was finished:

What I'm telling you is that the culture hasn't changed. The same culture people said caused the tragedy of STS-107 is still in place. It's really a sad story. The vehicle is only as safe as the people that are servicing it, preparing it, and analyzing it. It's up to the technical leaders and the program leaders to allow an environment where people are allowed to raise their hand and speak up. Unfortunately, we do not have a psychologically safe culture here at NASA. That's my opinion, based on what I've experienced. I've tried to write reports on this, and I'm not allowed to for whatever reason. Kind of shocking, isn't it?

The next logical question, then, is this: If a flawed culture remained in place, the same culture that helped send the crews of *Challenger* and *Columbia* to their graves, how was it possible for so many other flights to return to Earth safely? Camarda's response was nothing short of chilling. "By the grace of God," he concluded.

Hale and Camarda would appear to have been on completely opposite sides of the NASA fence. Yet when it came to the matter of the ethos in place at the agency, to a certain extent, Hale did not necessarily disagree with the former astronaut. Hale theorized that NASA's overall character was determined not so much by turf wars between the agency's various centers, or among clashing groups of scientists, astronomers, and astronauts, but more so by simple personality traits. Some people needed more data to make decisions, while others did not. Some were comfortable with risk, others were not. The danger came when one side or the other became unwilling to take into account any other viewpoint:

One of the things that we have to recognize is that we need to be aware of that arrogance. We need to be aware of those bad cultural decisions. I am here to tell you, despite all of the work we have done to improve the culture at NASA, some of that still exists. It serves as a reminder to me of what we need to continue to do at NASA to be as safe as we can be, because we always run the risk of doing something dumb because we believed we were smarter than we were. That's the only defense I have. I would certainly agree that some of the old NASA culture is still there. It's part of who we are. We have to recognize that and use that old

NASA culture where it's good, and be very aware of it when it could lead you to make bad decisions.

Hale went on to say that he welcomed "the worry warts." He wanted them to stand up and warn against unnecessary risks. "The bad old NASA culture would say to those people, 'You don't know what you're talking about. You're not an expert. You don't have the standing to make comments like that,'" he continued. "I would hope the new NASA culture would say, 'You know, even though you may not be the expert in the area, you've got a good point. We've got to make sure we do our work to cover that concern.'"

There were differing camps within the NASA hierarchy, that much was true, Griffin added. Important to remember, however, was a sense of perspective. There remained in place throughout the Space Shuttle program that same "can do" attitude that had saved the crew of *Apollo 13* way back when. Those were the better angels of the agency's nature, available at a moment's notice to carry the day. However, Griffin continued, NASA had also operated "in an atmosphere of scarcity" ever since those very same days of the Apollo era. "Not enough resources were made available to the space program to do the things that NASA was really set up to do and that the country expects from its space program," the former agency boss said. That resulted in centers that eventually handled about half the amount of work on an annual basis during the shuttle program as they had during the Apollo program. The end result was this:

You've got more agency than you need for the amount of space program you have. That's just a fact. It's also a fact that it is either impossible or next to impossible to close any federal facility, whether it's NASA or not. So you have a bigger NASA. You have a NASA which was designed to service the peak of Apollo, and you can't close any of it down. So in that atmosphere of scarcity, of course, you get competition over turf and scarce resources. That is absolutely no different than what happens in any other agency or, frankly, in the corporate world.

The ball—or rocket, presumably, in this case—was in Griffin's court. Coming off the *Columbia* accident, he developed a plan to shake the cobwebs out. Each center had a role to play, and that, he said, "immediately reduced some of the turf competition and got rid of a lot of what you call the 'bad culture.'" The transition was not an easy one. "The leading roles were from Johnson and Marshall, our two biggest centers and certainly the two premier manned spaceflight centers in the world," Griffin said. "There was a certain amount of residual jealousy on the part of all our centers over Johnson and Marshall's roles, and then there, of course, was some—I would honestly say—minor amount of bickering between Johnson and Marshall themselves. But, honestly, once we formulated the Constellation plan, most of those symptoms really receded into the void because everybody was busy."

Constellation would have taken NASA back to the moon and then on to Mars, but it was canceled before the last shuttle ever left the launch pad.

8. A Kick in the Pants

For millions of kids around the world, it never seemed to take much to derail dreams of becoming an astronaut.

The birds and the bees.

Sports.

Cars.

School.

Mike Fossum tried. He tried to shove the dreams he had once dreamed of flying in space into the deepest recesses of his mind, if not out of it altogether. Such talk was crazy, he figured. Those kinds of things happened for other people, not a regular guy like him. All the way through college—Rough! Tough! Real Stuff! Texas A&M!—and even after he signed up for the air force, he ignored that childhood passion. That kind of thing was silly, was it not? He would make a career out of the military, and leave it at that. There would be no fantastic adventures in space for Mike Fossum.

And then the air force loaned him to NASA, and sent him to work at JSC, just down the hall from the astronaut office.

Whether he had originally believed it or not, these people did not use phone booths to change into superhero capes after all—they were actually not all that different from him. Fossum could do this, too. So he applied a first time in 1985. And when he got turned down, he applied again, and this time he got as far as the interview process. Melanie, Fossum's wife, was expecting and very near the due date. Fossum delayed his NASA interview as long as possible, but Melanie's labor was finally induced for the birth of their second son. Such commitment still did not put him over the top. Then came a third attempt and a fourth. After each round, he got the dreaded phone call: "Sorry, Mike. It was close, but you didn't make it this time." Once the fire was rekindled within Fossum, it became more than just a professional struggle. This was something with a far deeper impact:

I can't begin to describe the emotional, and at times, spiritual struggle that I had through the years. I felt in my heart that this was a calling for me, that this was really God's intent for my life that I pursue this. It was an incredible struggle for a lot of years. For me, it was a spiritual journey as much as anything else. Honestly, I struggled with God after going through this over and over and over again. Either get me in the door or take this dream away from me, please. I talked to my pastor about it, and really believed that it was His intent for my life for me to pursue this. I prayed about it. Am I going to get there? It was quiet. All I knew was I needed to be working in this direction, but I never heard the guarantee, "Patience, my son. I'll grant your wish in My time." I never heard that part of it.

After at last kicking the door down in 1998, it was another five years before Fossum was assigned to a flight. It was Columbus Day 2003, a little more than nine months after the *Columbia* accident, and he was hard at work in his garage when the call came. "It was still hot, and I was drenched in sweat and grease from working on cars," Fossum remembered. "My wife is a teacher and my kids were all in school, so I was at home alone and the phone rings." Kent Rominger, then the chief of the astronaut office, was on the other end, and he started off with some small talk. What's up? Fossum wondered to himself. What did I do wrong? At last, Rominger got to the point and asked if Fossum would be interested in joining the crew of STS-121 as a mission specialist. "It overwhelmed me," Fossum said. "After so many years of dreaming, working, and praying, I literally fell to my knees in the middle of a concrete floor covered in tools and stuff a guy with four kids is going to have stacked up his garage, and just said, 'Thank you, Lord. I can't believe it's finally happening.'"

Fossum had known Rick Husband since meeting at Edwards' base chapel. When he heard the future STS-107 commander's rich tenor voice, Fossum could not help but introduce himself and ask why Husband was not in the base choir. Husband was not the only close friend Fossum had lost in spaceflight—*Challenger*'s Ellison Onizuka had been instrumental in steering him toward the air force's test pilot school in order to beef up his application to the astronaut office. Fossum was reconciled to the inherent dangers of aviation in general and spaceflight in particular. All that was left to do was tell Melanie and their four children—Carrie, who was twenty-one years old when STS-121 flew; Mitch, then a sophomore at the United States Air Force Academy; John, fifteen; and Kenny, nine—of his selection. Would NASA be able to solve the problems that befell the STS-107 crew? Could shuttle missions be made safe enough to allow him to look his family in the eyes and assure them that everything was going to be all right? Fossum sat down with each and explained as best he could exactly what was taking place:

I gave each of them the opportunity to talk and if they wanted to ask questions about it, I was open and clear. I did tell them in general and not getting too technical, "We've done the following things." First, what makes us different from Columbia *is we've made changes to this confounded foam on the External Tank. We've added a new ability to inspect the wings and heat shield on the shuttle. We haven't solved all of it, but we have some ability to repair things ourselves. That was actually part of our mission, testing out these tools and techniques to do some repairs in case there were to be some damage. The final thing that makes it a little more safe is the fact that we were going to the space station and dock. If the shuttle was damaged to the point where it was not safe for us to fly it home, plans were in place for us to actually stay on the space station and wait for another shuttle to come up and get us in a couple of months. That would be kind of a dicey situation, but it was safer than flying a crippled bird home. I could tell my wife and kids, "Things* are *different. It's not without risk, but we're being smart about it."*

After the highly charged mission-management briefings concerning continued foam loss, the decision was made to proceed with a launch date of 1 July 2006 for STS-121. Bad weather scrubbed liftoff that day and the next, but after a stand-down of yet another day, *Discovery* stood ready to take Fossum and the rest of the STS-121 crew to orbit on Independence Day in America—4 July. It would be the greatest fireworks display ever. Going to pad 39B was an emotional event for Fossum. Every other time he had been here with the orbiter, it had been quiet. The vehicle was quiet, with attendants scurrying around almost like ants hauling toolboxes and so forth all over the place. The vehicle was always surrounded by scaffolding for both protection and access, but not now. Now, steam was pouring off the shuttle. Venting made for a cacophony of hisses and snorts. Things creaked and liquid nitrogen dripped out of lines as they were being flushed. She was awake—and

26. For STS-121 spacewalker Mike Fossum, being an astronaut was more than just a job. It was a calling. Courtesy NASA.

almost nobody else was around as the elevator hit the 195-foot mark. "After all these years, it's like, 'Hey, we're really here. We're really on the pointy end of this thing,'" Fossum said. "We're very excited about it, but honestly, a little nervous, too. Anybody who's not a little nervous strapping on four million pounds of explosives doesn't understand what's going on."

Harnessed into his seat on the flight deck, just behind pilot Mark Kelly, the nerves did not last for Fossum. "I have thousands of friends that are praying for us right now, and I know that," Fossum remarked. "In fact, it was almost surreal how calm I actually felt sitting out there. I was right with God. I was ready for whatever He had in store. I felt so comfortable that I actually took about a fifteen- or twenty-minute nap while we were waiting for the countdown to progress." After the quick snooze, Fossum got down to the business of getting the bird off the ground, and at T-minus ten seconds, he twisted just enough to look through the window at the flame trench underneath them. He saw the main engines light and smoke begin to billow, and felt the stack sway. It was time to get straightened back out in his seat, because he and the six others on board were about to go for a ride. "I got settled back looking straight ahead and the solid rockets kicked in," he continued. "Oh, boy! Oh, my goodness, what a kick in the pants that is!" The liftoff was widely documented, with more than one hundred high-definition digital, video, and still cameras pointed at *Discovery* during her ascent.

Fossum was having the time of his life, but his family? Maybe, just maybe, not quite so much. He said:

My wife actually described the whole process of going through the launch; to her, she felt like it was like giving birth. It's much *harder on the families. It was* much *harder on her than it was on me. I'm in the middle of this. I've got a job to do. I've trained for years. This has been my dream and not necessarily one she would wish I had, and so she equated the launch itself to the pain of childbirth. It was something she dreaded, something she knew was coming and felt in a way that it was inevitable. She had to get through that, because she knew the kind of joy that was on the other side of it.*

Each of their four children was at a different stage of understanding. The older ones, obviously, knew full well what could happen to their father but accepted the risk as part of his job. Fossum would discover later, however, that his youngest son, nine-year-old Kenny, did not react well to the liftoff. "My little guy, Kenny, he was taken aback by it all," Fossum admitted. "He was pretty upset when he saw the launch, just by the fact that Dad was sitting on top of that column of fire, smoke, and noise."

In addition to delivering German Thomas Reiter to the station for a long-duration stint, *Discovery* also ferried to orbit 4,600 pounds of cargo in the *Leonardo* logistics module. Best of all for Fossum, the flight plan also called for him and Sellars to conduct three spacewalks, mostly to test various means and methods for heat-shield repair. There was never enough time to fully appreciate the view on any EVA during the shuttle era, much less the initial three of Fossum's career. But two and a half hours into his first spacewalk, Fossum had a quick moment to catch his breath and whisper a prayer of thanks for an opportunity relatively few humans would ever have. "I think I was mentally prepared for this kind of view," Fossum said. "But to see the earth spinning in the blackness of space—and we could see the moon out there at the same time—was His creation laid out in front of you, in a Technicolor, big-screen kind of way." That was not all. After undocking from the station, the orbiter's flight path took it over the Holy Land. Fossum was struck by what he saw. "We were coming over Africa and I see Egypt coming up," Fossum remembered. "You could see the Nile River, the Nile delta, and the Red Sea. We passed right over Israel. I was just there in silence, watching, as we're coming right over the Bible lands. You cannot see the cities, but there's the Sea of Galilee. I was just stunned to silence, just looking at it and thinking, 'Wow. Here's where it all began.' It was just so profound to be passing right over the top of that part of the world, to see it all laid out."

Fossum would call the moment the most spiritually impactful moment he had during the flight. Contrast that, however, with a media session about an hour later:

We weren't in touch with world events while we were up there. We were doing a press conference and the interviewer made reference to the new hostilities that had broken out across the Middle East. I was just deeply saddened. Just a profound grief came over me. It is so beautiful from space. One of the coolest things about looking at the earth from space is you don't see those imaginary colored lines, those borders, that we draw on the map and put on our globe. It's one Earth. It's our home planet. From two hundred miles away, it looked very peaceful and beautiful.

With that, Fossum became the latest in a long line of people with the solution to world peace. Bring the world's leaders up on the shuttle, high above the surface of the earth, where there were no barbed-wire fences, no tanks, no signs of any discord whatsoever. Let them figure out a way to live together in harmony.

Joan E. Higginbotham got the call from NASA in her dorm room at Southern Illinois University Carbondale, and she thought it was a joke. She had applied to a lot of places and faced many rejections, but she had *not* sent anything to the agency. The school had sent her resume in, not her.

She had a job offer on the table from IBM, where she had interned for a couple of summers, but it was in sales due to a hiring freeze on engineers. Higginbotham had a decision to make. "I really knew nothing about NASA, and the gentleman agreed to fly me down," she said. "It was during spring break, and I left Chicago on a miserable, foggy, dreary day and landed in sunny Florida." Higginbotham was taken on a tour of the OPF and the launch pad. This might not be so bad after all.

She went with NASA, and for the next nine years, participated in one capacity or another in fifty-three shuttle launches. She was satisfied, never thinking that she would ever actually fly herself. "I got to walk to the shuttle every day, work on the electrical systems, launch astronauts into space, and get them home safely," Higginbotham said. "That was a fantastic job. How many people get to do that? I was really very content with what I was doing."

Her boss, Jay F. Honeycutt, encouraged her to apply as an astronaut. The numbers did not seem particularly encouraging to her, those who applied versus those who were accepted. Honeycutt persisted. "It's very interesting to see what people see in you that you may not see in yourself," Higginbotham said. "So he obviously saw something in me that I wasn't necessarily seeing in myself." She made the cut on her second try, becoming a member of the very large 1996 "Sardines" ASCAN class. She was originally assigned to the crew of STS-117, but in the wake of the *Columbia* accident was moved to the flight just before it, STS-116. It would be a full ten years from the time she was selected as an astronaut to the ignition of *Discovery*'s SRBS on 9 December 2006. Was there ever a time she wondered if she would ever get off the ground? "There were several," she corrected.

Higginbotham was on the mid-deck for the launch. A. Christer Fuglesang was seated to the left and slightly behind her, next to the hatch, while Sunita L. Williams, who was being delivered to the station for a long-duration stay, was on Higginbotham's immediate right. The Swedish Fuglesang became an ESA astronaut in 1992; Williams became a NASA astronaut two years after Higginbotham. "We had a lot of experience on the mid-deck alone," Higginbotham said. "When the boosters lit and we took off, the three of us kind of joined hands—I don't know, but all three of us also said something like an expletive."

Along with Williams, STS-116 brought to the station its P5 truss. During four EVAS, the P5 segment was installed along with a rewiring of the station's electrical system to a permanent power grid. Higginbotham operated the robotic arm during the spacewalks, a task that she admittedly found to be daunting. "We have simulators on the ground, and they're all pretty good, but none of them simulate the heart palpitations you get the first time you put your hands on the controls in space," Higginbotham said. "You're thinking, 'Oh, with one wrong move, I could put a big ol' hole in the side of the lab, and that just would not be a good day.' It was very interesting the first time I took the controls of the arm. I was just nervous. I have to be honest with you. I was just plain nervous."

It was nerve-wracking enough to manipulate the P5 segment on the end of the arm, but with a fellow crew member out there? "I had Christer on the end of the arm, and I'm thinking, 'His wife wouldn't be too happy with me if I slammed him in the side of the lab,'" she said with a laugh. "I could kill a man and I can do damage to the $20 million module. No pressure."

She was also in charge of moving more than two tons of supplies for the Expedition crew onto the station, and replacing them with used equipment and experiments for a return to Earth. It was a "huge" assignment, almost like a game of Tetris in getting everything situated just so. "The stowage master has to be a really detail-oriented person, and so they gave probably the most anal-retentive person on the crew that job, and that's me," Higginbotham remarked. The job included managing a massive checklist that was sent up on the second day of the flight. "I was kind of stressed out waiting to get that notice," she continued. "My whole morning was basically looking at the configuration and putting stuff in order."

STS-116's first landing opportunity on 22 December was waved off due to weather concerns at KSC, and gusty winds erased any chance of coming in at Edwards. When conditions improved back in Florida for a second try, Higginbotham was on her way back to Earth. "The second rev was really iffy, and they waited literally until the last minute to tell us the targets to come home," she began. "Because they waited so late to tell us, we pulled a heck of a lot of gs coming in."

For reentry, Higginbotham was on the flight deck, seated just behind pilot William A. Oefelein. As gravity once again began to take effect, a small video monitor attached to the back of Oefelein's seat by a piece of Velcro fell to the floor in front of her. "It just falls, bam, on the floor, and it yanks the video cord out of the camera," she remembered. "So I'm no longer recording anything. I'm thinking, 'Okay, thirteen days in space and the last thing I have to do, I'm going to screw up. This is not good.'" When she bent down to pick the cord up, Higginbotham experienced yet another unexpected problem. She could not lift her head. "It was just too heavy," she explained. "All of a sudden, I just burst out laughing. Finally, I lifted my head with both hands. *That's* how heavy my head was." She snared the video recorder with her feet and lifted it just enough to grab it with her hands so she could replace the wayward cord. Higginbotham's first and only spaceflight concluded with her in a sweltering reentry suit, the result of her deadlifting her own head and finagling the monitor, cord, and camera back into working order.

The glamour of spaceflight.

Jim Reilly rode a column of flame and smoke three times to orbit, and three times, he had a commander with a different approach to control of a crew. Two marines—Terry Wilcutt and Frederick W. Sturckow of STS-117—bookended an air force officer, Steve Lindsey. Reilly had expected one thing out of Wilcutt because of his leatherneck background, but found himself with a more collaborative give-and-take leftseater. Sturckow was a typical marine, and Reilly also called him one of the smartest people he had ever met.

Although "C.J." Sturckow was more of a hands-on coordinator than Reilly's other commanders had been, he allowed Reilly to oversee the payload all but autonomously. *Atlantis* had the S3/S4 truss segment and two more solar arrays to deliver, and it was Reilly's job to make sure they arrived and were installed on the station in good order. "He knew that I was fairly experienced," Reilly said of Sturckow, who like Lindsey, had been a 1994 ASCAN classmate of his. "He told me to take care of the payload and all the paperwork, and he would take care of the shuttle and mission requirements. We were able to divide and conquer quite easily using his philosophy. It was one of the more surprising aspects, how well he managed the team and how well the team performed."

Just as the team was hitting its stride in training, a severe storm on 26 February 2007 pounded KSC, and hail left thousands of divots in the External Tank and the orbiter's heat-shield tiles. In the aftermath of the STS-107 tragedy, it had been determined that a tiny eighth-of-an-inch pellet of ice liberated at just the wrong time could cause enough damage to a wind-leading edge to cause it to fail during reentry. With so many spots damaged on the tank, concern over ice formation was real.

Reilly and fellow spacewalker John D. Olivas had just finished a run in the NBL to prepare for their first EVA of the upcoming flight, and they were jazzed, ready to roll on a flight that was just around the corner. As they prepared to debrief with their trainers, another STS-117 spacewalker, Patrick G. Forrester, checked a nearby computer.

Oh, no.

Forrester told them of the hailstorm, and together they were able to track down stills captured from video of the tank. It was not pretty. Almost immediately, dialogue was set into motion about what could be done to either replace the tank or repair it. Reilly remembered:

Could they repair it? What could they repair it with? How could it be tested? There were a lot *of discussions that went on about that. I have to admit that I was probably one of the biggest skeptics about the repair capabilities and how well they worked. I had some disagreements with the tank management on that—not arguing, just philosophical disagreement with how the tests were run and whether we were answering all the potential problems we could see where ice might form and if it could be liberated. If it was there in the foam, then it most likely* would *be liberated, and then it's all about the path it would take, how fast it would be going, and what time it would come out. All of that information was pretty subjective.*

Meeting after meeting, the call was made to patch the divots. *Atlantis* embarked on its 8 June 2007 ascent—using pad 39A for the first time since STS-107—with what Reilly called the ugliest tank ever flown. "By the time we flew, we had beaten that story pretty much to death," he added. "There was always still that risk. You just have to either get comfortable with the risk or step away from it. There were still some concerns, but we've done everything we can do at this point. This is going to be our solution."

Along with the truss segment, STS-117 also delivered *Expedition 15*'s Clay Anderson to the station to replace Sunita Williams, who had set a record for women by spending the last 194 days, 18 hours, 58 minutes in space by the time the flight returned to Earth. The hail-damaged tank ensured Williams's place in the record books, at least for the time being.

There were other issues to address once STS-117 got to orbit. A voltage spike took out Russian computers on the station that provided redundant attitude control, orbital attitude adjustments, and the life support system. The situation was serious. If the station had required any substantial attitude changes, they would have been provided by *Atlantis*. The problem had to be fixed before the crew of STS-117 could depart. "There was a lot of work going on on the ground that we only saw a little bit of," Reilly said. "The Russian crew members who were up there with us were spending a lot of hours working on repairing those computers. We were so busy, you just kind of knew it was going on in the background, but not really reacting to it in any great fashion."

A four-by-six-inch corner of white insulating thermal blanket on *Atlantis*'s port-side Orbital Maneuvering System pod also came loose during launch, and it was a serious enough concern to warrant rewriting the plan for the flight's third EVA, to be conducted by Reilly and Olivas. Planners went to work on figuring out how to accomplish the fix, and Reilly could not have been more pleased with their efforts:

When you're up there, you're just kind of divorced from everything that's going on on the ground. We had some folks that literally stayed up all night for about three nights to come up with the procedures and the tools for us to use out there. They did a beautiful job. By the time we got the procedures from them, they were very mature. They were ready for prime time. Team Four came up with the tools that we had on board, found out what would work for us, and then tested it. Even more critical, which was probably the harder thing, was to find the stuff that we could use inside to train ourselves to go out and do this task on the outside. They did a phenomenal job. They replanned the entire EVA for us, so it was completely different than what we had trained for. When we had a chance to look at video, we were like, "That's how we're going to do it." So we went out and did it.

On 15 June 2007, Olivas rode the shuttle's robotic arm to the damage site and spent the better part of two hours very carefully stapling the blanket back into place and using his helmet camera to beam pictures of the scene back to mission control. As he did so, Reilly got a head start on the EVA's actual planned tasks and installed an external oxygen hydrogen vent on the *Destiny* module. Some of what the two of them had trained to do got bumped to the fourth spacewalk of the flight, conducted by Forrester and Steven R. Swanson. "We didn't know whether we did a good job or a bad job," Reilly concluded. "We were kind of scrambling with it a little bit, but when we came home, we got almost a hero's welcome. We were just looking at each other like, 'I guess we did okay.'"

Days after the *Columbia* accident, Scott Parazynski sat in the Husband household with Evelyn Husband's parents, Dan and Jean Neely, talking about Parazynski's exploits as a mountaineer, and how he had climbed Challenger Point in Colorado.

The conversation set the wheels in motion for a nearby peak to be named in honor of the STS-107 crew. Six months later, the United States Geological Survey Board of Geographic Names officially christened the 13,980-foot Columbia Point.

After the accident, Parazynski also became the lead astronaut for shuttle inspection and repair techniques. "I spent two and a half years developing tools and techniques that we would be using to both inspect, and then if we had to, get out and repair the Thermal Protection System of the shuttle," Parazynski said. "It was a very, very intense phase of my life. I spent countless hours in the Neutral Buoyancy Lab, in vacuum chambers and on the KC-135 airplane testing out different tools and repair techniques. I'm very proud of the work our entire team did. It was several hundred people whose every waking minute was dedicated to this task."

Somewhere along the way, however, his passion for spaceflight began to waver to the point where he did not fully know that he wanted to fly again. "I was tugged, of course, by concern for my family and my kids in particular," Parazynski said. Going forward, he gathered strength from the *Columbia* families and Evelyn Husband especially. That Parazynski ever flew again was due in large part to "the importance she still felt toward the space program and the fact that we should all buckle down and keep going."

Parazynski returned to orbit just once more, knowing that STS-120 would be it for his spaceflight career. During his four previous trips on board the shuttle, he had made a total of three spacewalks, and on STS-120 alone, Parazynski was scheduled to do four to install the *Harmony* module and relocate the P6 truss segment and solar arrays.

"That particular mission was always the one that I wanted to go do," Parazynski admitted. "It was the ultimate space station assembly flight, because it was a big module coming up plus a very, very challenging combination of EVA and robotics to install the P6 segment and arrays. It was really, in my mind, the most exciting assembly mission of the entire sequence." Basically, the truss had to be unbolted and disconnected from the top of the station and moved all the way out to the very end, an undertaking he called "a pretty audacious thing."

In charge of the mission, which launched 23 October 2007, was Pam Melroy.

Before her work on the *Columbia* reconstruction team and crew survivability report, getting a command of her own had been "a huge deal." Although she still very much wanted a flight of her own, Melroy knew full well that there were many other ways to contribute to the success of the agency. She explained:

Working on [the Columbia Crew Survival Investigation Report*] was really different than what most of my colleagues were doing in the astronaut office at the time. I had the opportunity to work with people from all over the center in lot of different areas—engineering, safety, and other parts of the center that I didn't get a lot of exposure to in my other jobs. So, I had a very different perspective. There's a whole lot of ways to contribute to flying safely in space. I didn't have to command a mission. I really wanted to, because I wanted to go fly in space again. I wanted the opportunity to land the shuttle, and I just really, really enjoy leading teams. It's the thing that I find most challenging and rewarding. I wanted to very much, but it would have been okay if, for reasons beyond my control, I couldn't have flown again.*

Douglas H. Wheelock, assigned as the partner for three of Parazynski's four planned EVAS, liked Parazynski right off the bat. Still, "Wheels" found it hard at first to get to know the veteran astronaut because of his own apprehension. Wheelock knew that he was a rookie. Was he asking too many questions? "He was probably kind of checking me out," Wheelock said. "He's the lead spacewalker, and I'm EV-2. It was not really in his job description to be my mentor, but I needed to learn from him, so I needed to be all ears."

Remember all those old Looney Tunes cartoons featuring Chester, the young pup who would do anything to impress Spike, his bulldog of a buddy? That was these two, so much so that it is who they became during the lead-up to STS-120—Parazynski's Spike to Wheelock's Chester, which eventually evolved into the more alliterative Spike and Tike. "I was just listening, watching, and asking the right questions, trying not to bother him so much, but just learning from him," Wheelock continued. "When something would come up, where I'd see him do it with ease in the pool and I was having difficulty with it, I would just sit down with him and say, 'Hey, I watched you do this and you made it look really easy.' He would give me tips, and so with time, he became the ultimate mentor for me."

Wheelock took to the tutelage so much, when he subsequently commanded the *Expedition 25* crew during a five-month stay on the ISS in the latter half of 2010, he gave briefings for a contingency EVA based largely off the notes he had taken from Parazynski. Two of the most important lessons the New Yorker took away from his time alongside Parazynski were meant specifically for EVA, but applied very much to life in general—learn from mistakes and expect the unexpected:

Scott told me early on, "We're going to make mistakes, so when you make a mistake, 'fess up to it and then move on. Leave it behind. Don't even think about it anymore. Don't look back." That is really hard to do, but that's probably the biggest gift that he taught me. Also, when you're in direct sunlight up there, you have a problem. When you're in cold soak and you're in shadow, you have problems because things just don't work like they do on earth. We don't know how to machine a piece of hardware that can withstand the 600-degree change in temperature in forty-five minutes. Here on earth, we can't replicate that. I could only do it from Scott's words. He said, "Listen, you're going to get out there and find that something doesn't work, so take a look around you. Is the sun shining on it? If it is, get your body between the sun and the object, let it cool down a little bit, and then try it again."

When Melroy docked *Discovery* with the ISS to hook up with the crew of *Expedition 16* led by Peggy A. Whitson, it marked the first time that two female commanders were in space at the same time. Joining Parazynski, Melroy, and Wheelock on the crew of STS-120 were pilot George D. Zamka, and mission specialists Stephanie D. Wilson and Paolo A. Nespoli of the ESA. Hitching a ride up to the ISS was Daniel M. Tani, with Clayton C. Anderson returning to Earth from a five-month stay. Parazynski's first three spacewalks—the second of which he did with Tani—went almost like clockwork. At first, the worst that could be said was that Wheelock discovered a small tear in his right glove, located on the webbing between his thumb and forefinger, after he and Parazynski made their way back into the airlock following their second EVA together on 30 October. Although he did not know for sure, Wheelock suspected that the tear was caused by the fabric of the glove catching in a locking mechanism of a foot restraint. No problem. Wheelock simply switched to a backup set that had been fitted based off Parazynski's hand measurements. "I gave him a hard time that he wasn't fully evolved, because his thumb was a little bit shorter than mine," Parazynski quipped.

The news was about to get a lot worse.

After getting out of their bulky EVA spacesuits following the 30 October EVA, Parazynski could remember vividly the glum looks on the faces of his fellow astronauts. He and Wheelock had been satisfied with their day's work, because they had accomplished tasks most had thought would be very, very difficult. "Then, we looked at the video feed and we knew that the character of the rest of the mission had just changed," Parazynski concluded.

As the work outside had finished up, an attempt to redeploy the solar arrays did not go well at all. Damaged by a piece of space debris, strands of the steel guide wire keeping the panels aligned jammed. The thin material was caught just enough to rip lengthy gashes in the outermost right-side array. With about 80 percent of the array extended, motion was stopped so the process of figuring out what to do next could begin.

The implications were huge. If the array was out of commission, would it be able to handle the physical loads of *Discovery*'s undocking, much less future dockings? Would it rip apart further, or possibly even damage the orbiter or the station? Adequate power was also needed to support the new European and Japanese ISS laboratories that were about to be launched. If the flapping of the torn array on the arm continued, the station stood a very real chance of losing attitude. In a worst-case scenario, would the station have to be abandoned altogether? It was a serious dilemma. "One of the solutions was, 'We'll go out and just throw away a billion-dollar piece of space hardware,'" Parazynski said. "Then, we wouldn't have been able to support those new laboratory modules. It was really a dramatic effort to find a way to go out and repair that thing." He continued:

It was extremely urgent. It was a process similar to what I had been involved in on several prior missions. There are three teams in mission control, but then there's the fourth team that figures out what to do if things go awry. For EVA, *I was always a part of those efforts to support other missions. But being on orbit, there were other people that supported us. A bunch of brilliant, unsung heroes spent seventy-two hours working around the clock, looking at different ways to get an astronaut out there to repair, and then what we would do once we finally got out there.*

The fourth EVA was delayed a day and a fifth optioned outright until after the crew of STS-120 returned home. Although Whitson was respectful of the fact that it was two of Melroy's shuttle crew members who were going to be performing the dramatic spacewalk, there was never any discussion as to who was actually in charge. Too much was on the line to get into a debate over semantics. "We haven't really ever gotten to a place where anybody's had to say who's in charge of what, at least not that I know of," Melroy said and then added:

I would say that it took so much work to do so many things—getting the tools ready, constructing the cuff links—if you tried to do it with just shuttle crew, you'd never have gotten there. There were too many pieces that had to get dealt with, and so everybody had a role. Everyone *played a part in the spacewalk. That was actually very cool. You don't go your own separate way, but for example, the station crew will help you find tools and get them set up for a spacewalk. They're very helpful in suiting people up, because they do it a lot and know where stuff is. But then when it comes to actually directing the spacewalk out the door, the shuttle crew has been practicing that on the ground, so the station crew goes off and minds the station and tends to experiments. So it was a really interesting experience for all of us to drop everything and have our world be consumed by the same thing. That was a little bit unusual. I think it worked out extremely well. I think Peggy would say the same thing.*

There was a noticeable amount of silence from the ground, so Wheelock felt relatively secure in the fact that folks back on the ground in mission control had hit the afterburners trying to come up with a solution. "We kind of kidded that it was going to be like *Apollo 13*," he said. "They were going to come up with some very, very inelegant way of doing this, but it's going to work. I would say we probably experienced every point in the spectrum of crew morale." As it turned out, the solution was inspired in its apparent simplicity. Parazynski would cut out the frayed piece of guide wire and then sew the rip together with five insulated wire "cuff links."

If that seemed easy enough, it most certainly was not in either principle or practice. To get them where they needed to go, the orbiter's inspection boom was coupled onto the station's robotic arm so that Wilson and Tani could maneuver Parazynski 165 feet down the truss. Parazynski was then moved ninety feet up to reach the damage site, while Wheelock remained at the base of the solar array. So far out were the spacewalkers, it took forty-five minutes just to ride the arm out to the work site. Most of the way, Nespoli reviewed a long assortment of warnings, some standard and some specific to a procedure in which they stood the very real chance of getting electrocuted while working in such close proximity to the highly charged arrays.

Melroy knew the EVA would be risky, and when she heard the plan for how it was going to be accomplished, her eyebrows raised even further. "I thought, 'So I guess that they are okay with standard practices and operating procedures being completely thrown out the window,'" she said, able to chuckle about it years after the fact. Along with going much further from the airlock than normal, coupling the boom to the station arm and having no heaters or power source going out to the boom sensors, Melroy knew that the crew would be "doing crazy things with the arm, where you're putting the arm in all these weird spots. The fact that we were buying into a lot of difficulties for ourselves going forward was interesting to me." At times, the arm would be a mere foot to eighteen inches from the inboard solar array.

Melroy situated herself in the commander's seat on *Discovery*'s flight deck to get as good a view as possible of Parazynski and the damaged array. She had made the call to personally direct the spacewalk, rather than Nespoli, because she wanted it to be her hand on the microphone with Parazynski and Wheelock so far out on a limb. Earlier on STS-112, Melroy was the intra-vehicular astronaut charged with coordinating EVAs, a job that went a long way in preparing her for what she was about to do.

"My strategy was about 100 percent visibility," Melroy said. "It's fun for me sometimes to think back now, because I spent a lot of time getting my nest ready." The perch included a video camera set up to record Parazynski's every move, with a row of tapes ready to swap out. There were a couple of still-photograph cameras, with a couple of lenses each, as well as binoculars and her microphone. She could also lean her head back and view monitors featuring video from cameras on Parazynski's helmet and another on the arm itself. "I spent the entire time crouched in that little spot right there, watching him the whole time," she reminisced. The concern, she said, was very real:

The concerns I had were very focused about Scott. I was very concerned about his personal safety. I was worried that, because he was so dedicated and committed, we might not find out that we had pushed him too far until it was too late. I felt a lot of personal responsibility for his safety. The doctors told us that if he had gotten shocked by the solar array, it was not expected to be a lethal dose. I was very, very concerned about potential damage to the arm and to his spacesuit. I thought those were very real risks that we were taking, but with good cause. It was pretty clear that we had to do something if the outcome was going to be big.

The "what-ifs" for this spacewalk were daunting. With the inspection boom mated to the robotic arm, there was no way of knowing for sure the dynamics of how it was going to react with Parazynski situated out on the pointy end of a shaky spear. Because of concern over how much the configuration would flex and whether or not it could possibly strike the station's structure itself, the arm was placed maybe six to eight feet above the station truss. To take his place, Parazynski was forced to actually climb up Wheelock's back. Asked to make sure Parazynski's heels were securely in their restraints, Wheelock was left with no one else to use as a step ladder. Just over there was the earth, the station, and the shuttle. Above Parazynski, however, there was nothing but deep space. "Wheels" could not reach the boom by hand, but he could not leave the restraints unchecked either. He felt that he had only one option:

The only thing I knew to do was just undo my tether and push to grab his feet and the boom. It's the thing we're not supposed to say to NASA, that we did a free float. I didn't have to go far, but I made sure I could reach the handle on my SAFER jet backpack in case I missed. I had my safety tether still on, so I had that to count on. I just turned toward Scott and tried to get my body lined up, and I just pushed off and went floating toward the arm. I wrapped my arms around his feet, and grabbed on like I was climbing on to a tree or something. Sure enough, his left heel was not all the way locked in.

For Wheelock, the moment was an exercise in orbital mechanics, physics, and the power of fervent prayer—and not necessarily in that order. With Parazynski's feet checked and secured, it meant that his helper had to figure out a way to get back down to the truss. It was Parazynski's turn to become an impromptu jungle gym. "I said, 'Hey, Scott. Just stretch your left arm down toward the station,'" Wheelock said. "He's got long arms, so I crawled up his body and down his arm. I got to where I was only about two feet from the station, so then I could just hang onto his left hand and I could reach out and grab the station truss from there. It was another moment of sheer terror."

The Spike and Tike acrobatic act completed, they got down to business. Parazynski marveled at the view of Earth during the ride out, exclaiming at one point, "Oh, man! Words can't do this justice. No way, at least not mine." Still, the undertaking was far from a sightseeing trip. There was risk aplenty, Parazynski remembered:

It was a forty-five-minute ride out and another forty-five minutes to get back to the safety of the truss, and then whatever it would take going back to the airlock. Typically, you're to be no further away from the airlock than thirty minutes. That's what our secondary oxygen system is designed to support. If you were somehow to lose all of your primary oxygen, you would, in theory, have thirty minutes of oxygen supply to get back to safety. We were well *beyond that. Secondarily, we were working on a fully energized solar array panel. The shock potential was real. All the tools that I worked with had special insulation on them. Even the metal wrist disconnects on my suit had to be insulated, so that there wouldn't be any arcing between the solar array panel and my suit, which had 100 percent oxygen in it.*

Adding an even greater degree of uncertainty to the EVA was the fact that no one knew quite what to expect until Parazynski actually reached the damage itself. It had been inspected as much as possible with the telephoto lens of a camera and with binoculars, looking through one of *Discovery*'s windows. Even then, it was an oblique view, in difficult lighting. "It was just really, really hard to get a full sense of what the job would be like once we got out there," Parazynski continued. "There was a certain pucker factor on the way out, not knowing whether or not we'd be able to actually get the job done."

As the arm brought Parazynski closer to the damage, Melroy remarked, "We've got an excellent view of your close-up of the snarl."

"Isn't that amazing?" Parazynski responded.

"Oh, that's just ugly."

"Yeah."

Later, Parazynski continued, "I think that the guide wire has become frayed. It's almost like it's been stripped."

"Oh, wow," Melroy concluded. "Oh, that's really frayed!"

With the damaged array in front of him, it took the full extent of Parazynski's reach to get the job done:

I was on the end of this ninety-foot boom, with an extender at its full length. I was fully stretching out to my six-foot, three-inch height to do some of these repairs. It was pretty sporty. I remember at one point mission control saying, "Hey, we need to start thinking about wrapping this up." I don't think we had actually even installed the first cuff link yet. I was thinking, "Maybe I need to have Dan and Stephanie reorient the arm, to try and do this a different way." They said, "We could probably do it, but we'll have to pull you out and then shoot you back in a different way." I knew that would take more time and probably make it impossible for me to get it in during the seven and a half hours that I had, consumables-wise, in my suit. I ended up using another tool to pull the solar array carefully towards me—not too *close—but enough to where I could put in the farthest cuff link first.*

Almost immediately after removing the frayed guide wire and installing the first couple of cuff links, Parazynski was confident that the solution was going to work. Meanwhile, at the base of the array mast, Wheelock was dealing with risks of his own. As Parazynski cut the frayed wire, it was to retract into a spool right in front of his partner's helmet. Wheelock remembered, "The engineers said, 'Well Wheels here's the deal. That thing retracts at ten feet per second. So when Scott cuts that knot out of there and it releases the cable, we're afraid it's going to come flapping down the array and slice through the lower panels and then come out like a whip and hit your suit.' I was like, 'Excuse me?!?' It was just so intricate. It was just this little dance that Scott and I had to do with the release of that cable."

Later, he added:

One of the biggest worries was arcing, so we taped all of our metal tools. We were worried about arcing to the control unit on the front of the suit and killing the suit. If there are solar cells out there and the sun is shining on them, they're going to collect energy. They're going to send the energy into the power strip that comes down into the mast where I was hooked onto. You can't turn off the collection in the solar array. The power strip is on the very end of the array and it carries 210 amps of power. A fraction of an amp will kill you. I was actually grabbing that guide wire with these needle-nose pliers, about four inches away from this power strip that was carrying this huge amount of current.

The whipping of the cut guide line and deadly electrical arcs were theoretical issues—*if* the problems actually took place, *then* the EVA astronauts would have been in a world of hurt. To minimize the amount of power coming through the array, the station was maneuvered to keep the array in as much shadow as possible. That took care of one potential problem, but it also placed Wheelock in frigid shadows at the base of the solar array, where temperatures bottomed out at minus 300 degrees, for hours on end. "I was more naïve than I was concerned, initially," he admitted. Although his gloves had heaters, there was no warming element within the suit itself. There was either cooling or no cooling, and it could be turned off for only thirty minutes before suit performance began to degrade.

Wheelock became so cold that he could feel neither of his hands as the EVA progressed. For the first time in his career, he was—Wheelock did not actually admit that he was scared—but the fact was, he could not close his hands. He went to bypass on his suit's cooling unit, and by the time he reached its thirty-minute limit, Wheelock could no longer get his fingertips on the dial to its control mechanism. Though trepidation was growing back on the ground over his situation, there was virtually no need for him to say anything about it—it was almost possible to hear Wheelock's teeth chattering over the communication loops.

As Parazynski carefully placed the last cuff link, Wheelock began to wonder how he was going to make it back to the safety of the airlock. He swapped tethers back and forth to get to the base of the array, and he now had to collect all his tools. The array was deployed to its full extent, a process that took about fifteen minutes. "I didn't know how I was going to get back, because I couldn't hang on to anything," he recalled. "How am I going to get warm enough to make it back? I had been cold during training in Canada and Russia, but it was nothing compared to what I was in there. I had no way to warm up—I was starting to get a little bit alarmed."

Luckily for Wheelock, he had gone to the opposite end of the temperature spectrum and very nearly scorched his hand on his first EVA of the flight. The memory gave him an idea.

Working outside the station's *Destiny* laboratory, Wheelock had grabbed the reflective surface of a handrail that was exposed to direct sunlight. The seven layers of his gloves protected him for a time, but after a minute, his hand began to cook. It stung almost like an ant bite, or still worse, an electrical shock, and he pulled back from the railing. When he got back inside, there was actually something similar to a sunburn on the back of his hand. The experience came back to him in the super-cold shade of the mast:

I noticed at the very top of the mast, there was a sliver of sunlight. I thought immediately back to that first EVA. *This was a reflective surface, and I thought, "You know what? I can stick my hand into a 400-degree oven right now!" Because I couldn't close my hands, I had to actually use like the butt of my hands and tried to shimmy like I was on a tree or something up that mast. I reached up with my left hand, and I put my left hand in that sliver of sunlight, right against that metal. It took probably two minutes before I could start feeling my hand thaw out. As I calmed, I thought, "That feels so good." I just left it there, kind of hugging onto the mast. Once I could move my fingers, then I reached down and I brought my cooling out of bypass because I was starting to get fogged on my visor. I was able to move my left hand, and so then I held on with my left hand and stuck my right hand up there and got my right hand warm. Then, I was warm enough to actually make my way back.*

The array repaired, Melroy was not ready to relax just yet. The cuff-link fix had worked, yes, but Parazynski and Wheelock were still outside and under a tremendous amount of time pressure. When the overall mood became rather festive as Parazynski and Wheelock finished their work, Melroy lowered the boom. "I got on the radio," Melroy recalled. "I said, 'Hey . . . there's plenty of time for celebrating when they're back in the airlock. They're still a long way away. We've got a robot arm to maneuver that we've never done before. Let's talk about it when we're inside.' It was funny, but their reaction was definitely kind of like, 'Awwwwwwww, Mom.'"

Wheelock remembered Parazynski also shushing the celebration:

We got the sutures into the solar array. It's locked down and fully deployed. I understand that there was like a standing ovation in mission control, but we've still got to get back inside. We're exhausted. I was freezing cold. We still have work to do, and it's still not safe. I remember Scott, he said, "Guys, guys, too much chatter." Mission control was calling, "Congratulations, guys!" I remember Scott cutting them off and saying because Houston was talking, "Guys, guys, break, break. Guys, we're still outside. Let's finish well." That was one of his mantras—let's finish this thing well. We're not quite to the barn yet.

Parazynski could not have repaired the array by himself. With Melroy in her observation post on the flight deck and Nespoli reading from the checklist, Wilson and Tani were working the robotics. The cold notwithstanding, Wheelock acted as something of a spotter for his spacewalking partner and the robotics operators. A solar array is basically an enormous sail with a mind of its own, which presented all kinds of challenges. "If I pushed on it and let it go, it would float away from me five or six feet and then start come right back at me," Parazynski said. "'Wheels' would tell me to get ready. I had a tool that I called a hockey stick that I put out in front of me, and it protected me from any direct contact." Wheelock was a "real star," Parazynski said, as were Derek Hassman, flight director for the ISS Orbit 2 shift; Dina Contella, the lead EVA flight controller; and Sarmad Aziz, the lead robotics flight controller. "These people were just brilliant, the way they kept their cool to manage that risk," Parazynski exulted. "They did it very, very well. It was a great achievement for NASA to pull all that off."

Finally, after an EVA that officially lasted seven hours and nineteen minutes, the hatch of the station's *Quest* airlock was closed and sealed behind the two completely wrung-out spacewalkers. It was very much a sense for Wheelock of having been rescued from the abyss. He felt as if he would simply hang limply in his suit, every fiber of his being exhausted. For hours on end, his mind had been racing with the tasks at frozen hand, his body struggling to keep up with the stress load. Parazynski and Wheelock let the rest of the shuttle and station crew, as well as those back in Houston, know that they were secure and went about the business of powering up and pressurizing the airlock. When they were at last able to reenter the ISS itself, it was the biggest feeling of accomplishment and relief Wheelock had ever known.

With his helmet and gloves off, Wheelock faced one more order of business. Melroy—a recently retired *air force* officer, of all things—was about to induct him into the Order of Saint Michael, established in 1990 as a joint venture between the Army Aviation Association of America and the United States Army Aviation Center in recognition of individuals who have contributed significantly to the promotion of army aviation. Wheelock had just helped slay a dragon every bit as dangerous as the one depicted on the order's medal, and the significance of the moment and the relief of having finished the arduous EVA began to wash over him in waves. "I was so exhausted, I just started weeping," Wheelock confessed. "I was like, 'Why are you guys doing this to me?' It was just such a feeling of every kind of emotion you

27. Scott Parazynski (left) looks on as STS-120 commander Pam Melroy reads a proclamation inducting Doug Wheelock into Army Aviation Association of America's Order of Saint Michael. Courtesy NASA.

could ever imagine. It was like a vortex of emotion. Pam's reading this citation to mission control and to NASA TV, and I'm just staring at Scott, who was on the opposite wall. He goes, 'Are you okay?' I said, 'I don't think so.'"

When *Discovery* landed on runway 33 at KSC at 1:01 p.m. EST on 7 November 2007, Parazynski had just concluded the last and longest shuttle flight of his astronaut career. He stepped onto the tarmac, a veteran astronaut with more than 1,381 hours—more than eight weeks—of spaceflight experience to his credit. He had traveled more than twenty-three million miles over the course of 902 orbits of the earth, and during seven EVAs, had spent more than forty-seven hours in the vacuum of space.

His childhood dream of becoming an astronaut had come true. Now, Parazynski was determined to make another one happen. This particular goal was every bit as dangerous as, if not more than, anything else he had ever attempted.

Bob Behnken was just happy to be on the crew of STS-123. If he worked with one EVA partner, two, or however many the final number happened to be, he was fine with that. The primary task at hand—installing the Canadian-made, two-armed *Dextre* robotics system on the station—was what really mattered to the rookie.

Behnken's first time out, on 18 March 2008, he and Hubble veteran Rick Linnehan dressed *Dextre* out with a spare-parts platform, cameras, and tool-handling assembly. Years before, Linnehan had been the very first experienced spacewalker with whom Behnken had ever done a suited run in the NBL. He had been in the pool before, but with fellow newbies from his 2000 ASCAN class. They were as green as he was.

Behnken was on his way to becoming a lieutenant colonel in the air force and Linnehan a veterinarian, so they did not always speak the same language. "Rick is a great, outstanding, fun guy, but we didn't come from the same backgrounds," Behnken admitted. "We had to learn, a little bit, how to communicate most effectively. I'm a little bit more of the military mindset and he was a little bit more free-spirited."

Behnken's next two times out the hatch were alongside Michael J. Foreman, and they had much more in common professionally—Behnken had been the top flight test engineer and navigator in his class at the USAF Test Pilot School, while Foreman was a graduate of the United States Naval Academy in Annapolis, Maryland, who excelled at Pax River, the navy's test pilot training ground. The two of them were Cape Crusaders together, so Behnken had spent many a flight hour in the backseat of a T-38 with Foreman at the controls. Linnehan and Behnken were friends, but Behnken and Foreman shared a bond. "We could look each other in the eye, and just by where my eyes went next, he knew what the next task to go off and do was," Behnken said. "I don't think Rick would feel bad if he heard this. Mike and I are close friends, and it was just great to be out there with such a close friend."

It was sometimes easy for spacewalkers to develop a strange déjà-vu-all-over-again sensation as they headed out of the airlock, especially if they were veterans. Back in Houston, they had trained countless hours for just this moment in time, working on an ultra-realistic mock-up of the station. The work site in the pool looked very much the same as the real thing, except for the bubbles and with a much different view enveloping it.

As familiar as it might have been, a spacewalk could never have been routine for Mike Fossum, now a veteran and the lead spacewalker on STS-124. No way. The smart EVA astronaut treated this unforgiving beast with the utmost in care, or else, and Fossum had very wisely developed just that kind of respect. "A little bit of concern always stays in the back of your mind," Fossum remarked. "There's a little element of fear associated with doing a spacewalk, and I maintain that's healthy. If you don't have a little bit of fear, you don't belong outside." Working outside was a physical challenge, but it was also, Fossum knew, very much a mental game as well.

Gene Cernan once described his perilous *Gemini 9* spacewalk to a group of schoolkids in this way: Imagine yourself connecting two garden hoses and turning on the water full blast. Then, use only one hand to unscrew the hoses and do it with your eyes closed. For good measure, he told the kids, wear a couple pairs of extra-thick gloves. Before setting out to do all this, run a mile so that you happen to be good and tired during the task. Finally, stand on your head during some of the tasks. *That* was what it was like to do an EVA. Fossum would take the analogy a few steps further. Go through all of Cernan's steps, but to add even greater degrees of difficulty, do all those things while trying to stay on an oh-so-strict timeline, following a pages-long checklist and trying to make sure to not make any mistakes. All this took place on STS-124 while wrestling the Japanese *Kibo* module, as well as a Japanese robotic arm, into place.

Piers Sellars was Fossum's lead spacewalker on STS-121, and Sellars's dry British sense of humor helped take the edge off long and arduous test sessions and then during the flight itself. Ronald J. Garan Jr.—Fossum's EVA partner on STS-124—approached his duties with the precision of the test pilot that he just so happened to be. Garan and Fossum men were veterans of the air force, and both had been through the air force's flight test school at Edwards, Fossum as a backseater and Garan up front. "There were more similarities perhaps there, but in this case, I was the teacher and had the chance to share a lot of things," Fossum said. "We had a little bit of fun tying back to my experience with Piers. We followed kind of a British terminology. Instead of saying 'counterclockwise,' we would use the phrase 'anticlockwise' just because that's what I learned from Piers. Of course, for an American, that sounds all wrong. I think having a little bit of fun like that helped keep the ground team on their toes, too."

Garan replaced Stephen G. Bowen on the STS-124 crew, after Bowen was bumped to a subsequent flight in order to make room for an ISS crew swap—Gregory E. Chamitoff would go up on *Discovery* and Garret E. Reisman would come down. Garan had been hired in the 2000 ASCAN class as a pilot, one of the seven who came into the astronaut office that year. Asked if he would consider flying as a mission specialist, the native New Yorker told his bosses that he would fly wherever they needed him to fly. If that meant NASA wanted him to be a frontseater, great. If he was needed as a mission specialist, he would do that, too. They took him up on the offer, and onto the crew of STS-124 he went.

Aside from the Japanese Space Agency's Akihiko Hoshide, each of the STS-124 members was eventually slated to return to orbit. But with the clock ticking on the shuttle program with each and every launch, would they go on an American orbiter or a Russian Soyuz? Shuttle mission or long-duration? Commander Mark Kelly and pilot Ken Ham led subsequent shuttle crews, while Chamitoff and Reisman were named as mission specialists. Karen L. Nyberg was initially assigned to Ham's STS-132 crew, but was bumped due to her pregnancy. She was then named as a flight engineer on the crew of *Expedition 36*, a mission that left for the station in May 2013. That left Fossum and Garan, and while it was not necessarily the path they would have chosen for themselves, they ended up doing long-duration stays of their own together on the station.

First up was Garan. He had agreed to serve as a mission specialist once, but he very much wanted to make his second visit to space as a shuttle pilot. "That's what I trained for," Garan said. "I was a pilot my whole life. Honestly, personally speaking, I would've gladly taken a chance on never flying in space again for the opportunity to fly in the right seat as a pilot on the Space Shuttle." That was not to be. One by one, those mission slots went to other people. What was available was a slot as a flight engineer on *Expedition 27/28*. During a meeting with his boss, Steve Lindsey, Garan was told of the opening.

"You realize if I take this mission," Garan responded, "I'll never fly as a pilot. When you guys asked if I'd be willing to fly as a mission specialist, when I said yes, I didn't really realize that I would never fly as a pilot."

Lindsey's reply was simple and to the point.

This is what we need you to do.

As disappointed as he might have been, Garan accepted the assignment at face value. "At the time, there was no shortage of people to be Space Shuttle pilots," he said. "Just based on the timing and what crews were on what training flows, there was a big need to have people fly as long-duration crew members. I thought long and hard about it, and said, 'It doesn't really matter what I want, it's the needs of the office and space program that count.'" Garan agreed to the assignment, and he would later insist that he was glad he did. He consoled himself over not getting a chance to serve up front in the shuttle with the knowledge that "I was able to make a contribution, probably more than I would have been able to make flying in the right seat of the shuttle."

Fossum was in a similar spot. Sure, he would have wanted to make another shuttle flight, but he had also been taking a long, hard look at a long-duration stay on the ISS. He was not jumping up and down vigorously waving his hand, hoping to be picked to go into the space station training flow. As a teenager, he had once pictured himself living on a space station, but that was long before he had a wife and children. "The reality of the space-station flow is an extensive amount of travel and the Russian language requirements," Fossum began. "It's a tough grind. You're basically on a pretty tight leash for two and a half years prior to the flight."

First came a couple of test jabs from Lindsey, trying to gauge Fossum's interest. Finally, he told Fossum the very same thing he had told Garan—I need you. Fossum was hurtled back in time, to his childhood dreams, to the four rejection notices he received before landing his coveted gig in the astronaut office. He was going to do his duty. "How can you complain about a grind like that, or being away from home for six months?" Fossum said. "I have military buddies who are deploying to combat zones for six months to a year or more, and they're being shot at. That was part of it, but coming to grips with it, this is a calling. I really believe I was meant to do what I'm doing. I couldn't turn my back on that calling just because the road got tough. If this is what they need me to do, let's go."

Fossum and Garan were both on the station when *Atlantis* brought history's final shuttle crew for one last visit to the ISS. Garan returned to Earth on 16 September 2011, and after assuming command of *Expedition 29*, Fossum followed a little more than two months later.

Dale Earnhardt was one of the biggest superstars the world of NASCAR ever produced.

The ultimate rags-to-riches story, Earnhardt was the son of a racer who spent his earliest years in the sport banging around ramshackle race tracks in the southeastern United States and trying to make ends meet. Sometimes they did, and sometimes they did not. He was dirt poor, but once he caught his big break and made it to the sport's highest level for good, Earnhardt wasted little time in proving himself. He won his first championship in just his second full-time campaign, and before the end of his life, the man they called "The Intimidator" captured six more titles and untold millions of dollars.

Despite such dizzying success, one prize eluded him for years—a win in the Daytona 500, the sport's biggest race. He had lost it in virtually every way imaginable. He crashed. He ran out of gas. He cut a tire down on the last lap. He flat out got passed. Nineteen times Earnhardt tried to win the Daytona 500, and nineteen times he failed. Finally, at long, long last, Earnhardt held on to claim victory in the 1998 Daytona 500—the fortieth running of the race, which also coincided with the beginning of NASCAR's fiftieth season. After he took the checkered flag, members of every opposing team lined up on pit road to shake hands with the victor. In the end, it was a remarkable show of unprecedented respect.

In many ways Doug Hurley could relate to Earnhardt. He had waited nine long years after joining the astronaut office to fly. An avid NASCAR fan, Hurley took DVDs of two of the sport's most famous races with him on his rookie flight, STS-127 on board *Endeavour* in July 2009. One was of the 1998 Daytona 500 and the other was the 1979 Daytona 500, the first five-hundred-mile race broadcast live on television from start to finish. That legendary race featured a last-lap crash involving leaders Cale Yarborough and Donnie Allison and a subsequent post-race scuffle between Yarborough, Allison, and Allison's brother, Bobby.

With a few precious minutes to spare during the sixteen-day flight, Hurley popped the 1998 Daytona 500 DVD into an on-board computer and watched as Earnhardt steered his way to victory. "I'd waited a long time to fly in space, and so that just seemed to be the appropriate one to watch," Hurley said. "That picture of him driving down pit lane after he won, and everybody's out there giving him a high five or waving to him or slapping the car . . . I can definitely understand how he was feeling that day. It was kind of how I felt when the SRBs finally lit."

Hurley's journey to orbit had not been an easy one, even after being named to STS-127. The first two launch attempts were scrubbed before

Hurley and his crewmates—who also included commander Mark Polansky and mission specialists Christopher J. Cassidy, Julie Payette, Thomas H. Marshburn, Dave Wolf, and Timothy L. Kopra—ever made it to the suit-up room. During tanking in the wee early morning hours of 13 June 2009, a misaligned Ground Umbilical Carrier Plate allowed an unacceptable leak of gaseous hydrogen. Strike one. Four days later, on 17 June, the very same problem happened again. This time, the launch was waved off for nearly a full month because of range scheduling issues—NASA's Lunar Reconnaissance Orbiter lifted off the very next day—and to allow engineers adequate time to work the problem with the carrier plate. Strike two. The next try didn't get so far as the tanking stage after lightning struck pad 39A on the afternoon of 10 July, the day before the planned launch, forcing a check for damage in and around *Endeavour*. Hurley and the rest of the crew made it all the way to the pad on 12 July and were strapped in, ready to go, when approaching thunderstorms encroached on NASA's twenty-mile safety zone. Strikes three and four. For good measure, more bad Florida summer weather the very next day, on 13 July, meant yet another delay. Five attempts, five scrubs. It wasn't a record. It just felt like one.

Finally, on 15 July 2009, Hurley made it to orbit. It had been a long time in coming:

There were a lot of different events over the last nine years that were tough to deal with, and one of them, obviously, was the Columbia *accident. It makes you think about why you're here and what you're doing. The flight itself was great. The crew, you develop such a close relationship, it's almost like you're a family. The mission was one of the busiest, most complex missions they've ever thrown together for the Space Shuttle program. I'm sure I'll get in trouble for this, but it was probably one of the most complex space missions we've put together, period. We had our curveballs, but we made it work. It's a real, real, real tremendous sense of satisfaction. To do it with six or seven of your favorite folks to spend time with, plus the ISS crew as well, the wait was worth it, for sure.*

Hurley had never flown, and neither had Cassidy, Marshburn, or Kopra. In fact, there had been a bit of discussion leading up to the flight concerning an impending milestone for human spaceflight—one of the freshman fliers would become the 500th person to fly in space. Because Hurley was seated in the second-best seat in the house during launch, just to the right of Polansky on the forward-most part of the flight deck, he was considered the 499th. With Cassidy seated directly behind Hurley on the main flight deck and Marshburn and Kopra down below on the mid-deck, Cassidy got the nod as the 500th spacefarer. Not that it really mattered to the rookies or anyone else on the crew. They just wanted to fly, and they wanted to do so as quickly as possible, milestones notwithstanding.

One unknown awaited every rookie in the history of human spaceflight, not just the ones on STS-127. How would their bodies react to weightlessness? Few had ever come close to the sickness encountered by United States senator Edwin J. Garn, known as "Jake," during his trip on STS-51D. He had even coined a new measurement—the Garn Scale, which noted just how sick astronauts became in space. "I'm sure Mark was very, very concerned about flight day one, because he had four rookies, four guys who could've been in the fetal position in the corner throwing up. That is the unknown with somebody brand-new to spaceflight," Hurley said. "There's no analog on Earth that, A, tells you how physiologically you're going to react, and then, B, how mentally and emotionally you're going to react in space. It's just such a drastic change in your reality emotionally, mentally and, of course, physically. You just don't know."

Priority number one was to deliver Kopra to the space station, where he replaced Koichi Wakata on the latest Expedition crew. After docking with the ISS, for the first time in the station's history, there would be a total of thirteen people on board. Once there, the astronauts of STS-127 joined *Expedition 20* commander Gennady I. Padalka and flight engineers Koichi Wakata, Michael R. Barratt, Robert B. Thirsk, Roman Y. Romanenko, and Frank De Winne. Never before had so many people been in one place in outer space, but Hurley nevertheless felt he had room to spare. "The space station is so big now, you *never* felt crowded," he said. "I mean, never. Not even close. Frankly, it was really nice. We had a few nights where everybody's schedule allowed us to all have dinner together in Node One. It was really enjoyable. I think it just makes it more fun and more interesting. There's always a hand to help if you need it with six extra people on board." There was another reason for the sense of spaciousness, Hurley continued, that those of us destined to remain Earth-bound might not be able to quite grasp. If and when he or anybody else on board felt the least bit crowded, he could move—up, down, backward, forward, sideways, it didn't matter. They all could just . . . move. "On orbit, you can take advantage of every bit of space, because you can float anywhere . . . the floor, the ceiling, the walls," Hurley continued. "If there's a crowd of people that are down around the floor, you can float above them to the ceiling, and get out of people's way and not feel like you're being crowded."

A handful of those on board almost always seemed to be outside on yet another EVA. Five spacewalks were conducted to install and activate the last components of the *Kibo* Japanese Experiment Module Exposed Facility. Wolf, Cassidy, Marshburn, and Kopra were kept busy replacing batteries, installing cameras, removing thermal tile covers, making electrical configurations, and so forth. Hurley said:

We've never done a spaceflight of more than five EVAs, and we've only done a handful of flights with that many. A combination of that with the first time you've had a six-person space station crew and the amount of robotic operations we had, it kind of makes it uncharted territory. All the coordinated robotics in combination with all the EVAs, it was one of those things where the flight was filled from end to end. Everybody had to do their jobs. It wasn't a case where it was just a couple of people that were key to this mission. Everybody who was up there greatly influenced the success and/or possible failure, if you didn't do your job. We had multiple robotic operations with three different robotic arms, and it was all operational-type uses. We weren't doing checkouts of an arm. We weren't doing basic, minimal stuff. We were doing multiple EVAs where we had robotic support. We had guys on the end of the arm. We had a pallet on the end of the arm, supporting the EVA. I know everybody says they do a complicated mission, but I'm going to let the facts speak for themselves. I'd argue that you'd have to dig pretty deep into the books to find a more complicated mission.

The flight came to a stop on KSC's shuttle landing strip on 31 July 2009. Hurley was a rookie no more.

Faye Wilmore spoke, and her son Barry listened. That is the way the world should work, even for future astronauts.

After doing his undergraduate work at Tennessee Technological University in Cookeville, about an hour east of Nashville, Wilmore went to work on a master's degree in electrical engineering at the school. That allowed him to get in a final year of football, and, all along, he was considering joining the navy and going to flight school. He got a class date to Aviation Officer Candidate School, just after he completed his first year of studies for the master's degree. "I didn't know at the time, but I could have put a hold on that for a year and finished my master's degree, but my recruiter didn't tell me all that," Wilmore said with a laugh. He went to flight school, was commissioned an officer in the navy, and eventually took part in twenty-one missions in support of Operations Desert Shield and Desert Storm in Iraq. After returning stateside, instead of resuming his studies at Tennessee Tech, he planned to go after a master of science degree in aviation systems at the University of Tennessee, a much larger school in nearby Knoxville.

That is precisely when Faye Wilmore put her foot down.

"Barry, when you put your master's on hold at Tennessee Tech, *you promised me* that you would finish it," she told her son. He tried in vain to protest.

"Ma," and yes, Wilmore used that very word, a southern boy to the core, "you don't understand. It's been seven years. Things have changed. There is no way. I *cannot* do it." Long story short, Ma won out.

If Faye's request seemed unreasonable, consider that the purple-and-gold colors of Tennessee Tech ran very deeply in the Wilmore family. Not only did Barry Wilmore attend the school, so did his wife, Deanna; his father, Eugene; his brother, Jack; Jack's wife, Selena; as well as Jack and Selena's children, Lucas and Elizabeth Ann. Once he agreed to go after the dual degree, 1994 became the year that Wilmore never slept. By that time, he was an instructor at the navy's test pilot school. He was courting—yes, that is again the term he used, courting—his future wife and working on two master's degrees. "Oh, man, it was a rough year," Wilmore said, as if he really needed to point that out. In May 2012 Wilmore received one of the first two honorary doctorates ever awarded by the school.

As with his astronaut contemporaries and those who had gone before them, the roads Wilmore happened to be taking were leading to Houston. He was a 2000 ASCAN "Bug," and even though it took him more than nine years to finally fly on STS-129 in November 2009, he felt fortunate in that some of his classmates waited even longer than that. The flight did not take an Expedition crew member to the station, but it brought one home—Nicole M. Stott—the final crew rotation of the shuttle program. In addition, *Atlantis* lugged some fourteen tons of spare parts into orbit, pieces of

28. STS-129 pilot Barry Wilmore (right) had all the respect in the world for his commander, Charlie Hobaugh (left)—who kept the reason for his nickname "Scorch" a closely guarded secret. Courtesy NASA.

everything from the electrical, communications, and robotics systems to plumbing and air conditioning. It was one of the most intricate—and expensive—house calls ever made, and to Wilmore, at least, every astronaut who felt that his or her particular mission was the most important of the program was completely off base. "They are *all* wrong, because STS-129 was the most important, most detailed, most whatever, ever flown," Wilmore began, again with a laugh. Turning serious, he continued:

They don't allow the pilots and commanders to do spacewalks. But I was suiting the guys up. I was working the robotic arm. When you take a vehicle into space, you don't just float around inside. There's work. There's a lot you have to do, changing filters, doing water dumps, and on and on and on. The details of what you have to do to keep that spacecraft in prime operating fashion are incredible. The pilot, at least for my mission and many other missions, is the most heavily scheduled during training and the most heavily scheduled on the mission because of the various things that fall into our basket to do.

Charlie Hobaugh had flown twice before as a pilot, and even though STS-129 was his first command, it did not show to Wilmore. If "Scorch" ever actually had to learn anything about flying in space or being a shuttle commander, his pilot never saw it.

The marines prepared Hobaugh for leadership, Wilmore said, and he had worked to prepare himself for his tasks as a leftseater. "I've said this many times, and I'll tell you the same," Wilmore said. "If I could build a commander, I can't say that I'd do anything different than Charlie Hobaugh." He explained:

Charlie is a marine who comes across as a marine in attitude, yet he was compassionate. But you don't think of a marine as being compassionate. There were a couple of occasions where he could and should have probably said, "Butch, what are you thinking? You shouldn't have done that that way. You need to think about this. You need to do that." Instead, he would be like, "You know, Butch, if you need to work on that, we can reschedule this debrief. We can make it to where it better fits into your schedule." It was not belittling me, which he had every right to in a couple of instances. He didn't say it sarcastically. He said it as if he meant it. I realized that he actually meant, "Hey, Butch. Put your PDA *down during the brief and listen." The good Lord blessed all of us on that mission by making him our commander.*

When Wilmore returned from the flight of STS-129, with just six more remaining, he knew the chances of landing on another shuttle crew were slim to none. So he did the next best thing—the *only* thing, really—and began Russian language training. He was named to the *Expedition 41/42* long-duration station crew in January 2013.

9. The End of an Era

Although Milt Heflin was one of a select few people who had worked at NASA since the days of Apollo, he had never actually seen a Space Shuttle land in person. He had almost always been back in Houston, behind a console in mission control. The flights he did not work, he was somewhere other than here at KSC or out at Edwards Air Force Base in California when the birds came in to roost. He could have carved out some time at some point along the line and gone to a landing, but for one reason or another, he never had. Until now, there was always going to be another chance at some point down the road.

This was not just another mission. It was 21 July 2011, and the flight of STS-135, the last shuttle mission of them all, was drawing to a close. Decades earlier, Heflin found himself on the prime recovery ship following *Apollo 17*, the final lunar landing, as well as for the splashdown of the Apollo-Soyuz Test Project. And here he was, on the landing strip for the end of another American manned spaceflight epoch. No other human could lay claim to such a distinction—being present for both the final Apollo and shuttle landings. Not the most famous astronauts of them all—the twelve men who walked on the surface of the moon. Not John Glenn, Chris Kraft, or Gene Kranz. Not any American president or NASA administrator. Only Heflin could, and he was rightfully proud of it.

Those predawn moments were bittersweet for Heflin as he watched *Atlantis* descend out of the darkness and then come to wheels stop on the runway. The flight had been as very nearly perfect as it was possible for a shuttle mission to be, a textbook case for what the program could—and many figured, *should*—do. *Atlantis* had lifted the *Raffaello* Multi-Purpose Logistics Module to the station, and those on orbit spent the better part of the mission transferring 9,403 pounds of supplies and equipment—including

2,677 pounds of food—out of it and onto the station, and then replacing it with 5,666 pounds of items no longer needed. Commander Chris Ferguson, pilot Doug Hurley, and mission specialists Sandy Magnus and Rex Walheim took the ride in one of the most poignant spotlights NASA had ever known, and they did it while standing on the shoulders of hundreds of thousands of people who had over the years poured their lives into the program. This machine had been used over the years as an orbiting platform upon which to conduct a mind-boggling assortment of scientific research. Planetary probes and telescopes sprung from its payload bay to study the cosmos, and when one, the Hubble Space Telescope, turned up with a bad mirror, another crew blasted off to save the day and repair it. Without the shuttle, the International Space Station would have remained nothing more than a theory.

Heflin was there through it all. It was unsettling to him and, he said, to the orbiter itself that things were ending like this: "It was like *Atlantis* was saying, 'All right, you guys. Take a look. We're rocking and rolling. I just showed you what I can do and what my brothers and sisters can do as well, and you're just shutting us down. You're walking away from us.'" The sense of abandonment bothered Heflin greatly. Yes, he knew that NASA had been stuck in low-Earth orbit for far too long. Yes, he knew the shuttle could no longer realistically serve as a taxi to and from the station. Heflin's emotions roiled within him. "When *Atlantis* came out of that darkness, she was strutting her stuff and pounding her chest and saying, 'Look at me!'" he said, and then paused with what seemed to be a catch in his throat. "It was a big deal . . . really a big deal. I was *extremely* reflective on the side of the runway, very, very reflective."

For Heflin, the shuttle's conclusion might very well have been merely a symptom of an even larger issue for the nation he had served during six different decades:

You need to have that Kennedy moment again, I think. I'm just concerned as a country, we've kind of lost that pizzazz. We don't need to apologize for being exceptional at something. We don't need to apologize at all. We need to be exceptional, and we need to be known *as being exceptional. Even with our blemishes, I think we are a country that wants to find a way to improve life on this planet, to live peacefully and to do all these wonderful kinds of things. I think that's what we're all about. We're damn good at what we do, and we certainly make it look easy. In a way, that's a shame. It's too bad we don't make it look difficult.*

As hard and fast as the memories were coming, Heflin simply could not fully wrap his mind around what it all meant. What kind of legacy was the Space Shuttle leaving behind? That was a good question, and one that he was not prepared to answer. Not yet, at least. "I will tell you that I don't think we know yet," concluded Heflin, who had also attended the STS-135 launch, one of only two shuttle send-offs for which he had been present. "That's my answer. I do not think we know yet, because I don't think we have been able to appreciate what was accomplished in that program. It's almost like what happens to some presidents. A decade or decades later, you look back and say, 'Damn, that's pretty good what he did.'"

Heflin announced his own retirement from NASA in an e-mail to friends and colleagues on 8 February 2013.

Once Doug Hurley got back from STS-127 in July 2009, he wanted nothing more than to get right back in line for another trip to orbit. The problems he faced in doing so were big ones—there were only two flights remaining at the time for which crews had not been selected and, more important, Hurley had plenty of company in going after a spot on either of them. Even so, Mark Polansky, his commander on STS-127, was all but certain Hurley would fly again on either STS-133 or 134. "I thought I had a pretty decent chance," Hurley said. "In fact, Mark had said, 'You're a shoo-in. You'll be on one of those flights.' I thought I'd done pretty well over the years, but you just never know."

The native New Yorker did not have to wait long for the shoe to drop, and the news was hard to swallow on any number of fronts. Up first was the naming of the STS-134 crew, and Hurley was not on the list. Not only that, but the very same 11 August 2009 announcement included word that his wife, fellow astronaut Karen Nyberg, was being replaced on the crew of STS-132 due to a "temporary medical condition." She was, in fact, expecting the couple's first child more than three months prior to the launch date. Nyberg and her STS-132 commander Ken Ham came up with a plan that would have accommodated her pregnancy and kept her on the crew, but management dropped her from the flight. That was despite the fact that

NASA medical requirements stated that only births inside six weeks prior to a flight would be disqualifying.

Hurley and Nyberg were both chagrined, and little more than a month later, the door to another shuttle flight for Hurley appeared to have closed altogether. STS-133, then slated as the shuttle's swan song, was going to fly without Hurley on board. The 18 September news release stung the STS-127 pilot, and no doubt the rest of the astronaut office whose names were not on the final crew lists. Asked if there had been an elbow thrown here or there to get to the front of the line, Hurley responded without a moment's hesitation, "Oh, you bet, I would say unequivocally."

Serving at the time as chief of the astronaut office, Steve Lindsey landed the plum role as the commander of STS-133. In his place as chief went Peggy Whitson, a veteran of two long-duration stays on the ISS. Such a move was certainly not without precedent. Then astro-boss Al Shepard bumped his way to the front of the line to eventually walk on the lunar surface during the flight of *Apollo 14*. It had also taken place a number of times during the shuttle era. Said Hurley:

It's been done before, but it's pretty interesting when the chief of the office gets a prime assignment. That created a fair amount of discourse—that would be the best way to describe it—when he was assigned as the commander of 133. There have been various explanations as to who, what, and where, but the bottom line is that I personally don't think that the chief of an organization [should fly]. To fly in space is the ultimate goal of anybody who is here, then that person is not only your boss and has ultimate authority and power, but then ends up with an assignment. Like I said, it's not unprecedented, but it hadn't happened in many, many years.

Hurley continued, expressing frustration over the selection of the STS-134 crew as well. "There were definitely some eyebrows raised about 134, too," Hurley said, although he would not discuss what the issues might have been. "Whether they were above-board, below-board, or purely on merit, I would argue of those three flights, our assignment [as the crew of STS-135] was probably the fairest."

When engineers opted to extend the service life of STS-134's Alpha Magnetic Spectrometer primary payload, it was bumped to last on the manifest in April 2010. For a few months, it looked as though Mark Kelly would be the last shuttle commander instead of Lindsey, along with Gregory H. Johnson,

E. Michael Fincke, Italy's Roberto Vittori, Drew Feustel, and Greg Chamitoff as its final crew. All along, yet another flight was in the rumor mill. "It wasn't long after she took over, Peggy called me into the office and said, 'If there's a 135, you're going to be the pilot,'" Hurley remembered. "So I knew if and when that would happen, I knew very early on in January 2010 that I would be on it. The flight was kind of on and off, on and off, on and off."

The chips finally began to fall the right way for STS-135 in mid-July 2010, when a Senate committee passed a reauthorization bill allowing for NASA to add one more flight to the shuttle's manifest. The full Senate approved the act a couple of weeks later, and while President Obama signed it on 11 October, yet another appropriation was needed to actually foot the bill. In the meantime, Ferguson, Hurley, Magnus, and Walheim were officially announced on 15 September as the crew of STS-335, a rescue mission that would fly only in the event of an emergency on STS-134. Then if *Atlantis* was not needed for any save-the-day heroics, STS-135 might just have a shot at getting off the ground.

Not since STS-6 way back in 1983 had a mission been flown with just a quartet of astronauts on board, but the reasoning behind naming only four people for this flight was practical. With the remaining shuttles on the way to being retired, no other orbiter would be available to launch and rescue the crew of STS-135. So if *Atlantis* had been somehow damaged to the point of no return, Ferguson, Hurley, Magnus, and Walheim would have been stuck on the station until various Soyuz ferry ships could bring them home one by one.

Planning for such an emergency was a moot point if the mission did not take place. For the first half of the crew's nine-month training flow, Ferguson gave STS-135 at best a fifty-fifty chance of actually taking place. Right before Christmas in 2010, William H. Gerstenmaier—NASA's associate administrator for space operations—sent out a note to those involved in the shuttle program saying, in effect, that STS-135 was actually going to happen. Doubt remained in many corners, and around February, Ferguson sat down with John Shannon, manager of the shuttle program office, to get the scoop. Ferguson described the meeting:

I asked to talk to John Shannon for a little while, just to get his read on where this was all headed. His story was very consistent on the likelihood of flying as a scheduled flight. The money was there, but it had to carefully walk through the minefield associated with Congress to ensure we were not overtly scheduling something that was not previously approved, to make sure they were comfortable NASA *had the money to fly it and that we intended to do so. I don't want to say we were concerned—maybe cautiously optimistic is the way to put it—about the commercial supply chain that would exist after the shuttle program was over. The story was very consistent, even though we didn't know what the answer was going to be—we had the money to fly it. It was the intent of Charlie Bolden to fly it. Gerst wanted it. Everything was lining up, and the fact that* NASA *could basically execute the flight using internal shuttle funding, they wanted to make sure they were not stepping on any congressional toes by doing that.*

Despite the fact that he was buoyed by the chat with Shannon, Ferguson was still not quite sure his small crew would fly. "If you had to ask me specifically when the moment was where we were fairly confident we were going to fly, I don't think that moment *ever* really came," Ferguson admitted. "Nobody said this was a done deal. I watched the news like everybody else did." He continued, "Once 134 went, I was fairly confident we were going. There was no way to turn around gracefully at that point. But, of course, if we'd had a maintenance issue, perhaps a pad abort, I could easily see somebody coming back and saying, 'That's it. We're not flying. We can't support the engineering effort it'll take to fix this problem and the delay in the layoff of the workforce to actually pay for this flight.' So, ultimately, I knew we were going when the SRBs lit and not before that."

Once in a great while, on those rare occasions when there was a break in the training and a chance to breathe, certain doubts had to have crept through the crew. *Is this flight ever really going to take place? Is there a chance that we are doing all this for nothing?* Early on especially, there were almost constant comments to the effect that they were never going to fly.

Despite the awkward circumstances brought about by the uncertainty surrounding the end of the Space Shuttle program in general and more specifically STS-135 itself, Hurley was proud of the fact that no one on the flight ever let up on the work at hand. It was just the four of them, their bond forged almost from the moment they were assigned. Each knew what the other was going through. They focused, trained, and put out the effort as if there were never any concerns whatsoever. "*Nobody* showed up at any sim or any training event and said, 'We're not going to do this. I'm just going to phone it in,'" Hurley said looking back. "That just never happened."

Other flights had some wiggle room when it came to learning the ropes, but not the shuttle program's last mission. The STS-135 commander had already commanded, the pilot had already piloted, and the mission specialists had already specialized. Each of them had already been to the station. All four had been with the agency for a decade or more, so there were few if any ropes left to be learned. Not that it was old hat by any means, because every flight featured its own unique challenges, but the four astronauts of STS-135 were experienced enough to know the process for figuring out what to do when those surprises cropped up. With just four people strapped into the crew compartment, it was vital that each knew his or her job in and out, and, to a certain extent, his or her crewmates' tasks as well. Magnus remembered:

One of the more challenging things is that on a typical shuttle mission, you train to be a very narrowly focused specialist in the areas that you are in charge of. For example, if you're a robotics person, you're going way deep into robotics and you might not pay attention to what the photo/TV person or what the science payload person is doing. That's not your task. But what we found with a four-person crew is that we had to do a lot more cross-training, which, to me, felt a little bit more like the training you do for a long-duration mission because you tend to become a jack of all trades. We had to back each other up all the time, because there were only four of us.

Personalities that meshed well were also important, and the crew's leader helped make sure the mix was just right. Any number of factors played in the decision, according to Ferguson. They all knew who they got along with well in the office, so interpersonal relationships were important. With a smaller crew, each member would have to execute a plan and do it very quickly. This flight especially, the astronauts involved would have to accept a level of media interest with an appropriate level of decorum. The spotlight could not be chased, but it could not be shunned either. "We needed people who were perhaps not overly interested in communicating with the media as much as possible, or on the other hand, unwilling to do it," Ferguson admitted. "We had to pick the right kind of people, the right kind of personalities, the ones who would be willing to do things from a public relations standpoint on the last flight, without it going to their heads, for lack of a better term." It was crucial to find just the right mix of people, and for a variety of reasons, Ferguson explained:

Crew chemistry is really important. It's important in any selection. Did I have an input? Yeah, I did and I think that helped. We really had to depend on one another a lot more so than some crews that have—I don't want to say an extra person. For example, we all had to know exactly how we were going to feel right after launch. Some people don't feel good the entire first day. Some people don't feel good all the way through the second day. We couldn't spare a person. When you take a rookie into space, you don't know how they're going to feel because they *don't know how they're going to feel. We had four experienced people, four people who knew how they were going to feel the first couple of days. They were fairly confident that they were going to get through a very aggressive flight plan to get us set up for rendezvous on the third flight day.*

Next, it was Hurley's chance to turn the tables on his commander. "How he conducted and handled himself, I can't think of anybody in the office who could have done a better job than he did, publicly and within the crew," he said, and then continued, crediting Peggy Whitson for her input on the crew selection:

I was told by way *more than one person that Peggy got the crew assignment right for* STS-135. *Peggy did a* great *job of matching personalities and skill sets on our crew. We'll be friends for life. We had a great time together, and obviously, Chris's leadership was a huge help. But by the same token, we got along great from the start and became very good, close friends. I think you saw, you had to, that we enjoyed being around each other and worked well and very effectively as a group. Chris drew on our strengths, and we did a good job of covering up each other's weaknesses. We built a bond fairly quickly. I give a lot of credit to Chris, but I also give a lot of credit to Peggy. She could've picked anybody. There were way more than four people who were eligible, capable, and wanted to be assigned to that flight, trust me.*

This was a group of four people with distinct personalities, and, again, Ferguson laughed when he was asked to describe each separately. "They're all sitting here, so I can't talk about them too much. I have to lie," he said, still chuckling.

29. STS-135 commander Chris Ferguson, Rex Walheim, pilot Doug Hurley, and Sandy Magnus share a laugh during training for the highly anticipated and emotionally charged flight. Courtesy NASA.

The joker in the crowd, Ferguson began, would have to have been Hurley, although Walheim figured that he had actually taken that role at one time or another. "I was a little bit of the jokester at the time, trying to keep things light," Walheim noted. Ferguson concluded that Magnus was the detail person, while Walheim added that she was the flight's certified smart person. "She uses a word to describe her level of attention to detail that I'll reserve for your imagination," Ferguson joked, and he would not budge on what that word might have been. "She's *very* detail oriented, which is exactly what you needed to manage the cargo going in and out of the station." Walheim was the straight shooter. "I called him the canary in the gold mine," Ferguson continued. "When everyone was getting tired, I'd just look at Rex. If Rex is tired, *everybody's* tired. If Rex thinks we should do something, we should probably do it. He kept us honest."

If Walheim was the straight shooter, he also became a bit of a target during training. Naval aviators Ferguson and Hurley took great delight in giving him a good-natured pounding about his service in the air force. A civilian, Magnus was able to stay out of that particular line of fire, but she remembered the exchanges well. "I'm not as good on my feet with jokes as the guys are," she said. "It was a lot of fun to listen to them. We had the whole military rivalry thing going with Rex getting beat up by the navy. He stood tall, though, survived it, and got a few jabs in for the air force."

Hurley did not miss a beat, and readily admitted to the grief he and Ferguson gave Walheim. In fact, he seemed to relish it. "True!" Hurley proclaimed. "We often reminded him that we had T-shirts and underwear older than the air force. Also, we often called to his attention all the space firsts and lasts with naval aviators—basically all of them!" Naval aviators, after all, could lay claim to a vast number of spaceflight achievements:

NASA's first suborbital spaceflight was made by Alan Shepard.
Marine John Glenn became the first American to orbit the earth.
Wally Schirra served as commander of the first Apollo mission to fly.
The first man to walk on the moon, Neil Armstrong, was a former navy flier.
Apollo 12 featured Charles Conrad Jr., Richard F. Gordon Jr., and Alan L. Bean, navy men all.
Gene Cernan was the last person to step foot on the lunar surface.
Command of all three *Skylab* flights went to the navy's Pete Conrad, Al Bean, and Gerald P. Carr.
John Young and Bob Crippen composed the crew of STS-1, the first shuttle flight.
The first ISS commander was navy SEAL Bill Shepherd.
The last Space Shuttle flight featured none other than Chris Ferguson and marine Doug Hurley at the controls.

Wait just a second, though. Yes, the navy's astronauts had some important achievements in space, but so did those from the air force. For instance:

Three of the Mercury Seven pioneer astronauts—Gus Grissom, Leroy Gordon Cooper Jr., and Donald K. Slayton—were air force jet jockeys.
Ed White did the first EVA by an American.
Neil Armstrong might have flown for the navy during the Korean War, but his other two *Apollo 11* crewmates—Buzz Aldrin and Michael Collins—were air force.
The entire crew of *Apollo 15*—David R. Scott, Alfred M. Worden, and James B. Irwin—was composed of air force officers. Another, Charles M. Duke Jr., was an *Apollo 16* moonwalker.

Guion S. Bluford Jr. was the first African American in space.
The first Hubble servicing mission was led by Dick Covey.
Jerry Ross was the first person—and one of only two—to fly the shuttle seven times.
Eileen Collins was not only the first female shuttle commander, but she also led the first flight following the *Columbia* accident.
Collins was the first woman to lead a shuttle mission, and Pam Melroy was the second.
Kevin P. Chilton became a four-star general, the highest rank ever achieved by a former astronaut.

Surely, such lofty accomplishments by the men and women in blue were enough to even out the ledger. Not so, at least not for Hurley. "Not even close!" he barked, albeit with a smile. "Nice try, though." As comebacks go, Walheim's response was decidedly mild—if not measured. "They certainly never missed an opportunity to give me a hard time about my air force background," he said. "They had quite a lot of fun teasing me whenever they could. We'd have to give it right back to them when we could. It was all in fun. We had a good time with it."

Despite all that camaraderie, Hurley was never quite able to convert any of his crewmates into NASCAR fans. This is how much the guy loves his favorite sport—for years, Hurley held season tickets at Texas Motor Speedway in Fort Worth. When STS-135 launched with a speedway flag on board, Eddie Gossage, the track's president, and Mike Zizzo, its vice president of media relations, were in the family viewing stand atop the Launch Control Center as Hurley's guests. Hurley was fully and without reservation on board as a NASCAR nut, but the rest of the crew? Not a chance. The most any of them would allow was Walheim. "There was one night we were at Kennedy Space Center, and there was a big NASCAR race on," Walheim related. "We all got around and watched it. It was pretty complicated, so we were asking all these questions, 'Why is he doing that? Why is he doing this?!?' We were probably bugging him too much. He was probably thinking, 'I wish these guys would go away so I can watch the race in peace.'" Magnus did a public-relations appearance with Hurley at another track, Chicagoland Speedway, and let us just say that she evidently came away with a passion for the sport that was something a bit less than overwhelming. "That was my first experience with NASCAR, and it was . . . it was very interesting," she allowed very mildly. "It's a whole world I hadn't been exposed to yet. I don't know if I'm a fan, but I found it very . . . interesting."

The joking about the navy and the marines and the air force, the back-and-forth debates over the merits of automobile racing and who knows what else were simply ways to pass the time and blow off a little steam. There was serious work to be done, and it sometimes had to be done amid emotionally difficult circumstances.

In the months following the flight of STS-135, Mark Kelly never mentioned to Ferguson anything about knocking him and his crew out of the history books. Kelly was too busy to worry about such matters.

Kelly's spaceflight career was over, but another chapter of his life had begun to unfold several months before. He was deep in training for STS-134 when wife Gabrielle "Gabby" Giffords was critically wounded on 8 January 2011 in a shooting near her native Tucson, Arizona. Giffords, a Democratic member of the United States House of Representatives, was meeting with constituents when Jared Lee Loughner opened fire with a Glock 19 semi-automatic pistol. Nineteen people were injured, six of whom died. Giffords, who was shot in the head, was initially reported as one of the dead.

Kelly rushed to her side, leaving Houston and preparations for the flight temporarily in his wake. In the months afterward, while he did not end up with command of the historic last flight, Kelly indeed became one of the most recognizable astronauts of the shuttle era. "Of course, Mark's life has changed dramatically just in the last year or so," Ferguson said. "He's got challenges he's going to have to face. Whoever went last, I don't know if it's really going to make a difference in a couple of months, I really don't."

For better or worse, the attention garnered by the tragedy kept a majority of the spotlight off the STS-135 crew. As Giffords's condition very slowly improved, the story became the stuff of almost fairy-tale proportions—the gallant warrior-astronaut standing ever faithful alongside his wounded public-servant spouse. At one appearance soon after the shooting, Kelly held hands with a supportive Michelle Obama as the president himself gave an update on Giffords's condition. Kelly was interviewed by Diane Sawyer on a special edition of *20/20*, and Sawyer then scored another press coup when she sat down with Kelly and Giffords for an interview aired on ABC News on 14 November 2011. That chat and an appearance on the cover of *People* magazine were in support of the book *Gabby: A Story of Courage and Hope*. In a press release announcing the project, Giffords and Kelly were called "one of the most admired and beloved couples in recent American history." The fanfare did not end there. Separately, Kelly graced the cover of the esteemed *Esquire* magazine in December 2011 as one of its Americans of the Year. "Prior to 134, we were not so much on the horizon," Hurley figured. "Mark and Gabby were very much at the forefront of the news, him flying and her being there. Once those guys landed, that's when it really picked up for us. The media was like, 'Oh, 135 is the last shuttle flight.' It went from there."

That it did. Smiley N. Pool, chief photographer and photo coach at the *Houston Chronicle*, received permission from NASA to accompany the crew throughout much of its training. Pool was there when Ferguson, Hurley, Magnus, and Walheim did their final simulator run and when they entered quarantine. He snapped artsy shots of the last shuttle crew with the first, Young and Crippen. He captured, Magnus concluded, "a lot of really special memories," and she was glad that he did. The memories they could not purposefully gather for themselves were instead being collected by Pool and a horde of fellow media members.

Pool's was far from the extent of the flight's news coverage. Among many others, camera crews from the Science Channel followed the crew and numerous other shuttle program participants throughout the training flow for the program *Last Shuttle: Our Journey*, which debuted the night of the launch. "We had to be so focused on figuring out how to do this mission that normally you have five to seven people for, instead of four," Magnus said. "Some of that extra-nice stuff, keep a journal, take a lot of photos, we did not have the capacity to do that. So to have the world around us helping us with that, turned out in retrospect to be a real bonus."

With STS-134 back on the ground, interest in the finale snowballed almost daily. Hurley marveled at his commander's ability to shield them from it all:

Chris did a great job of insulating the three of us from what had to be just a mountain of requests, demands, and directions, especially the last couple of months. That's when you hit a fever pitch with training and stress. All that stuff peaks one to two months prior to the flight, because that's when everybody goes, "Holy crap. I'm going to go fly in space here in a couple of months." From a crew standpoint, you're kind of on edge more so than you probably were a month after you got assigned. Then combine that with a reduced crew. Combine that with the demands privately and publicly of flying the last shuttle flight and being the commander of the last flight, I just thought Chris handled it very, very, very, very well.

The furor added a degree of difficulty to preparations for the flight that were already compressed to just nine months, and divided among just the four of them. Based on their previous experience, some of the more mundane training tasks were waived. When that happened, there was always a certain amount of doubt left behind as to whether it had been the right thing to do—what if, after all, that one small detail happened to be something that ended up impacting the success or failure of the mission? "Are we good enough that we don't have to sit through this three-hour class?" Ferguson remembered wondering. "Are we good enough to where we don't have to fly to El Paso this afternoon to fly a Shuttle Training Aircraft? Are we good enough? By and large, that's how we made our decisions. Those kinds of things worried me a little bit."

Overshadowing everything—the saga of Mark Kelly and Gabby Giffords, the historic nature of this being the last flight and preparing to fly it—was the cold, hard fact that thousands of people all around them were losing their jobs on a regular basis. As a result of the utterly depressing layoffs, the last shuttle astronauts also needed to serve as de facto cheerleaders. It was, after all, their lives on the line if somebody left a screw loose somewhere in the vast innards of the orbiter. Ferguson remembered:

We were making frequent and regular contact with the folks at the Kennedy Space Center. There were a couple of lines in the sand that we drew. We went out there and watched the vehicle roll from the OPF over to the VAB, and then again when it went from the VAB out to the pad. We knew a lot of these KSC workers were going to be losing their job, and in our own little way, we needed to maintain a pretty close contact with them. We needed to make sure that they understood—as they had so many other personal issues to worry about, like the loss of their jobs and perhaps the moving of their family—that up until the very end, there were still live, living, breathing people on these Space Shuttles that they turn wrenches on day in and day out.

If there was one very best thing to be said about the shuttle workforce in the waning days of the program, it was that on no occasion did the crew of STS-135 encounter anyone who was openly bitter. Many workers chose to stay the course, even though they knew full well that a pink slip was waiting at the finish line. "There was just an unbelievably incredible professionalism, and I'm talking the trainers, the instructors, the flight controllers, and, of course, the people down at KSC, who probably bore the brunt of the layoffs," Hurley said. "There was story after story about these people who had been with the program since the beginning and wanted to finish it out." Thousands of people attended *Atlantis*'s rollout from the VAB to pad 39A—pad 39B had been in the process of deconstruction for quite some time already. Hurley admitted:

It was awe inspiring, it was sad, and it was emotional all at the same time. It was a vicious circle of this, especially the last four to six months, where it was someone—or a group of people—leaving and you were saying good-bye. You'd go to a training event, and it was the last one ever. This is the last time we're going to do this, this is the last time we're going to do that. It was the last class this person was going to teach. It was tough, but it's like everything else in life. Things end. Programs end. People change jobs. But it was not easy. At times, it was very, very difficult. But it was just amazing how the folks did it professionally, with dignity, and very much honored the legacy of the shuttle.

Emotions had indeed been rubbed raw many times over. Launch director Michael D. Leinbach received a standing ovation from teams in the Launch Control Center on 26 May 2011, following the last certification runs of the shuttle program, when he lambasted both the agency itself and the president for what he saw as their lack of leadership. Leinbach, in part, told those assembled:

What I'm about to say would not be appropriate on launch day, and this is our last chance to talk together. The end of the shuttle program is a tough thing to swallow, and we're all victims of poor policy out of Washington DC, both at the NASA level and the executive branch of the government and it affects all of us—it affects most of you—severely. I'm embarrassed that we don't have better guidance out of Washington DC. Throughout the history of the manned spaceflight program, we've always had another program to transition into—from Mercury to Gemini, and to Apollo to the Apollo-Soyuz Test Project, to Skylab and then to the shuttle. We've always had something to transition into. And we had that, and it got cancelled and now we don't have anything. I'm embarrassed that we don't. Frankly, as a senior NASA manager, I'd like to apologize to you all that we don't have that. So there you are. I love you all. I wish you all the best.

The program was ending with few solid answers about what was coming next, but that did not mean that shuttle workers were simply playing out the string. Far from it, actually, and never was that more evident than in the issues that plagued the outset of STS-133. A planned launch of 5 November 2010 was delayed initially by a significant hydrogen leak around the Ground Umbilical Carrier Plate—the same problem that plagued Hurley's STS-127 launch—and then again when an even more serious problem was discovered. Cracks developed in the structural ribs known as "stringers" on the External Tank, problems that evidently took place during the super-cold fueling process. After a rollback to the VAB for repairs, STS-133 did not fly until 24 February 2011. There was a time when other such issues might have been glossed over or ignored altogether—too cold to launch? No problem. Foam was liberated from the External Tank? That had happened on multiple flights. It has never hurt anything before, so it will not hurt anything this time. It is a processing issue, nothing more, nothing less. Debris hit the orbiter? The bird is tough. It can take a hit. The voices speaking out in protest? Do not worry about them. Such blasé attitudes were not the case this time around, not in late 2010, early 2011.

In many ways, that kind of pride in the program was a way for the shuttle workforce to thumb its collective nose at the powers-that-be who were showing them the door. "The stringer problem was a serious, serious matter," Hurley said. "They worked the problem. They went where the data took them. They gave it due diligence, and they solved the problem." Hurley continued, "The program did just an amazing job in solving the stringer-crack problem and modifying the tanks. Obviously, we fixed it correctly, because we didn't have an issue on 133, 134, or 135 with the tanks. Think about it. To have that kind of a technical challenge thrown at the team, all while the shuttle program is downsizing and there's layoffs, it said a lot about the program at that point. The way John Shannon managed the program should be a model. We completed the flights, and we did it safely."

Like Heflin on the runway tarmac at the end of the flight, such attention to detail in the midst of uncertainty left Hurley with the disquieting notion that the shuttle could have continued flying well past the sell-by date thrust upon it. "Unfortunately," he concluded, "the shuttle program hit its stride just as it ended."

One of the most striking series of photographs shot by Smiley Pool of the *Houston Chronicle* was taken as Ferguson, Hurley, Magnus, and Walheim prepared to enter quarantine at JSC late on the evening of Friday, 1 July 2011, before they headed to the Cape the following Monday for the launch itself. The pictures capture everyday moments being played out in the most extraordinary of circumstances. Magnus shares a scoop of ice cream with a friend. Ferguson and Hurley, both solitary figures in separate photographs, lug huge bags into the quarantine building.

But the most emotionally evocative one of them all, the one that truly is worth a million words or more, is one of Walheim saying good-bye to his twelve-year-old son, Jeffrey. They are hugging, but it is no half-hearted, quick, all-but-embarrassed, one-armed embrace. Instead, it is a full-on, rib-cracking bear hug, and one that is at once both tough and endearing to see. For thirty years, shuttle astronauts had said their farewells to family and friends, and for thirty years, it had been very challenging to do so. "It is hard," Walheim began. "You just have to keep focus on the mission, because if you dwell on it too much, it *will* affect you a little bit. You try not to let that happen, so you try and keep everything upbeat."

As hard as it was to say those good-byes, there was also relief with game time finally at hand. Friends of friends of friends had been coming out of the woodwork, hoping for an up-close-and-personal spot from which to view the launch, a process that only picked up speed as the big day approached. While each crew member was allotted 330 tickets for the grand event, it never seemed to be enough. That was just for the launch. To actually get something flown on the mission was just as big, if not bigger, a quest. Every flight carried a wide-ranging assortment of knickknacks—hundreds of small flags and patches, medallions, and so forth that were prized mementos back on the ground. Packed away on STS-135 was everything from a printed recipe out of KSC's crew quarters to a Joe Gibbs Racing cap (Hurley again!) and a banner for the Quail Park Retirement Village in Visalia,

California. None of those were particularly out of the ordinary, but it was just the sheer number of requests cascading in from every which direction that seemed almost overwhelming at times. Walking through the doors of the quarantine facility put it all behind them, so the crew could deliberate on issues that truly did matter. Hurley remarked:

I don't think I would be the first person to say this, but it was almost a relief to get into quarantine. We call it a white-collar prison, because you're kind of stuck there. But the facility is great, and you really get a chance to relax a little bit and to rest. Among ourselves, we just talked about, "Okay, this is what I'm worried about on the timeline." We were able to go through particular days or particular phases of the mission. I thought it was great. If I ever get to command a space mission, I would do it, too, to sit down and go through the checklists, the procedures, the choreography of these particularly busy days. As much as I hated being away from Karen and especially Jack—because Jack is so young, I basically couldn't have any contact with him until I got back and that was very, very, very difficult for me—it was nice because the quarantine allowed us to recharge our batteries a little bit.

The time for the final shuttle launch, the last crewed liftoff from American soil for who knew how long, had finally arrived.

As the famous silver astrovan made its way from the Operations and Checkout Building to pad 39A on 8 July 2011, one of its passengers had as many flights to his credit as the crew about to be launched had combined. Jerry Ross's astronaut career was older than the shuttle program itself, and so he had seen and done it all. Seven times he had flown the bird to orbit and back again, and as chief of the Vehicle Test Integration Office at JSC, Ross was riding shotgun with the three men and one woman charged with putting the finishing touches on the program. "The mood was not much different from any other flight, frankly," Ross said. "They were all anticipating the flight. They knew it was the last one, and so that was the elephant in the room, maybe, but it was not something that was focused on. It was the excitement of getting ready to go fly, and looking at the crowds and waving as we went by." His photographs from the brief journey provided yet another behind-the-scenes glimpse into the flight, and this time, he did not disembark at the Launch Control Center as he had a few other times. This time, he took the ride all the way to ground zero. "But they watched me to make sure I didn't get into the vehicle," Ross joked.

Thousands of people lined the nine-mile route, and the throng only thickened as the modified 1983 Airstream Excella motorhome made its way past the VAB and the press site. The trip took twenty minutes or so, and it was one that each of the STS-135 astronauts had made at least once before. No matter how much they might have tried to pretend otherwise, this time was different. "This was off the charts," Ferguson said. "We were all astounded. We didn't know there was room to fit that many vehicles at the press site. If we missed anything before that indicated how closely the country and the world were watching, we got it right then and there."

Media was in attendance from not only the United States but around the world also. A note from a buddy prompted Ferguson to wonder if there might have been an ulterior motive for the international media's interest:

It was not just the United States. It was global, a lot of media speaking different languages over there. I don't know quite how to take that. After the flight, I got a note from a friend who I'd gone through the navy's test pilot school with. He was a Swedish fighter pilot who ended up going to work for Saab Aircraft. He wrote and said, "It's so sad to see the United States end the Space Shuttle program. The rest of the world counts on the United States to always maintain the lead in these kinds of things. To see it take a step backwards is demoralizing." I thought about his comments and I thought about the presence of all the foreign media there for launch. You do wonder, does this mean that the rest of the world thinks we've given up, that we're either broke, unwilling, or unable to continue on from this point? Is this because they think it's the end? I don't know the answer to that question myself. I certainly hope it's not. The plans are encouraging, but the money isn't.

Given the weather forecast, it all seemed like just another dress rehearsal. Predawn drizzle gave way to heavily overcast skies and a minuscule 30 percent chance of favorable conditions for the scheduled 11:26 a.m. EDT blastoff. Hurley got out of the astrovan at the pad wondering what movie he and his mates were going to be watching that night back in quarantine, following the scrub for weather that was all but assured the way things looked.

As the cloud ceiling began to break up, each update got a little more positive. Problem was, it needed to get *a lot* more positive to get this par-

30. After arriving at the launch pad on the morning of 8 July 2011, the crew of STS-135 takes a few moments to check out their ride to orbit—the shuttle *Atlantis*. Courtesy NASA/Jerry Ross.

ticular show on the road. Walheim, on STS-122 in February 2008, had launched on another 70 percent no-go day. So maybe he had a shot this time as well.

Fellow astronaut Rick "C.J." Sturckow was responsible for flying approaches and landings at the nearby shuttle landing strip in the Shuttle Training Aircraft, checking to make sure *Atlantis* could safely land back at KSC in the event of an abort. The call was his. Visibility was acceptable, and so the launch would proceed as planned. "There are easy weather flights and there are difficult weather flights, and I'm sure that was a difficult weather call," Ferguson said. "If the guy on the scene says the weather's acceptable enough for an RTLS [return to launch site] landing, then the weather's good enough for an RTLS landing. We just take his word for it and go for it." As unlikely as it might have seemed that morning, the countdown came out of the standard hold at T-minus nine minutes and proceeded unimpeded through the sixty-second mark. Then, at T-minus thirty-one seconds, it stopped when indications showed the External Tank's beanie cap had not fully retracted. Within minutes, the issue was cleared and the countdown continued, the last few remaining seconds clicking away to the end of a long and storied era.

T-minus zero. *Atlantis* was off and running one last time. Almost from the time he had taken his place in the flight engineer's seat located just behind the center console on the flight deck, Walheim's face had been locked in an almost constant grin. If there were any nerves, he was hiding them very well and the smile grew even wider when the SRBs kicked in. "I just love the ride to orbit, it's too much fun," Walheim said. "My wife hates it that I'm always smiling. She's like, 'Can't you just show a little bit of concern?' It's an interesting thing that the guys riding on the rocket seem more calm than the folks who have to watch it."

Within seconds of leaving the pad, *Atlantis* punched through the low cloud deck, the tip of a spear atop a column of incandescent flame and smoke. A little more than a minute into the flight, a warning buzzed through the cabin signaling the possibility of a change in cabin pressure due to its stretching under the forces of launch. It was a known quirk of the shuttle structure, and CapCom Barry Wilmore told the crew just after he gave the crew the well-known "go at throttle up" message that no action was needed on the dp/dt (delta pressure over delta time) sensor alarm. This was the easy part. They had already been through the wringer many times over during training, Wilmore remembered:

I can tell you this. I've done many challenging things in my professional career, and right up there and equal to any other challenge was being an ascent-entry CapCom. During the missions, it's easy. But a typical mission without problems is not what we train for. We train for multiple, simultaneous *malfunctions. We do not have time to sit and mull it over. Literally, life-or-death situations are occurring every other second it seems, and you've got to make a decision* now. *The flight director has ultimate authority in what is done on ascent and entry because it's such a dynamic environment of flight, but the Cap-*

Com works hand in hand with them in transferring all the information to the crew. It's very challenging in the simulation environment. Preparing for the worst-case scenario, you can have seven major malfunctions in seven different systems simultaneously. We're shutting down an engine here so we can make it to a certain point in the trajectory so we can abort overseas and land in Spain. It's just very, very dynamic.

Fortunately, it was not necessary for *Atlantis* to land in Spain or any other contingency site. Those who walked NASA's hallowed halls—and anybody who wanted to imitate those who did—would have called it a "nominal" launch, just the kind this last journey needed. Within moments of main engine cutoff, Magnus was up and out of her seat to Walheim's right, already getting set up to take documentation photos of the External Tank that was soon to be released from the orbiter's underside. The post-insertion phase of the flight—from ascent through the crew's first sleep period—was one that had been greatly concerning. With just four hands on deck, Ferguson, Hurley, Magnus, and Walheim had to manage a workload normally handled by six and most times seven people until they docked with the station, where they were to meet up with *Expedition 28*'s Mike Fossum, Ron Garan, Andrei I. Borisenko, Aleksandr M. Samokutyayev, Sergey A. Volkov, and Satoshi Furukawa.

It was in precisely these moments after launch and before meeting up with fellow spacefarers that their previous experience on orbit paid off. Their bodies adapted faster. On STS-127, it had been two or three days before Hurley really had an appetite. This time, he felt ready to chow down from the word go. There had been cases where rookies got to orbit and were unable to do almost anything as they battled space sickness. This crew did not have that kind of luxury. "The concern of not sleeping well or not being able to adapt just wasn't a big issue for us," Hurley said. "We needed to have all four of us operating at capacity to just get the cockpit set up for the next day, which was the inspection day." They got their work done and got in their racks for some shut-eye, basically on time. "Flight day one, in a lot of cases, really sets the tone for the mission," Hurley continued, and then added:

If you have a screwed-up flight day one—if people are sick or whatever—you don't get a bunch of stuff done. That puts you behind on flight day two, and then, flight day three, you're into the rendezvous with the station. If you're be-

hind going into the rendezvous, trust me, you may end up rendezvoused, but it won't be pretty. On 127, we went right from rendezvous on flight day three to EVA-1 on flight day four. So, you think about it, it can domino so easily on a mission. That doesn't even take into account if you have a significant systems issue, which can also put you way behind. It can definitely be a challenge.

The crew was very much relieved that things had gone as well as they did from the outset. That was certainly a good thing, because the flight plan was about to get that much busier as it headed toward a 10 July docking.

Mike Fossum was as crazy busy as he ever wanted to be while getting ready for his friends on STS-135 to arrive. Fossum rocketed out of Russia on 7 June 2011, and he had been prepping for *Atlantis*'s docking almost ever since. To prepare "the road to 135," as Fossum called it, meant sprucing up the place to a certain extent, but there was also a lot of grunt work as well— prepacking items to be transferred into *Raffaello*, configuration of various systems, and so forth. It went without saying, for instance, that it would be the last time a station crew set up photography of an orbiter's heat-shield tiles. Then, once the docking took place, it was a reunion and homecoming rolled into one. "They were a wonderful crew, just wonderful," Fossum said. "I knew them all as friends already. They came on board, got to work but were very respectful of us, our place, and the fact that we're sleeping right there near where they entered the station." Fossum especially enjoyed Magnus's reaction to returning to a home she had known for four months. She rolled in, remarking about various changes that had been made here and there, much like someone coming back to a childhood home years later. "She says, 'Hmm, I wonder why you have it that way?'" Fossum said with a laugh. "Ron and I looked at each other. We don't know. It was that way when we found it."

Magnus once figured that her spaceflight career was finished, but that was before she received a phone call from Whitson asking if she would be interested in joining the STS-335/135 team. Detailed to NASA headquarters at the time, Magnus headed back to Houston to begin yet another journey to orbit. Once she got there, the sense of awe and familiarity was striking. "I don't think you can really appreciate how incredibly amazing and huge that station is until you're rendezvousing and docking with it," Magnus exulted, and then added, "I felt like I was coming home. This had been my home, and I was just coming by to visit. It had changed a little bit. As soon as we opened the hatch, it was like, 'Oh, I'm home.' I found bags with my handwriting still up there. A lot of the stowage was where I remembered it. Some of it had moved. We had new modules, of course, and I had to get used to that. But I immediately felt like I was in my environment."

She was back in her element, but a long-duration stay aboard the station and a relatively quick visit there as a member of a shuttle crew are two entirely different beasts. Magnus was the payload commander on STS-135, and she simply could not ever afford to work at a leisurely pace as she directed five tons of equipment out of *Raffaello* and another three back in for a return to Earth.

A long-duration mission is a marathon. You know you're going to be there next week. There's no panicking. There are no twenty-seven contingency plans if a piece of equipment breaks, because you don't have *to get it done tomorrow. The way you plan and execute a long-duration mission is way different. On a shuttle mission, it's a sprint. Consequently, you only have a finite amount of time to get a certain amount of work done. You have to be ready instantly to execute contingency plans. If this piece of equipment breaks, you've got to have work-arounds ready to spring into action. The disadvantage of a shuttle mission is that you're just so busy, you can't take in what it's like to actually live in space. It's a lot different than just visiting.*

The crew worked as fast and as carefully as it could, always keeping an eye out for what was supposed to take place next in order to stay on the timeline—or, at the very least, as close to it as possible. "It is draining sometimes," Walheim admitted. "It's not only that you're doing a lot of stuff, you're doing a lot of *important* stuff. It's stuff that can't be messed up or it can really cause a problem. It's a little bit of mental strain trying to do all this stuff, and get it done without making a mistake. It was definitely the busiest flight I'd been on." Ferguson put it this way. If the crew was awake and not eating or going to the bathroom, its nose was to the grindstone. Add into the mix appearances prescheduled by NASA's public affairs office (PAO), and things tended to get still more complicated. There were interviews with television stations as far apart as Binghamton, New York—near Hurley's hometown in Apalachin—and San Francisco, and as many as six on any given day. Then, on 15 July, a series of interviews was followed by a call from President Obama.

The media sessions provided breaks—if not entirely welcomed ones—in the daily grind. "Here we are five minutes before a PAO event where we're supposed to be either on national television or talking to the president," Ferguson said. "You're still trying to rustle people together to get settled into their positions. Then, as soon as the event's over, everybody breaks right away and they go right back to what they were doing."

The whole spiel was quite unlike the talk-show appearances the crew did following the flight, where they would show up five hours in advance for makeup, pre-interviews, and so forth. Primping for such a job took up the better part of half a day, but not on orbit. There was no time for such luxuries. "Up there, you just kind of drag a comb through your hair, make sure your shirt's tucked in, and then you stand there in front of the camera," Ferguson concluded.

Walheim was the only STS-135 crew member with an EVA to his credit, having completed a total of five spacewalks between STS-110 in April 2002 and STS-122 in February 2008. He had been outside for nearly thirty-six and a half hours, not nearly a record in the grand scheme of things, but that was thirty-six and a half hours more than any of his crewmates could claim. When an EVA to retrieve a failed Pump Module was shoe-horned into the STS-135 flight plan, planners figured it prudent not to use *Atlantis*'s astronauts due to the short training flow and reduced crew. That meant that Walheim had to settle for planning and coordinating the spacewalk from the orbiter's flight deck while Fossum and Garan did their work, which also included installing a couple of experiment packages and repairing a base for the station's robotic arm. Hurley and Magnus operated the robotic arm for the final EVA of the shuttle era.

In the hours before the spacewalk, Walheim was greeted with a traditional wake-up call featuring the song "More" by contemporary Christian music artist Matthew West. "I wanted to pick a song about how beautiful the Earth and Creation are," said Walheim, who also flew one of West's CDs despite the fact the two of them had never actually met. The tune itself simply meant that much to him. "That song kind of describes what it looks like from space—look at the mountains and the oceans and the deserts and how they spread out for thousands of miles," Walheim added. "It's a neat description."

There had been little time for Walheim, Magnus, Hurley, Fossum, and

Garan to train together, once again making previous experience in the roles they would play all the more critical. Walheim had coordinated other spacewalks; Hurley and Magnus had worked with robotics on other flights; and, on STS-124, Garan and Fossum had done three EVAs together totaling more than twenty hours. Sticking to the basics, the five of them went over plans for how they would communicate and so forth the night before. Hurley credited Walheim—even if he was an air force guy—for ensuring that the EVA turned out as well as it did. "Even though he wasn't going to go outside, Rex really had that EVA suitcased," Hurley said. "He really spent a lot of time with those guys when we got into orbit, going through everything. Frankly, Rex was probably the biggest key to success on orbit, because he knew that EVA inside and out, frontwards and backwards. It showed."

The station's long-duration crew rarely had the time to train as thoroughly for an EVA as the spacewalkers who were brought up and then right back down again by the shuttle. Shuttle-based spacewalkers had the luxury of countless runs in the pool, virtual-reality lab, and table-top sessions to prepare for a trip outside. "The space station, you just don't have that ability," Fossum admitted. "So you have to adapt and work to figure out how to get it done." This was to be a "moderately complex" spacewalk, Fossum continued, so "I felt comfortable keeping the pressure turned up on the two of us and our team, because we were experienced out there." Garan added:

We had very, very little time to train for that spacewalk. When it came time to do the spacewalk, there were things that we had never *trained for that came in later, after I was already in space. Space Shuttle spacewalks are* extremely, *in minute detail, choreographed to the second. The fourth spacewalk that we did together, when we were space station crewmates, was more—I don't want to say ad-libbed. We definitely had a plan. We definitely executed the plan, but there was definitely a lot more flexibility built into it because we didn't have a lot of time to train for it. Because we had spent hundreds and hundreds of hours together underwater in the NBL, because we'd done three spacewalks together, because we basically knew exactly how the other one thought and what they were doing and what their strengths and weaknesses were, the ground team was very comfortable to have us do this EVA.*

The six-hour, thirty-one-minute EVA was the 119th conducted out of an ISS airlock, and the 249th by American astronauts. Only one other pair of astronauts—Clay Anderson and Richard A. Mastracchio—had done as many as four spacewalks together, but in terms of total time elapsed as a duo, the record went to Fossum and Garan with a little more than twenty-seven hours to their combined credit. As their tasks grew to a close, the two of them took a moment to watch the sunrise. Fossum had done that on their last STS-124 EVA, and he remained glad to have had such an experience. "I had Ron stay outside," said Fossum, who wore the red stripes of the lead spacewalker on his spacesuit. "There was a sunrise coming in just a couple of minutes. The ground knew what I was doing, because I'd done it before. What's the hurry? Let's just catch our breath here and enjoy the view." He continued:

It's really phenomenal to see that through the helmet, where you've got your head in a fishbowl and you can really appreciate the sunrise as it builds. Finally, there's a blaze of color and it's a big band that stretches across the horizon. The sun comes up through the middle of that, in just a blinding white light. After Ron was inside, I paused for a moment just to look back. Atlantis *was hanging there not that far away. It's an iconic image of the Space Shuttle, doing what it was meant to do. For me, that was a really special moment, to just hang there for about thirty seconds and study* Atlantis *as she was perched on the front of the station, ready to head home in a few days for the last time.*

More than a week remained in the flight following the EVA closeout, and the majority of it was spent in transferring cargo back and forth. There was no getting around the fact that this was the last shuttle flight, because in interview after interview, they were constantly reminded of just that. It was the nine-hundred-pound gorilla lurking—or floating, rather—in the corner. Among themselves, it went mostly without discussion. "We talked about it occasionally, but quite frankly, we were so incredibly busy trying to get all that stuff transferred across the hatch," Magnus insisted. "We didn't have a lot of time just to hang out and contemplate life in the universe. Most of the time we were running around like chickens with our heads cut off trying to get everything done before we had to undock."

The gravity of the moment and what it truly meant to the history of the Space Shuttle program hit home when hatches between the station and the orbiter were closed and latched for good on 19 July. It hit Garan when he reached across the hatch to shake hands with Magnus. He realized that never again would such a scene be played out on orbit, at least not with a shuttle crew. The realization was depressing, and Garan afterward masked it with a bit of humor. He could talk on intercom to the shuttle completely unheard by anyone else, but to reply, those on *Atlantis* had to speak over a radio frequency anyone could listen in on.

Hey, can you guys hear us? Garan would ask.

Yeah, we can hear.

Do you miss us yet?

He repeated the ruse a couple of times. "I was kidding around, but it really was kind of sad to close the hatch and realize these guys were about ready to undock for the last time," Garan said. Fossum had a twinge of regret that the STS-135 crew could not stop for just ninety minutes and watch an entire trip around the planet through the windows of the station's cupola. There were other emotions. Who knew when another crewed vehicle would dock at this particular port? The hatches were closed hours before the shuttle actually undocked to allow time for leak checks and so forth, which obviously prevented any further personal interactions between the separated crews. "I never liked that, because I have buddies that are a few feet away and we could be having dinner, telling some stories and enjoying our time together up here now that the work is really all done," Fossum remarked. As *Atlantis* undocked and began to back away on 19 July, Garan rang a bell and announced its departure. For Walheim, it turned out to be an emotional moment. "It was like, wow. It kind of got me a little choked up for a second," he confessed. "This was it. We're leaving, and we're not coming back."

Earlier, Ferguson had been struck by a recoiling bungee cord on the corner of his right eyebrow, just to the side of the bridge of his nose. The injury, plainly visible in post-flight photographs, could have been far more serious had the cord struck Ferguson less than an inch lower. Some wondered if a more substantial wound would have given Hurley the opportunity to actually fly *Atlantis* to its final landing, but no little boo-boo was going to bring Ferguson down. He was going to have control of the stick and rudder, no ifs, ands, buts, or maybes about it. Hurley handled the undocking and fly-around of the ISS, just as he had done on STS-127. He had flown part of the rendezvous, and would also take the stick for a few brief moments during the approach to landing. "Chris was very generous about allowing me to do that kind of stuff," Hurley said. "Would it have been great to land the Space Shuttle? That would have been awesome, but you're a professional. You don't do that stuff unless there's a reason to do it."

On 20 July 2011, the day after undocking from the ISS, the shuttle released its 180th and final experiment into orbit—the breadbox-sized Pico-Satellite Solar Cell package. All that was left to do was one more round of interviews with the major American networks—ABC, CBS, CNN, Fox, and NBC—before beginning preparations for the next day's reentry.

As *Atlantis* plowed its way back through the atmosphere, Fossum captured on film its long and fiery path in some of the most remarkable photographs of the entire shuttle era. "I got lucky in several regards," Fossum said. "First was the presence of the cupola windows, our bay windows to the universe, if you will. I'm not sure there had been a reentry in darkness since we got our really good windows. We had a new camera that has some really good low-light capability, and I had been practicing. If I could see that plasma trail, I knew I could capture it." This was no meteor. There were living, breathing people in the middle of the fireball, and they were Fossum's friends. Not only that, but he had also been in such a chaotic scene himself twice before. Time stood still for at least the minute or so that the plasma trail was visible. "It was emotional," Fossum added. "I was a little bit stunned. I wasn't sure I was going to be able to see it, and there it is. They're dragging a plasma trail across the planet. I knew that we would not see that again. That was the last shuttle, and it is blazing a trail home right now. I just *have* to catch that on camera to share that. I took over two hundred photos, and I'm glad a few of them turned out."

Atlantis touched down on runway 15 at precisely 5:57 a.m. EDT, with the nose gear coming down twenty seconds later. At 5:57:54 a.m. EDT, the wheels came to a stop for the final time. A Space Shuttle would never move another inch while engaged in active flight operations. "Mission complete, Houston," Ferguson called just after motion ended. "After serving the world for over thirty years, the Space Shuttle found its place in history, and it's come to a final stop." Wilmore, back at the CapCom console in Houston, responded, "We copy your wheels stop, and we'll take this opportunity to congratulate you, *Atlantis*, as well as the thousands of passionate individuals across this great spacefaring nation who truly empowered this incredible spacecraft, which for three decades has inspired millions around the globe. Job well done, America."

When Neil Armstrong guided *Eagle* to a touchdown on the lunar surface in 1969, CapCom Charlie Duke had momentarily fumbled a word or two. Wilmore was not concerned that he might repeat the miscue, because he joked that he did such things *all* the time. "I'm used to it," Wilmore said with a laugh. "Certainly at rollout, I don't think I had it written down, but I was prepared for what I wanted to say. That stuff might be replayed down the road, and you don't want to sound like a hick." He paused, and then reconsidered the comment given his childhood in Mount Juliet, Tennessee. "Actually, I don't mind sounding like a hick. I *am* a hick. You want to sound like an intelligent hick."

After thanking Wilmore, Ferguson continued, "You know, the Space Shuttle's changed the way we view the world, and it's changed the way we view our universe. There's a lot of emotion today, but one thing's indisputable. America's not going to stop exploring. Thank you, *Columbia*, *Challenger*, *Discovery*, *Endeavour*, and our ship *Atlantis*. Thank for protecting us and bringing this program to such a fitting end. God bless all of you. God bless the United States of America."

Asked what the gravity of the moment meant to him, the normally jovial Wilmore turned serious and took a few moments to form an appropriate response. Finally, he began:

The United States of America came to a place where we don't have a vehicle we can launch into space ourselves. Many, many reasons combined to put us in that position. You knew it ahead of time, but I realized this was it. There's no more shuttle launches, and there's nothing else for the foreseeable future. I can tell you that the proudest moment of my professional career was when I got out of the van on November 16, 2009, for the launch of STS-129. I walked up and looked at the Space Shuttle sitting on the pad. It's steaming and creaking and groaning—all with that United States of America flag on the side. I can remember it vividly. Realizing that we don't have that anymore, it's obviously going to be a sad thing. You're eager and excited about the future, but the future is years down the road. We don't have access ourselves to the space station we put so much into to build. Certainly, you're going to feel a saddened emotion. There's a lot of pride in this nation coming together and working for a common goal. I've met and worked with many people across this nation who are passionate about human spaceflight who love their job. It's more than a job to them. We've got pa-

per models that we're working toward, but we don't have hardware that we can go out and see and sit in and launch in. That is a sad thing.

On the runway, the four astronauts looked at each other. The last few minutes were a blur as they blazed through reentry, and the hectic last one hundred thousand feet as Ferguson and Hurley very briefly exchanged control of the vehicle. Hurley remembered:

Right after wheels stop and just prior to folks starting to talk, you're looking around and looking at each other. There was that moment, and I think for me, then it kind of hit. It's emotional. It's exciting. You're very satisfied that you did the best you could. It's just a lot of different emotions. It's the cumulative efforts of thousands and thousands of people over thirty-plus years, and it's over. All those things kind of flood in for a few moments. I don't think anybody really said much, other than "Wow" and Chris saying the landing could've been better, but he always *says that. He's such a perfectionist. It was a moment I won't forget, ever. The sun is rising and you start to see the vehicles pull forward towards the vehicle. It was tough.*

Just like that, the Space Shuttle program was over. For the crew of its final sojourn, however, there were times in the coming weeks and months when it seemed as if the flight itself had been the easy part.

What Ferguson called an "aggressive post-flight schedule" was more like an outright tornado of nonstop, often grueling activity. There were highlights, certainly—a tour of the 9/11 Memorial with New York City mayor Michael R. Bloomberg, appearances by the crew on the *Colbert Report*, Ferguson's stint on the *Late Show with David Letterman*, and a meeting in the Oval Office with President Obama. "I think we all get along well—I think," Ferguson said. "We laugh. We complain. I think this aggressive post-flight schedule we have is probably going to be a challenge, but I think we've been around each other long enough to know when we're reaching our limit. We're all still talking to one another, which is a great sign."

There was an emotional letdown after the flight, which was nothing new. There were debriefs to handle, and the processes of getting used to life back on terra firma and getting reacquainted with their families. Mix in a mind-boggling assortment of PAO events, and it was "very, very, very challenging,"

31. The four astronauts who flew the final Space Shuttle mission are welcomed back home to Houston during a 22 July 2011 ceremony at Ellington Field. Although the flight was over, there were times during the next few very busy months when it seemed their feet had not yet hit the ground. Courtesy NASA.

according to Hurley. The STS-135 crew visited every NASA center, sometimes twice. It made an Armed Forces Entertainment tour of Germany, Turkey, and Italy that required a monumental twelve flights in six days. From the day they were assigned to the flight to post-flight, the four astronauts had basically not a single day off. Personally, Hurley readily admitted that the grind was exhausting:

It was surprising that a lot of people didn't realize the toll that it took on us. I think it was more like, "Well, you got to fly the last flight of the shuttle, so you should be thankful." It really has nothing to do with that. I needed a break. The other part of it was my personal situation. It would be a challenge for anybody to have an eighteen-month-old son running around. But Karen had all her travel and the demands of her schedule and mine. Just to be able to bring the chaos to an end and reconnect with your family is something that should be considered, that I think is often overlooked by folks.

During the public-relations blitz, there was strength in numbers. The crew made many of its appearances together, so each knew almost exactly what the other was going through. They bucked each other up every chance they got, and Walheim was especially good on that front. Ferguson remembered that "if we got a little frustrated after an aggressive day on orbit or with perhaps a long line of people to sign autographs for, we always leaned on Rex a little bit because he kept us moving forward. He always reminded us that we have our place, and our place right now is to smile and do our job." For a year, this was a closely knit unit that had formed a bond that would have been hard to break by anything other than maybe time and distance. "We liked each other," Hurley said. "I think we'll be friends for life. Frankly, it was very, very difficult when we finished our post-flight tour, I think, for all of us. Chris said, 'We'll get to hang out and see each other every ten years, anyway. They're bound to have reunions.'"

After everything the crew of STS-135 had experienced together, that somehow did not seem good enough.

10. A Magnificent Machine

What happens next?

At the conclusion of the 135th and final Space Shuttle flight, the longest concerted human spaceflight effort in NASA's history had come to an end. Just two years separated the first and last manned flights of the Mercury program, with another two and a half encapsulating Project Gemini. It took five years to complete Apollo's historic journeys to the moon and back—eight if the Apollo-Soyuz Test Project (ASTP) is taken into account. In all, fourteen years went by between Al Shepard's suborbital Mercury hop and his buddy Deke Slayton's lone jaunt into space on ASTP. They were turbulent years in American history—Vietnam; the assassinations of President John F. Kennedy and his brother, presidential candidate Robert F. Kennedy, in addition to civil rights activists Malcolm X and Martin Luther King Jr.; the hippie movement; and Watergate—but they paled in comparison to the shuttle in terms of sheer length. STS-135 marked exactly thirty years of flights. The program had been filled with some of the utmost triumphs NASA had ever known, and its greatest tragedies. And, in the predawn hours of that July day, it was all over.

For the vast majority of those directly connected with the human spaceflight effort, the end of the shuttle program would have been perfectly acceptable if only something else definite had been on the horizon. There were assurances from Charlie Bolden, the former shuttle astronaut turned NASA administrator, that America was in good shape to maintain its status as a global leader in human spaceflight. On 1 July 2011, exactly one week before the launch of the final shuttle flight, Bolden spoke before the National Press Club in Washington. He told the audience:

Some say that our final shuttle mission will mark the end of America's fifty years of dominance in human spaceflight. As a former astronaut and the current NASA administrator, I'm here to tell you that American leadership in space will continue for at least the next half-century because we have laid the foundation for success—and for NASA, failure is not an option. Once again, we have the opportunity to raise the bar, to demonstrate what human beings can do if we are challenged and inspired to reach for something just out of our grasp but not out of our sights. President Obama has given us a mission with a capital "M" to focus again on the big picture of exploration and the crucial research and development that will be required for us to move beyond low Earth orbit. He's charged us with carrying out the inspiring missions that only NASA can do, which will take us farther than we've ever been.

Targets included landing on an asteroid by 2025 and Mars at some point in the 2030s. Companies from the private sector such as SpaceX and Virgin Galactic stepped up in an attempt to fill the void created by the end of the shuttle program, but only one had ever very briefly put a human beyond the accepted threshold of space. In September 2011 NASA announced that its next heavy-lift launch vehicle would be a $35 billion monstrosity, the "largest, most powerful rocket built," according to NASA's exploration and operations chief, Bill Gerstenmaier. None of that, however, meant much to the NASA and United Space Alliance workers who were being laid off in droves.

Michael Griffin, the NASA administrator whom Bolden succeeded, spent almost four years in the trenches as the head of the agency. He ascended to the position in the aftermath of the *Columbia* accident, and he was there when President George W. Bush announced his Vision for Space Exploration on 14 January 2004. The move signaled the end of the shuttle, but also the start of a Constellation program planned to take NASA back to the moon and on to Mars. It was a plan to get the agency from point A to point B and back again, and Griffin felt that it was on the money:

This came about because the Columbia *accident board indicted, in their words, thirty years of failure by presidents and Congress to provide NASA with a strategic mission. So President Bush said, "Okay," and the White House put together a team of people and said, "Craft a strategic mission for NASA." I think they did an extremely good job at that, I really did. Did I have something that I wanted to do that was different from that plan? No, not really. I think the quality of that plan was further evidenced by the fact that in 2005, a Republican Con-*

gress codified it into law, and in 2008, a Democratic Congress did exactly the same thing. In those years, NASA *was not a partisan issue.*

Not just the moon but Mars as well was within reach. Griffin felt the agency was ready to take on just such a challenge. "I was definitely confident, if we kept on the strategy and the budget," he continued. "Getting to Mars, from 2005, was a three-decade proposition. First, we had to finish the space station, retire the shuttle, rebuild the capability to go to the moon, establish a lunar base, and then we'd be ready to move on to Mars. I thought that the people who laid out that strategy did a remarkably good job, that it was a good strategy, that it was worthy of being supported and sustained."

Although he had already left office by then, Griffin's confidence was backed by the Review of United States Human Space Flight Plans Committee—also known as the Augustine Commission after chairman Norman Augustine, a former chief executive officer of Lockheed Martin. In a series of meetings in the summer of 2009, a panel that included former astronauts Leroy Chiao and Sally Ride met in order to determine a series of recommendations for the country's future in space. Announced on 8 October 2009, the Augustine Commission put forth three courses of action that it felt held the most potential.

There was the Mars first option, in which NASA would make an all-out push to our closest planetary neighbor. There it was, Mars, first and foremost, the supreme prize. The commission made its case:

A human landing followed by an extended human presence on Mars stands prominently above all other opportunities for exploration. Mars is unquestionably the most scientifically interesting destination in the inner solar system, with a history much like Earth's. It possesses resources which can be used for life support and propellants. If humans are ever to live for long periods on another planetary surface, it is likely to be on Mars. But Mars is not an easy place to visit with existing technology and without a substantial investment of resources. The committee finds that Mars is the ultimate destination for human exploration; but it is not the best first destination.

A second and more likely scenario was to start out by heading to the moon, with lunar surface exploration focused on developing the capability to explore Mars. "What about the moon first, then Mars?" the report continued. "By first exploring the moon, we could develop the operational skills and technology for landing on, launching from and working on a planetary surface. In the process, we could acquire an understanding of human adaptation to another world that would one day allow us to go to Mars."

Last was the suggestion that NASA take a flexible path to inner solar system locations, such as lunar orbit, near-Earth objects, and the moons of Mars, followed by exploration of the lunar and/or Martian surface. It was not difficult to figure out that the Augustine Commission favored this third possibility, if not some sort of hybrid of the second and third:

The flexible path represents a different type of exploration strategy. We would learn how to live and work in space, to visit small bodies and to work with robotic probes on the planetary surface. It would provide the public and other stakeholders with a series of interesting "firsts" to keep them engaged and supportive. Most important, because the path is flexible, it would allow many different options as exploration progresses, including a return to the moon's surface, or a continuation to the surface of Mars. The committee finds that both moon first and flexible path are viable exploration strategies. It also finds that they are not necessarily mutually exclusive; before traveling to Mars, we might be well served to both extend our presence in free space and gain experience working on the lunar surface.

President Barack Obama visited KSC on 15 April 2010, and he brought with him a major policy shift. Constellation was behind schedule and over budget, and its *Ares 1* launch vehicle had already been canceled, soon to be followed by the rest of the program. President Obama accepted none of the Augustine Commission's three recommendations in full and, instead, opted to have the country rely on Russia and eventually launch vehicles developed and constructed by private aerospace companies to get to an ISS whose lifespan was being expanded by five years or so. The Orion crew module would be revamped for use as an escape pod from the station in the event of an emergency. He also announced that designs for a Heavy Lift Launch Vehicle would be ready by 2015, with construction to begin shortly thereafter. In his speech at KSC that day, the president did not appear to be the slightest bit concerned about privately held companies taking the lead in human spaceflight:

32. Concept art for NASA's proposed Heavy Lift Launch Vehicle. When and if the project would ever become a reality were the subjects of heated debate as the Space Shuttle program reached its conclusion. Courtesy NASA.

I recognize that some have said it is unfeasible or unwise to work with the private sector in this way. I disagree. The truth is, NASA *has always relied on private industry to help design and build the vehicles that carry astronauts to space, from the Mercury capsule that carried John Glenn into orbit nearly fifty years ago, to the Space Shuttle* Discovery *currently orbiting overhead. By buying the services of space transportation—rather than the vehicles themselves—we can continue to ensure rigorous safety standards are met. But we will also accelerate the pace of innovations as companies—from young startups to established leaders—compete to design and build and launch new means of carrying people and materials out of our atmosphere.*

Elon Musk was just the kind of entrepreneur Obama was looking to in making the announcement. Born in South Africa, Musk founded PayPal and Tesla Motors before turning his attention to Space Exploration Technologies Corporation—SpaceX, for short. Musk, as the chief executive officer and chief technology officer of SpaceX, released a glowing statement in support of the president's plan:

We can ill afford the expense of an "Apollo on steroids," as a former NASA *administrator referred to the Ares/Orion program. A lesser president might have waited until after the upcoming election cycle, not caring that billions more dollars would be wasted. It was disappointing to see how many in Congress did not possess this courage. One senator in particular was determined to achieve a new altitude record in hypocrisy, claiming that the public option was bad in healthcare, but good in space! Thankfully, as a result of funds freed up by this cancellation, there is now hope for a bright future in space exploration. The new plan is to harness our nation's unparalleled system of free enterprise [as we have done in all other modes of transport], to create far more reliable and affordable rockets. Handing over Earth-orbit transport to American commercial companies, overseen of course by* NASA *and the* FAA, *will free up the* NASA *resources necessary to develop interplanetary transport technologies. This is critically important if we are to reach Mars, the next giant leap in human exploration of the universe.*

Buzz Aldrin, who was a little more than a week removed from being booted off the American reality show *Dancing with the Stars*, was also there in support of President Obama's plan. He had long been a proponent of commercializing space travel:

The truth is that we have already been to the moon—some forty years ago. A near-term focus on lowering the cost of access to space and on developing key, cutting-edge technologies to take us further, faster, is just what our nation needs to maintain its position as the leader in space exploration for the rest of this century. We need to be in this for the long haul, and this program will allow us to again be pushing the boundaries to achieve new and challenging things beyond Earth. I hope NASA will embrace this new direction as much as I do, and help us all continue to use space exploration to drive prosperity and innovation right here on Earth. I also believe the steps we will be taking following the president's direction will best position NASA and other space agencies to send humans to Mars and other exciting destinations as quickly as possible. To do that, we will need to support many types of game-changing technologies NASA and its partners will be developing. Mars is the next frontier for humankind, and NASA will be leading the way there if we aggressively support the president's plans. Finally, I am excited to think that the development of commercial capabilities to send humans into low-Earth orbit will likely result in so many more earthlings being able to experience the transformative power of spaceflight. I can personally attest to the fact that the experience results in a different perspective on life on Earth, and on our future as a species. I applaud the president for working to make this dream a reality.

Aldrin was on board with the president, but that was *not* the case for his *Apollo 11* commander Neil Armstrong. Not long after President Obama visited the Florida space coast, Armstrong and fellow Apollo legends Jim Lovell and Gene Cernan released an op-ed in which they blasted the policy shift. Yes, they said, Constellation was behind schedule, but because it had been underfunded. Yes, various entrepreneurs could claim to offer low-cost alternatives to NASA, but at the possible expense of safety and reliability—criticism that brought Musk to tears in at least one televised interview. President Kennedy had boldly presented a challenge to the American people in the 1960s, and they had accepted it en route to becoming the most successful spacefaring nation ever known. The trio of lunar commanders charged President Obama with frittering away that global leadership. "America's leadership in space is slipping," they wrote. "NASA's human spaceflight program is in substantial disarray with no clear-cut mission in the offing. We will have no rockets to carry humans to low-Earth orbit and beyond for an indeterminate number of years. Congress has mandated the development of rocket launchers and spacecraft to explore the near-solar system beyond Earth orbit. But NASA has not yet announced a convincing strategy for their use. After a half century of remarkable progress, a coherent plan for maintaining America's leadership in space exploration is no longer apparent."

A year later, Armstrong and Cernan went before the U.S. House of Representatives committee on science, space, and technology to continue their defense. Armstrong was vexed that America would not be able to launch to the ISS from its own soil for the foreseeable future, calling the situation "lamentably embarrassing and unacceptable." The brain-drain taking place due to cutbacks on manpower throughout the aerospace industry was worrisome, Armstrong charged, stating what countless believed. A significant trend downstream was that with a declining pool of jobs in the industry from which to choose, capable American students were choosing fields other than engineering, science, technology, and mathematics. "Our choices are to lead, to try to keep up or to get out of the way," Armstrong said in his written testimony. "A lead, however earnestly and expensively won, once lost, is nearly impossible to regain."

Although he was not a reclusive hermit, Armstrong had long been known for his reticence in the public eye. Far and away the most famous astronaut of all, when Armstrong spoke with such ferocity on President Obama's plans for America's future in space, it was very much like one of the faces on Mount Rushmore had suddenly opened up and started shouting from the mountaintops.

Armstrong had never focused on the shuttle during his time at NASA, but it had been Wayne Hale's life. The thoughtful e-mails he distributed among his brethren in the spaceflight community became almost required reading, and they morphed into a popular blog. His posts were in-depth, at times moving due to their eloquence, and at times maddening because they were *so* on the mark. He would later joke that he became "addicted" to the process. Before retiring from NASA, Hale concluded one post about the end of the shuttle program with the line "That horse has left the barn." Three years later, after leaving the agency, Hale surmised on his personal blog that not only had the horse left the barn, but "now the barn has burned down." He continued:

It is simply heartbreaking to wander the halls of the NASA human spaceflight facilities and listen to the empty echoes. Yes, the ISS continues on; and the other parts of NASA—science, aeronautics—they continue on. But the exquisite dedicated professional team that made human space flight look so easy is dismembered, dispersed, and nearly completely dismantled. The plan, such as it is, consists of looking for the entrepreneurial heirs of Henry Ford to produce the Model T. The hope is the genius of free enterprise will move us from the horse and buggy era to the gasoline alley era of space exploration. That is a good hope, but in the meantime personally, I would have kept the horse until the automobile appeared. But as they say: hope is not a plan.

Hale was not alone among former NASA higher-ups chafed by what they deemed as the administration's lack of vision. After leaving the agency, Griffin was named an eminent scholar and professor of mechanical and aerospace engineering at the University of Alabama in Huntsville. "The issue is, 'Are we going to have a U.S. government human spaceflight program or not?'" Griffin concluded. "This administration has made the decision that we're not. Certainly, we have retired hardware before. I'm not particularly saddened by retiring one generation of hardware. I am *very* saddened by retiring a generation of hardware with the Space Shuttle with nothing coming after."

It was announced in December 2011 that Griffin would serve as a board member of Stratolaunch Systems, a project developed by Burt Rutan and financed by Microsoft cofounder and billionaire Paul Allen. The company planned to construct a carrier aircraft weighing more than 1.2 million pounds with a wingspan longer than an entire football field, as well as a 120-foot multistage booster capable of reaching low-Earth orbit. To Griffin, the move represented neither a change of heart nor a case of "if you can't beat them, join them." He had always supported private companies' involvement in human spaceflight, just not as it was presented by President Obama:

I don't see how anyone could glean from my actions and statements that I do not favor "commercial space." However, it is true that I am not a big fan of the Obama administration's policies, which are mislabeled as being in support of commercial space. It's not "commercial" if you're giving out money in advance to preselected contractors in a private-public partnership. "Commercial" is when people develop their own products on their own money and, if successful, put it on the market—any market, government or otherwise—for whatever the traf-

fic will bear. That's not what is going on today at NASA. The administration is not developing "commercial space". They are providing development money to a different set of contractors, under different rules, than has been the case in the past. It is true that I do not support their plan. That is not at all the same as not supporting "commercial space." Further, to support policies designed to promote commercial space ventures—something of which I am strongly in favor, if done properly—is emphatically not the same as saying that NASA should not be allowed to develop a successor to the shuttle, or that NASA cannot provide astronaut access to space. To support commercial space is one thing; to be hostage to the market for a strategic national capability is quite another, and is a policy to which I object.

Before *Challenger*, the official designation for all things shuttle was the National Space Transportation System Program. It was truly a space truck for hire, with a plan for almost nonstop service to low-Earth orbit. The shuttle was to be a means by which an American payload would be launched, and it would take those originating in other countries—along with their personnel and participants—into orbit as well.

After *Challenger*, it was simply the Space Shuttle program. That meant no more foreign participants. No more congressmen in space. No more teachers in space for the time being. The drastic moves cut down dramatically on the shuttle's marketing base, and, ultimately, it meant that the program could not sustain a high enough flight rate to keep costs for each mission low enough. Still, Bruce McCandless would always be amazed at the technical sophistication of the vehicle, what with its five computer systems, redundant thrusters, and engines and all. "It's proven to be a marvelous platform for doing things like serving the Hubble Space Telescope," McCandless responded, and then added what countless were thinking. "I think that we'll look back and wish we still had the thing flying, even at a low rate or as an auxiliary capability."

David Hilmers flew four shuttle missions in all, two of which were DoD flights he could not discuss to much extent if any. He was on the first bird to fly following the sadness of *Challenger*, and he eventually served on an almost countless number of medical mission trips due in very large part to the things he had seen from on orbit.

Hilmers had been selected as an astronaut in July 1980, nine months before the shuttle's first flight, so he had been around since its infancy. The program had not turned out the way many expected, in some ways. It was going to be a space truck, which it was, but it did not fly nearly as many times as originally predicted. Twenty-four missions or more a year? Those were the days, all right. As unrealistic as those particular expectations were, Hilmers knew that the spacecraft had far exceeded others:

You see the types of things that we developed during the course of the Space Shuttle program. I don't think we ever envisioned that we could build a space station like we did. Some of the things that were done on spacewalks, I think, were beyond what we thought. Rescuing satellites, I don't think, is something we thought we could do. It turned out that the Space Shuttle had an incredible amount of flexibility in what it could do and what we could make it do. Despite the two accidents that we had, it's a pretty good safety record for something as complex as the Space Shuttle. It flew a lot of people into space very safely. So on one hand, we were disappointed with the number of flights we were able to fly, partly because of the accidents we had, partly because of budget constraints, and partly because it turned out to be more complex than maybe we thought. But on the other hand, it turned out to be magnificent in the sense of what it could do, what we learned, and the type of innovations that we developed.

Hilmers joined the chorus of those frustrated with the lack of a plan on NASA's part. Yes, there were promises, but it quite simply is not possible to fly a promise into space. "I'm not sure that the decisions about the directions of the space program are being made by people that do it for the right reasons," he surmised. "Maybe it's driven more by budget, rather than what makes a logical goal."

Upon taking office in February 2009, President Obama made headlines with bailouts of automobile manufacturers Chrysler and General Motors and a massive effort to reinvigorate the country's economy. They were some of the most controversial decisions in the history of American politics, and Hilmers was not so sure they were the right moves to make, especially at the expense of throwing decades' worth of spaceflight experience out the window:

I'm terribly frustrated, too, by seeing a lot of money being spent on economic stimulus, for example, that doesn't have a lot of long-term benefit for anybody. What can be a better use than something that's going to pay benefits for generations? If we need to spend money to stimulate the economy, why can't we use it for a program that everyone has seen a benefit from and that really excites young people? We could be throwing away all this wonderful experience that we've had over the last thirty years of the Space Shuttle program and before that, and losing personnel that have been there and that have all this experience. We could just be throwing that away. If you think the space program is expensive, try doing it from scratch and basically restarting it with new people, with brand-new equipment, after years of not having a manned space program. I don't see a good outcome of this. I think many of my peers and people that are in the astronaut office now probably feel the same way—very frustrating.

Hilmers left NASA in 1992, and Charlie Hobaugh was about to when *Atlantis* lifted off for the final time. Hobaugh carried himself with an air of supreme confidence. Tall, square-jawed, and perfectly postured even when he was sitting down, he came across as the very definition of what it means to be a member of the United States Marine Corps. As he strode through the KSC press site on the morning of the last shuttle launch ever, it was Charlie Hobaugh's world. Everyone else just happened to be living in it.

The shuttle's legacy, he began, was one that would be unsurpassed for the foreseeable future:

It was a vehicle ahead of its time. It'll probably not be matched in capability any time soon. The payload and multi-role capability of the shuttle is something that will never *be realized again in the near term. We would be able to bring up fifty thousand pounds or so of external hardware. We could deploy satellites. We could deploy telescopes. We could repair equipment that had been in space. We could retrieve things and we could bring large items back and analyze them, whether it's an experiment or a failed piece of hardware that you want to understand why it broke. I think the Space Shuttle had capability—spacewalks, robotics, everything else—that's going to be hard to beat.*

Three times, the native of Bar Harbor, Maine, flew to the ISS—twice as pilot and the final time as commander. Those were the tasks he had been given, and he took to them with great aplomb. Like many others in the astronaut office, however, he had hoped to go even further—out of low-Earth orbit. "I think [the shuttle is] still a fantastic vehicle that will never be equaled in the near term," Hobaugh said. "I think it's important that we rededicate ourselves to going beyond low-Earth orbit, which is hopefully what [the end of the shuttle program] will allow us to do. It'll be a shame if it doesn't, if we just end the shuttle program and don't get out of low-Earth orbit. Ultimately, that's where I hope we end."

Was it a goal Hobaugh felt could be reached? It had been done in the past, during the glorious days of Armstrong, Aldrin, and Apollo. "We know we can do it—it's just that we haven't done it in a really long time," he continued. A launch vehicle needed to be designed and constructed in order to re-create the incredible Saturn V. He hoped that low-Earth orbit and commercial competition would be used as stepping-stones to get where he felt NASA needed to go, at the same time as it developed the kind of crew vehicle needed to get there. The goals were obtainable, but it was not going to take place very fast:

Time is going to be the thing that's going to be most hurtful to us, in that as we stand the shuttle down, we have no vehicle to fall back on, to just immediately step into and go into low-Earth orbit. That was something that was recognized a long time ago. This is not a surprise. It's an unfortunate outcome of not having a big enough budget to be able to do simultaneous programs. We're basically having to stand one down to be able to bring another one online. Unfortunately, that gives us that gap, where we don't have anything.

Sitting in a small press-site conference room at KSC, Hobaugh was a few months shy of his fiftieth birthday. "We're looking forward to hopefully stepping out into the future, and ultimately get beyond low-Earth orbit, which was something I hoped would've happened during my prime years as an astronaut," Hobaugh said with a smile. "I feel like I'm starting to get up there as I broach fifty. I don't know. Maybe fifty is the new forty." He would never fly the shuttle again, certainly, so how long could he reasonably afford to wait for President Obama's space policy to shake itself out? He retired from NASA in September 2011, just one of many spacefarers who left the astronaut corps as the shuttle program was ending. They were not laid off like so many of the talented workers who had trained them and prepared their vehicles, but the number of flights dwindling and hitching a ride to the ISS an "iffy" proposition at best, why stick around?

The shuttle itself was never far from the public's conscience during its three decades of flight. In the news, the program's exposure ran the gamut—from fairly positive to quite negative to downright sensationalized. Coverage was nonstop in the days and weeks following the loss of its two crews in flight. John Glenn's return to orbit received generally favorable reviews—although his role in the flight did have its fair share of critics.

Two other astronaut-specific storylines were complete polar opposites as well, one becoming one of the most bizarre incidents NASA had ever faced while another was one of its most heartwarming. The saga of STS-134 commander Mark Kelly and his wife, Gabrielle Giffords, was reported by countless unlikely sources. It was hard not to be touched by the story, just as it was difficult not to have an opinion on Lisa Nowak.

Nowak, an STS-121 mission specialist, was arrested at the Orlando International Airport on 5 February 2007 and subsequently charged with the attempted kidnapping and attempted murder of Colleen Shipman, a romantic rival for the attentions of STS-116 pilot Bill Oefelein. Nowak and Oefelein later became the first two astronauts ever fired. More than two years after her arrest, Nowak pled guilty to charges of felony burglary and misdemeanor battery and was sentenced to time already served and a year of probation. At the time a captain in the navy, Nowak retired in 2011 with an "other than honorable" discharge and was docked one pay grade. The whole episode was sensationalized in gossip magazines and the insatiable twenty-four-hour news cycle and became fodder for punch lines around the world.

Hollywood framed the shuttle as only Hollywood could, in heroic yet most of the time utterly unrealistic terms. The middling 1986 flick *Space Camp* featured five teens and an instructor who were accidentally launched into orbit on board the shuttle and somehow miraculously returned safe and sound. Released less than six months after the *Challenger* accident, it was one of those low-budget affairs where the actors moved and talked slowly in order to simulate weightlessness. Twelve years later, the Bruce Willis–Ben Affleck–Liv Tyler blockbuster *Armageddon* cost an estimated $140 million to make, and while spectacular, its portrayal of the shuttle was almost laughable in its unreality. That two shuttles were launched within minutes of one another was one thing, but then one actually landed on the big, bad Texas-sized asteroid that was headed toward Earth. It carried a crew of misfit drillers and must have had a gravity generator as well. The list could go on. In terms of thrills a minute, it was the greatest of popcorn action adventures. Nothing more, nothing less.

On television, the shuttle generally fared much better in terms of the respect it was given. NASA in general and the shuttle in particular was a running theme on the classic American television series *The West Wing*. In the last episode of its first season, White House communications director Toby Ziegler, played by Richard Schiff, had a brother who was an astronaut stranded in orbit on the shuttle. Although the always-unseen sibling was able to land safely, he later committed suicide. In one of its biggest storylines, Ziegler was fired in the show's seventh and last season after disclosing to a reporter the existence of a military Space Shuttle. In the series finale, Ziegler was pardoned.

On the 1990s *Home Improvement* situation comedy, the crews of STS-61 and STS-73 visited the fictitious *Tool Time* set after astronaut Ken Bowersox and lead actor Tim Allen developed an off-screen friendship. "I know people know that there have been astronauts on *Home Improvement*—in fact, Ken Bowersox went back a couple of times, I think," STS-61 commander Dick Covey said with a laugh. "But I don't think they knew who we were, though. I think we're most known for fixing the Hubble." In another episode, Allen trained to become an astronaut himself. On *The Big Bang Theory*, a sitcom that debuted on American television about a decade after *Home Improvement* went off the air, geeky engineer Howard Wolowitz was sent into a panic when he discovered a flaw in his design of a space toilet the shuttle was about to deliver to the ISS. Wolowitz was launched to the ISS himself later in the series, alongside real-life astronaut Mike Massimino.

In the United Kingdom, the fantastic BBC *Top Gear* car review show got in on the action when presenters James "Captain Slow" May and Richard "Hamster" Hammond attempted to fashion a shuttle—complete with External Tank and side boosters and remote-controlled landing system—from a three-wheeled Reliant Robin. The launch went exceedingly well, the boosters detaching on cue, right up to the point that the vehicle refused to separate from an External Tank that had to have been every bit of twenty feet tall. The stack crashed into the ground in a huge ball of flames, leaving presenters May and Hammond in comedic disbelief.

Mainstream documentaries presented the shuttle in a more realistic light, though some were more objective than others. One could almost not help but stand and cheer following a viewing of *Hail Columbia!*, *Destiny in Space*, *Blue Planet*, *Mission to Mir*, *The Dream Is Alive*, *Space Station*, or *Hubble 3D*.

Television news legend Walter Cronkite narrated one, while A-list actors Leonard Nimoy, Tom Cruise, and Leonardo DiCaprio handled voice-overs for others. Such high-profile talents were just one sign of the finely honed productions that these projects became. Each played up—for lack of a better term—the "golly, gee whiz" wonders of human spaceflight, with little or no mention of its real-world dangers. They were not exactly NASA "infomercials," but they were close.

Filmmakers Renee Sotile and Mary Jo Godges pulled out all the stops for their unsurpassed 2005 project *Christa McAuliffe: Reach for the Stars*, which was eventually aired on CNN. Actress Susan Sarandon provided narration and singer/songwriter Carly Simon composed songs, becoming in the process examples of the impact McAuliffe's story had on the public at large. In making the film, Sotile learned that on board *Challenger* was a cassette tape of Simon's music, which was some of McAuliffe's favorite. She sent Simon's management a letter in which she explained the teacher's love for the artist's music, a contact attempt Sotile called "totally a shot in the dark."

A month or so later during a late-night editing session for the film, they received a call from none other than Simon herself. "It was unbelievable," Sotile said. "She not only wrote an original song for this film, but she gave us other music as well, all from the goodness of her heart." And then there was Sarandon, whom Godges and Sotile met during a screening of one of their other projects at the Harlem International Film Festival. "She began reminiscing about where she was at the time of the tragedy and asked us who was narrating it," Godges reminisced. "I said, 'How about you?' She gave us her telephone number and three weeks later, we were recording her audio. The only thing she asked of us was to give Christa's family her love."

The Traipsing Thru Films effort became one of the most riveting ever produced on a shuttle-era astronaut. It captured to perfection McAuliffe's delightful exuberance during training contrasted with the stark reality of her untimely death and its impact on her family. Like so many others, Sotile called the *Challenger* accident her "JFK moment" and years later wondered if there might be some sort of documentary on the famous teacher's life. There was not. Envisioning a tribute to McAuliffe when she and Godges began work on *Reach for the Stars*, the project became something much more than that. "It still clearly shows that she was an American hero and a pioneer ahead of her time," Sotile related. "I didn't even know that about her at first. I just knew she was a schoolteacher. It just turned into something much more emotional and bigger and real. The intimacy of the grieving came into it, that I didn't realize was still so prevalent in their feelings."

Sotile and Godges stayed with McAuliffe's mother, Grace Corrigan, during the making of the film, and they talked to her two sisters, Lisa Bristol and Betsy Corrigan. Steve McAuliffe, Christa's husband, and many in Concord, New Hampshire, the town in which they lived and worked, declined to be interviewed for the project. Sotile called the difference in how the two sides of the family handled their grief "the double tragedy in this story." Said Godges, "Doing research on it, there were some other documentaries and articles, but it always ended with, 'Her husband leads a quiet life.' They didn't really talk about her kids. One of our first questions to Christa's sister was, 'Did you become a second mother to the kids?' I would have assumed that would've happened naturally, and she said no. Then she started opening up on what happened. She was like, 'Steve didn't really need me.'"

In the end, *Reach for the Stars* was the crown jewel in the Traipsing Thru Films catalog, "the single most important thing we will ever do," according to Sotile. Godges added, "We had to create a movie that did justice to Christa's life, one that her family, friends, colleagues, teachers, students, and future students would say, '*That's* Christa.' I knew if we could set the tone of the movie to match Christa McAuliffe's, we would be on target. Even the title, *Reach for the Stars*, is something she always said. Her pioneering spirit was a guiding force from the get go."

Another guiding force in the five-year odyssey to make the film was Barbara R. Morgan, McAuliffe's Teacher in Space backup who became a full-fledged member of the astronaut office in 1998. She flew on board STS-118 in August 2007, with Sotile and Godges watching from the press site. Morgan was the very first person they had interviewed for *Reach for the Stars*, and it was a powerful thing for them to see their friend continuing what she and McAuliffe had started more than two decades earlier. "From the very beginning, Barbara Morgan was sort of our cheerleader," Sotile began. "On launch day, it was exciting and terrifying. We could just imagine her sitting in the mid-deck. We were surrounded by seasoned journalists who had covered the *Challenger* explosion. Needless to say, there wasn't a dry eye on the lawn that day. When the announcer said, 'Go with throttle up!' and *Endeavour* continued to soar was a moment we could breathe again."

Another Traipsing Thru shuttle-related release, *Astronaut Pam: Countdown to Commander*, had a far breezier tone. The filmmakers became aware of Pam Melroy long before *Reach for the Stars* ever saw the light of day. While watching television in California, they heard the twin sonic booms when Melroy and the rest of the STS-92 crew were on final approach to landing at nearby Edwards Air Force Base on 24 October 2000. "The TV said, 'That was the pilot, Pam Melroy!'" Sotile remembered, to which Godges added, "It was so bizarre. We all stood up and screamed. We were like, 'A girl did that, right?' I didn't even know who she was." Then, when they met face-to-face during the filming of the McAuliffe documentary, Godges and Sotile found Melroy to be "a very cool chick who proves that girls can do anything on and *off* the planet." She dazzled them with her ability to relate the most technical of details in ways they could very easily understand. Godges remembered:

The way she explained what she was doing in space and all these different projects she was involved with, I understood what she was talking about. A lot of people are missing out on what's going on in the space program, probably because they just don't get it. But the way she explained everything was so exciting and simple, but impactful at the same time. The bottom line was how she described things. She just made them so you could relate to them. One example, she said that on ascent, it feels like your cat's gained 400 pounds and is sitting on your chest. Everyone with a cat can relate to that.

Melroy was at first reluctant to give her consent for the project to proceed, because, she conceded, she does not consider herself to be a "big limelight kind of person." Her mother, however, stepped in to convince her to go ahead with it. "It felt a little bit it was going to be 'I love me' kind of stuff," Melroy said without the slightest trace of false modesty. "It didn't feel comfortable to me, that there was enough to make some big story about me." Did she have a point? What *was* there to talk about, except a career in the air force that included combat tours and three spaceflights that saw her become only the second female shuttle commander in the history of NASA? In the end, she was pleased with the outcome, in that it focused on the shuttle program as a whole and not just her story. "I think that they did a really good job of just using some of the story that I told to educate," Melroy said. "I was really, really happy with it, actually. I was very glad that my mom encouraged me to do it."

The finished product was thirty-nine minutes in length, and broken up into shorter "countdown" segments that could easily be shown to school groups separately or as a whole. Even after completing McAuliffe's story, there remained so much Sotile and Godges did not know about the space program. Working on the projects put NASA and its accomplishments into a much clearer perspective. Sotile admitted, "I didn't realize the importance and significance of the space program. Until I started making the Christa film, I was just kind of mad at NASA. Then, I saw the importance of it to our everyday life and our technology. I saw how much it means to all the people who are astronauts, including Pam. They're not just following their dreams, which is great, but they're risking their lives for the good of all mankind. But Pam and Christa were still exceptional people before they got into the space program."

Forget for a moment the cultures in and around the human spaceflight workforce that helped lead to two fatal accidents, or what NASA should or could have been doing in the nearly four decades following the end of Apollo. Never mind the fact that absolutely none of the pummeling NASA took in the wake of the discovery of the Hubble Space Telescope's incorrectly ground mirror was the fault of the flight crew that deployed it or the steed upon which they rode to orbit. Disregard the political posturing that took place over whether the shuttle should continue to fly or not. Recognize that to enjoy Hollywood's version of the shuttle required a definite suspension of disbelief.

Push all of that to the side and consider the Space Shuttle itself, an extraordinary conglomeration of millions of separate components. It most certainly had drawbacks, that much is quite obviously true. But as a mode of transport to low-Earth orbit, it had no equal. Perhaps the best person to explain just such a dichotomy was Michael Griffin, who in no way, shape, form, or fashion had his wagon hitched solely to the shuttle. He had actually gone so far as to criticize the Nixon administration for killing the Apollo program to concentrate on using the shuttle in low-Earth orbit, which he felt ruined a prime shot at proceeding from the moon onward to Mars. "Working in low-Earth orbit was not bad," he said. "Working *exclusively* in low-Earth orbit was bad. I spent some time analyzing what we could have done had we used the budgets we received to explore the capabilities inherent in the Apollo hardware after it was built. The short answer is we would have been on Mars ten or twenty years ago, instead of circling endlessly in low-Earth orbit." For Griffin, the winged spacecraft was never the ultimate end goal when it came to human spaceflight, nor should it have been. Instead, the shuttle was a means by which NASA could extend its capabilities before moving out further into the universe. He was fully aware of its limitations.

According to Griffin, "The shuttle was by no means as safe as its designers intended. That's just a fact." Second, "It was not able to fly as frequently as its designers intended." Finally, he reeled off a third and probably most important drawback when it came to what determined the shuttle's future. "It certainly was not as cheap to operate as its designers intended. Those things are all true," Griffin conceded, but almost in the same breath, he offered a decisive "*However . . .*" and a quick rebuttal to those concerns:

The shuttle was, if you will, a second-generation step in developing the art and science of learning how to fly in space. Exactly what did people expect? That we were going to figure it out perfectly, starting out in 1971 and '72, only a decade after the very first human spaceflight had been done? I mean, we've been flying airplanes for a hundred years and we still figure out subtle ways to kill ourselves with them. The human race has been doing open-ocean maritime voyaging for over a thousand years, and we still lose ships. Starting the Space Shuttle a decade after the first-ever human spaceflight, anybody who thought we were going to figure it out and get it just right was drinking too much Kool-Aid. When I look at the legacy of the shuttle, I look at it in a positive tone. I look at it and say that 90 percent of all people who have flown into space have flown on board the Space Shuttle. This is a vehicle that really allowed us to learn how to live and work in space. This is a vehicle that really allowed us to learn how to do EVA. It's a vehicle that allowed us to build the space station.

To those fortunate few who were able to ride a shuttle into orbit, the Space Shuttle was the most noble of beasts. "Even though some people say that it was a mistake, that it cost too much or it was risky," started astronaut Jean-François Clervoy, "for me, the main thing that will remain is that it will be the most amazing machine invented by humans. To say it another way, the intelligence of humankind will never be as concentrated in one machine as this was. It's an amazing machine." The shuttle was a launch vehicle. It was a satellite and space station in and of itself. It was a platform from which spacewalks, robotics, and rendezvous could all be conducted. It was a cocoon able to withstand the heat of reentry and an airplane on landing.

For Clervoy, although its technology was some thirty years old, it would remain for a long, long time as the best combination of good, old-fashioned know-how in any number of areas—electronics, computer and material science, mechanical, structure, propulsion, telecommunication, and physics.

Clervoy was on a roll. Quite possibly best of all was the shuttle's aesthetic quality:

When you see it landing as a glider, it's just beautiful. It's a piece of metal that comes back from space and touches the earth just as a kiss landing, and that is probably the strongest memory for me. It's how space lands on Earth—as a kiss. There won't be any ship like this for a long time. Suborbital flights will give access to space for a few minutes to people, but from a technological and operational point of view, it is far closer to aeronautics than to astronautics. It's closer to an aircraft than to a spacecraft. So it is not as amazing as the Space Shuttle.

One by one, the astronauts continued to answer the relatively simple question—in the long run, years after all of the orbiters have been retired to their respective museums, what will be the legacy of the Space Shuttle program? It did not take long for two major themes to arise. First was the fact that it was able to deploy, save, and then service the Hubble Space Telescope. It also became a high-tech transfer truck to take construction workers, equipment, and pieces to the International Space Station. However, the Space Shuttle was more than just those forty-three flights.

When the shuttle program ended, Bill McArthur was NASA's orbiter project manager—the second-best job he had ever had at NASA. Letting go of the shuttle was an emotional experience, but one that caused those in his office to ponder the program's contributions. Surely, future generations could point to its advancement of technology, the science that was gained, the medical research that was conducted on orbit, the Hubble Space Telescope, the International Space Station, all of that. McArthur, however, came to the conclusion that the true meaning of the shuttle era was even more fundamental. He was about ten years old when Yuri Gagarin flew and a new cadet at

West Point when Neil Armstrong and Buzz Aldrin walked on the moon. As incredible as those accomplishments were, they were *visits* to space:

Human beings would leave our planet and spend a few hours, a few days, or a few weeks in space and then return. The legacy of the Space Shuttle is that it marked the transition from people believing that human beings really didn't belong in space to us viewing ourselves as a spacefaring race. Since 2000, there has been a permanent human presence in space. If we are successful, there will never be another time in human history when we don't have people living off of our planet. I think that's the legacy of the shuttle. Kids today think that human beings belong in space, that spaceflight is almost routine. It's not. It is anything but routine, but the human race views itself as no longer confined to one planet.

Dick Covey took part in two of the highest-profile missions shuttle crews ever attempted. He piloted the agency back into orbit on STS-26, the first flight following the sadness and shock of *Challenger*, and Covey also commanded the first Hubble servicing mission during which he and the rest of the STS-61 crew saved NASA's keister. Later he became president and chief executive officer of United Space Alliance, and for Covey, the shuttle's place in the annals of aviation was assured by what it was as much as what it did. "It may be some time before we ever have another winged spacecraft," Covey began. "The Space Shuttle is the first winged spacecraft that flew 135 missions, went to orbit, and then came back and landed on a runway. Someday, people will look back and say, 'We flew spacecraft to a runway?!?' I think its uniqueness from that standpoint will be what distinguishes it."

Jeff Hoffman was the first person to spend one thousand hours of flight time on the shuttle, and he was a member of Covey's crew that corrected Hubble's flaws. He was a shuttle astronaut, and while he had "very warm feelings about the shuttle," he was also very aware of its limitations as well. The shuttle was instrumental in building the ISS, yes, but that was the way it had always been designed to work. From its earliest concept stages, the station was to have been built piece by piece, with the various components trucked to low-Earth orbit by the shuttle. However, NASA once had a massive Saturn V vehicle that launched the *Skylab* station as a whole. The shuttle "allowed us to do *extraordinary* things in low-Earth orbit," he enthused, but at the same time, "that's not where we want to end up." Hoffman added:

I think we're doing the right thing by retiring the shuttle now that the station is finished, because frankly, once the station is finished, we're going to go back and use the shuttle for scientific flights. We're not going to use it the way that we used to use it. To keep flying the shuttle just as a transport for people back and forth to the space station, I think would be problematic, given the limitations on NASA's budget and all the other things we'd like to do. I think it's the right time. It's harder and harder to keep doing it safely. It's getting older. I think as sad as we'll be on the day it makes its final return to earth, I think that's the right decision.

The time might have been right for the shuttle to sail off into the sunset for good, but like so many others, Hoffman appeared perplexed by a lack of planning for what was to follow:

I think what's a real shame is that ten to fifteen years ago is when we should have been planning for the shuttle replacement. We perhaps could have built a smaller version of a reusable shuttle that could have taken advantage of what we've learned from the current Space Shuttle, to develop a more affordable, more robust system that would serve the transportation purposes that we have now in the twenty-first century. But we didn't do it. I think we'll look back on the shuttle era as having been a very extraordinary time where we really developed a mastery of operating in low-Earth orbit. Whether or not we're going to want to re-create some of those capabilities remains to be seen. We certainly are going to be giving up a tremendous range of capabilities to do things anywhere. We can still do spacewalks. We can still use the manipulator arm. We can still do scientific experiments on the space station, and that's great. But as far as working other places, working on satellites and so forth, we'll be giving that up with the Space Shuttle. But I think on the whole it's the right decision.

After leaving NASA, Hoffman headed back to MIT, where he became a professor in the department of aeronautics and astronautics and director of the Massachusetts Space Grant Consortium. The roles kept him in close contact with NASA, so much so that the day after the interview for this book, Hoffman took part in the final stage of a joint venture between the agency, MIT, and local middle schools. Students from the Massachusetts Afterschool Partnership wrote programming for small robots that resembled the famous remote droid from *Star Wars: A New Hope*, which was loaded and then tested in real time on the station by *Expedition 25* flight engineer Shannon Walker.

It was precisely the kind of work Hoffman enjoyed doing on orbit, but because the shuttle did not fly as frequently as originally intended, that cut down on its scientific return. Although there had been fifty or sixty flights devoted to methodical research, relative few were focused on, say, life sciences. Therein was the problem:

In laboratory science, you've got to do things over and over, repeat experiments and be able to make changes. But at least we learned how to do science in space, the proper way to do it. It certainly set the stage for designing science experiments on the International Space Station. The difference with the space station, of course, is that once we put the equipment up there, it's very hard to change it. The advantage is that you're up there for a long time, so you can do experiments many times. Unlike the shuttle, where you can optimize something as you learn things and continue to make improvements, on the space station, what you have is what you get. There's a little bit of advantage in doing it each way.

Hoffman went on to theorize that one of the shuttle's most important innovations was to demonstrate the utility of interaction between humans and robotics. The New Yorker had been on the end of a robotic arm during spacewalks on STS-51D and STS-61, so he knew full well the extremely precise nature of what it took to be maneuvered around on one:

This, I can't emphasize enough. A lot of times, people talk about, "Should we send humans in space or robots?" I always like to point out that it's when humans and robots can interact that we'll get the best results. That's exactly *what we did in the use of the robotic arm in missions such as Hubble, for instance, or in the construction of the space station. We did things that astronauts on their own would not have been able to do, but the robotic systems on their own would also not have been able to do. Working together, we accomplished things that neither humans nor the robotics systems could do on their own. I think that is an incredibly important heritage of the robotic arm on the Space Shuttle, and now, of course, on the space station.*

Finally, there was Hubble. "In overall scientific impact, you can't do much better than to talk about Hubble," Hoffman continued. "Hubble would have been impossible without the shuttle. Every statistic that you look at shows Hubble as NASA's most productive scientific mission. More scientific papers have come out of Hubble than just about all other NASA science projects combined. That's certainly the thing that I'm most proud of."

Although Steve Robinson was an astronaut who spent nearly 1,200 hours and traveled 19.8 million miles in space during four shuttle missions, he was probably best known for yanking a small strip of gap filler from *Discovery*'s belly, much like Androcles gently pulling the thorn from a lion's paw. The importance of the shuttle was actually two-fold, Robinson postulated:

The most obvious one you can see up in the sky. The International Space Station could not have been built without the Space Shuttle. The Space Shuttle was designed for two reasons—to bring up large satellites and to build a space station. It did bring up a number of large satellites. I guess the one we're all most familiar with is the Hubble Space Telescope, but it was by no means the only one. But almost the entire space station was put together piece by piece after being brought up on the shuttle, with very clever use of robot arms.

The station was composed of modules that had been built all over the world, in many different languages, and somehow they all fit together. That was the technical legacy of the shuttle. Aside from hardware, its parts and pieces, Robinson marveled at the wisdom gained by putting them all together:

There's a legacy of knowledge that comes in maybe two parts. We learned to do all this difficult engineering stuff in space, and that's brought us to a state of technical expertise that we wouldn't have been at had we not done this. It's been a technology engine for us like no other. The other part of the knowledge legacy is looking forward, into the future. The space station is just now opening its doors for scientific investigation. There's never *been a laboratory on Earth built for the range, the variety of types of scientific investigations that can be conducted on the space station. In the history of humankind, it's absolutely unique and it's just getting started. So I think maybe the legacy of the shuttle is yet to be written.*

Andy Thomas grew up in Australia dreaming of becoming an astronaut. He watched Wernher von Braun in "Man in Space," a legendary episode of the *Disneyland* American television series that helped foster an interest in human spaceflight. The episode first aired on 9 March 1955, and even back then, the German rocket designer had conceived of a reusable launch vehicle. So, for Thomas, the shuttle was von Braun's final legacy in NASA. "I can remember him showing these big rockets that he had in mind, with space planes on the top of them," Thomas remarked. "That was the shuttle."

Aside from the idealism of the Disney program was the fact that access to space is both difficult and risky. "The shuttle never lived up to original expectations, but I think we now recognize that's not a shortcoming of the shuttle," Thomas said. "It's a shortcoming of those expectations, of not really understanding that those weren't realistic expectations—that it was a rocket you could launch every week or two, carry multi-ton payloads and crews, and do all these EVAS and these robotics and then land, dust it off, and fly again next week."

Eileen Collins, the first female commander of a shuttle crew and the leader of Thomas's and Robinson's STS-114 flight, agreed that the ISS and Hubble would always and forever remain large parts of the program's heritage. Along with that, however, was the fact that NASA figured out how humans could repeatedly gain access to low-Earth orbit. None of the three, she added, was more important than any of the others—they were too different to distinguish in such fashion. "We learned how to do it right and how to do it wrong," Collins said. "Despite the fact that there were two accidents, I'll say it over and over again that the Space Shuttle program was tremendously successful. In my mind, it was a test program. It will be successful in history if we learn from it."

She could envision students in the future taking courses on the history of spaceflight and the shuttle. "I contend even if you're going into finance, or banking, or teaching, or any profession, or if you're a parent, that you can still learn lessons from *Challenger* and *Columbia*," Collins said. "There's so much there to learn on human nature, not just on the technical side of it, but on how we make mistakes working together as people—maybe our ego and our arrogance. As much as I've read on *Challenger*, I'm still learning new things. People had the way that they were going to do something, and by God, that's how they were doing it. They were wrong."

More than anything else, astronaut Ron Garan felt that the international cooperation fostered by the station program was the shuttle's most important contribution. "History will show this," Garan began. "I truly believe that the International Space Station is the most complex, complicated structure ever built, and it was built in orbit by fifteen nations. A lot of these nations don't have a real good track record of friendship together. It's really remarkable from a political standpoint that we were able to pull this all together. People are people. It doesn't matter what your nationality is. If you're all working toward a common goal, then you find common ground. You collaborate. You share knowledge. You work together."

There were indeed many, many lessons to be learned from the shuttle era—in engineering, science, and business management. There would always be a right way to do things and a wrong way, and NASA had experienced them all over the thirty years that the shuttle flew. It was with a deep sense of nostalgia tinged with no small amount of sadness that crews in the orbiter processing facilities at KSC started preparing *Discovery*, *Atlantis*, and *Endeavour* for display at the museums that would serve as their final resting places.

Discovery was the first to leave, in April 2012, bound for the Smithsonian Institution's National Air and Space Museum Steven F. Udvar-Hazy Center in Chantilly, Virginia. The test vehicle *Enterprise*, which had been on display there, went to the Intrepid Sea, Air and Space Museum in New York City.

Endeavour was placed at the California Science Center in Los Angeles, after being towed through the city's streets to great fanfare. The event was even featured in a handful of Toyota commercials.

Atlantis remained closest to home, at the Kennedy Space Center's visitor's complex.

There was second-guessing aplenty concerning the locations—the Smithsonian and KSC were no-brainers, but how in the devil did Houston and JSC get shut out? Yet whether the choices were good ones, bad ones, or none of the above was irrelevant. There were no firm plans in place for what came next, so NASA's course of action was wide open for debate. Should the shuttle have continued flying, or was it time to shut it down and move on to the agency's next great human spaceflight era?

Only time would tell.

Epilogue

As long as human beings had strapped themselves to rockets, figuring out what to do next when they landed safely for the final time had always been a life-changing, crossroads moment. It was hard to imagine anything ever coming close to matching such an experience, much less going beyond. As much as they might have liked, there was no way to stay on orbit forever. The thousands of space workers who lived vicariously through the astronauts faced similar issues—how could anything ever top working on the Space Shuttle?

Charlie Bolden led the agency out of the Space Shuttle era.

Many went into NASA management, flying a desk rather than the shuttle. Some, like Tom Akers, retired to Small Town USA.

Dave Hilmers became a medical missionary in troubled locations all over the world.

Scott Parazynski reached the summit of Mount Everest a year after being forced to turn back due to an injury.

While waiting his turn to become an army general, Doug Wheelock flew in support of military operations overseas.

Their abilities were still intact, and they had to be channeled somewhere. It was time to move on to other challenges.

David Hilmers had always been very deeply moved by the view from orbit. How could something that looked so beautiful from so high up actually be home to so much dire misery? Hilmers would do his part and return to the dream of his childhood to become a doctor. After his third flight, STS-36 in early 1990, Hilmers had his game plan laid out.

Although he was offered a spot on a subsequent research mission, Hilmers turned it down in favor of beginning his career in medicine. His first night class came on 17 January 1991, the very night that American forces began leading a heavy aerial bombardment of Iraq's military infrastructure following the country's invasion of Kuwait a few months earlier. "I was driving to class, and I really remember thinking, 'My goodness, here I'm stepping off on this new career, and some of my marine buddies are probably up in those planes. Is this really what I want to be embarking on?'" Hilmers said. It *was* what he wanted to do, so he threw himself into studying for his medical school entrance exam, cramming before heading to work at NASA by day and taking classes by night. Then came 5 April 1991.

Sonny Carter had already been named to the crew of STS-42, which was to carry the International Microgravity Laboratory-1 to study the human nervous system's adaptation to low gravity and the effects of microgravity on other life forms such as shrimp eggs, lentil seeds, and bacteria. Within a day or so, Hilmers was asked to replace Carter on the *Discovery* crew. With his Medical College Admission Test just a couple of weeks away, NASA needed Hilmers to start training "right now." He accepted the flight assignment, placing him into the difficult routine of rearranging the training schedule to continue classes, with a day off to take the entrance exam. He would fly back and forth for training in Huntsville, Alabama, and classes in Houston. Accepted into medical school at Baylor University in Houston, STS-42 launched 22 January 1992, just three weeks or so after Hilmers's last premed final exam. He recalled:

Some of the classes, I remember, I had to take in the middle of the day. So I would either come early to do the training or stay late. I'd drive and do the class and come back. We made things work. If you would've said, "How can you do those things?" I myself probably would've said they're mutually incompatible. Of course, I had to learn a lot of stuff to do the scientific mission. I had to study a lot for that, too. It was a really busy time. I think it kind of gave me the feeling, "Well, if I can do this, I can make it through medical school."

The flight quickly became a series of lasts for Hilmers—it was his last time in crew quarters, his last astrovan ride out to the pad, his last launch, his last opportunity to see the splendor of Earth from orbit. "The whole astronaut experience is incredibly unique," Hilmers remarked. "It's not something you get to do anyplace else. Even though I'm many years away from what I did back then, it's never too far from your mind. I don't spend my days thinking about the space program, but there's a lot of times that something will just remind you of what happened back in those days—some picture you'll see, some reference to a place that you saw from space. You always have those pictures in mind."

Hilmers graduated with honors from Houston's Baylor College of Medicine in December 1995. By the time his residency ended four years later, Hilmers was already getting involved in the school's overseas pediatric AIDS initiative. He now serves on Baylor's faculty, and he has also made very nearly as many trips around the world as a medical missionary and researcher as he ever did while on orbit. Some of the efforts in which Hilmers has taken part include the following:

Mission to the World, the medical missions arm of the Presbyterian Church in America: Azerbaijan, Sri Lanka (after the 2004 tsunami), Iraq (after the start of the war), Belize, Haiti (after the 2010 earthquake), and Myanmar.

Hope Initiative, an organization of Vietnamese in Houston: Cambodia and three separate trips to Vietnam.

Rejoice, a charitable organization in Thailand serving AIDS patients: Thailand and Laos.

United Nations, working on malnutrition: Thailand, China, Vietnam, Pakistan, India, Sri Lanka, and South Africa.

Medical Research Council, the British equivalent of the National Institutes of Health: Gambia.

National Institutes of Health: Nigeria.

Centers for Disease Control and Prevention: Thailand and Burmese refugee camps.

Coca-Cola Corporation, working on nutritional drinks to use in the developing world to combat malnutrition: Botswana, Peru, Brazil, Colombia, and Mexico.

Local nongovernmental organizations, churches and volunteer trips: El Salvador, Nicaragua, Ecuador, Peru, Brazil, Costa Rica, Haiti, and Nepal.

Baylor College of Medicine education and teaching trips: Peru, Panama, Guatemala, Honduras, Sri Lanka, Israel, and Palestine.

Baylor Pediatric AIDS Initiative: Romania, Botswana, Mexico, and South Africa.

YUST/PUST Foundation: China and North Korea.

33. David Hilmers checks a Nepalese woman for signs of leprosy during a 2005 mission trip. Courtesy David Hilmers.

Hilmers is one of a very few people who could even begin to make being an astronaut seem almost like a fallback gig. He accumulated a wealth of memories in the mission and research field, just as he did during his four flights on board the Space Shuttle. There was an impromptu road trip across Iraq months after the start of the war in 2003. He befriended a witch doctor in Belize. He sat mere feet from the door of a tiny clinic in El Salvador when a gunman entered and started shooting. One person was fatally wounded, and the woman whom Hilmers had been treating was hit in the leg. He had just arrived in earthquake-ravaged Haiti when his father, Paul, passed away. With flights into and out of the country at a premium, Hilmers had little choice. He stayed in Haiti and finished out the rest of his two-week tour.

Through all these experiences and more, Hilmers asked himself one simple question.

Am I making a difference?

Dave Leestma was in the management pipeline before he began training for STS-45, having worked for more than a year as deputy director of flight crew operations. Seven months after his final "wheels stop" call on the KSC landing facility, Leestma was named director of flight crew operations. In essence, he went straight from the playing field into the front office, where for the next six years, he was in charge of the astronaut office and selection boards.

In 1971 Leestma had graduated first in his class at Annapolis. Yet without hesitation, he would call overseeing a group of highly competitive type A personalities with nothing more on their minds than getting a slot on a shuttle crew a much tougher assignment. "Director of flight operations—selecting astronauts, picking folks for flights, making sure that the missions themselves are being trained for right and going well—those were different and harder in a much broader sense," Leestma said. "Doing schoolwork is tough, but the other area is focusing on people—which I like—but it runs a real wide range of emotions."

What, exactly, was the secret to getting a spot on a shuttle crew? Surely, ability had something to do with it, but there was something to be said for being in the right place at the right time as well. Maybe, just maybe, mix in some political maneuvering along the way and an astronaut might be well positioned to fly. "It's a combination of probably a hundred different factors," Leestma began. "I would say the *least* of those is politics. Selections are made based on what the mission is, what the skills of the people are, how well they've been doing in their training, is it going to be a good mix of personalities. Then there's experience and inexperience." Leestma and the chief of the astronaut office would then pass their recommendations up the chain of command.

Surely, there was a good bit of sucking up along the way, and Leestma smiled broadly at the suggestion. "Oh, yeah, there's that. Without question, there's that," he said. "Everybody that isn't assigned to a flight *wants* to be assigned to a flight, and they want to be assigned as soon as they possibly can. That's just human nature. That's what they came to do. There probably wasn't as much overt sucking up, but if they could get face time or if they were given a little project and they had a chance to present that, they looked forward to that."

One of the hardest parts of the job for Leestma was easy enough to understand. Putting someone in harm's way—in many cases, someone he had helped select as an astronaut in the first place and then for a spot on a particular crew—always gave Leestma pause. "It's much, much harder to assign a crew to a flight and selecting astronauts and then watch them go fly than it is to be on the vehicle yourself," he admitted. "You're committing these people to a mission where there's a lot of risk, and I'd much rather be taking that risk myself, even though they're just like me. They want to go do this just as badly as anybody else. They're more than willing to do that. But putting them on the vehicle and then having to watch the countdown—I did that forty-seven missions in a row."

If Leestma could have done without that kind of doubt, actually informing someone that he or she had been selected as an astronaut was by far the better option. Someone else gave the bad news to those whose applications had not been accepted, but if it was Leestma on the line, the news was no doubt going to be good. He served as chairman of the selection board for Group Fifteen in 1994, Sixteen in 1996, and Seventeen two years after that. The forty-four members of "The Sardines" were packed into the office for the agency's upcoming International Space Station efforts. When Leestma called Mark Kelly with word that he had been selected, Kelly wanted to know if his twin brother, Scott J. Kelly, had also been selected.

"Mark's first reaction was, 'Hey, that's great. Yes. I can't wait to come, but did my brother get picked?'" Leestma remembered. "I said, 'I can't tell you, because we haven't talked to him yet.' He said, '*You've got to tell me!* Please!' He couldn't wait to hear. It was an amazing reaction, the way those guys cared about the other. I've never had that reaction before."

For the record, Scott made the cut as well.

After NASA, Byron Lichtenberg kept his head in the clouds—and beyond.

Between his two shuttle flights, he had helped start the Association of Space Explorers, and afterward, not only did he fly for Southwest Airlines but he was also a founding member of the X-Prize Foundation and the International Space University. Zero G, a company that offers weightless rides to the public on a specially outfitted Boeing 727, was also his brainchild. It all started with the view from orbit. "When you're on orbit and go around the entire planet, you see everything that we live in over ninety minutes," he began. "You realize we're just this little speck in this huge, almost infinite universe. You see all of civilization, humanity, life as we know it today right below you."

Perspectives change. A person is no longer a Texan or an American or a Russian. Lichtenberg had attended some meetings of the Association of Space Explorers, where he met Pham Tuan, a Vietnamese cosmonaut that he could have flown against once upon a time; Jose Hernandez, an astronaut of Mexican descent; Saudi prince Sultan bin Salman bin Abdul-Aziz

Al Saud, who flew on the shuttle before the *Challenger* tragedy; and American senator Jake Garn, another pre-*Challenger* shuttle participant. Their nationalities did not matter—as humans, their lives had been changed by spaceflight. "It really made us feel more like global citizens," Lichtenberg said. "Yes, I'm American and I'm very proud of that, but we're also citizens of Earth, and we really need to do good stewardship of our home planet."

Tom Akers left NASA in January 1997 to take on the role of commander of the air force ROTC Detachment 442 at his alma mater, the University of Missouri–Rolla. He retired from the military in 1999, at which point he joined the school's mathematics department. "I put in ten, eleven years teaching math there," he began, before quipping, "I really loved every minute of it—except the time I was grading papers. That's the one thing I don't miss."

Best of all, Missouri was home. It was where his family was. It had been his plan all along to go back. Life in school was not all that foreign to Akers; before he joined the air force in 1979, he had been a high school principal in his hometown of Eminence, Missouri. It was the stuff of *The Andy Griffith Show*'s fictional and idyllic setting in Mayberry. As of the 2010 census, the year Akers left the UMR classroom behind for good in favor of life in a small town, Eminence boasted a population of six hundred people. He had a small twenty-acre farm and served as trail boss for the local Cross Country Trail Ride. "There are not enough hours in the day to do everything you'd like, riding horses, helping neighbors, hunting," Akers said. "In fact, just this morning I had to go help a friend go move a hay elevator from one place to another and then get back home to meet with the farrier who shoes my horses."

In short, post-NASA and retired life for Tom Akers has been very good.

Although he knew it was at best a long shot, Jerry Ross had always held out hope for an eighth shuttle flight until *Columbia* went down and his friends were lost. The knowledge that he would never fly in space again hurt:

I found it hard to accept, but being realistic, I knew that was the bottom line. I actually went into a little bit of a blue funk after Columbia. *It was just a very depressing time for a period. It just took time to work back out of that and get back into the real world again. I wasn't clinically depressed or anything like that. I was just bummed out. I logically knew that was the end of it, but I didn't want it to be emotionally. I guess I had the goal in life of never having to grow up and earn a living.*

Ross stayed with NASA through the shuttle program's final flight, serving as manager of the Vehicle Integration Test Office. In one version or another, that group had been around since the days of Gemini, and its engineers were charged with the task of helping crews get ready to fly. That meant going all over the world to inspect and test hardware for such things as sharp edges that might pose a risk to a spacewalker, doing fit checks for tools, and if a prime crew member wanted to inspect hardware personally, Ross's group would handle the logistics. It also meant that he was in charge of setting up crew quarters—bringing in cooks, suit technicians, flight doctors, and so forth.

Ross eventually became what he called a "voice in the wilderness." Ross elaborated:

Basically, I was just trying to do the right thing. But when you get down to the Cape sometimes, there is this launch fever kind of thing that does kind of creep in. On several occasions, I was down there monitoring what was going on and I would call somebody on the phone and say, "Hey, are you really sure you ought to be doing this?" I would send e-mails in the middle of the night, and while I didn't always get a direct, personal response back from whoever I was sending things to, basically I think without exception the points I was trying to make eventually were the ones that were acknowledged and caused us to step back from decisions that could have gone a different way.

Ross's retirement from NASA was announced on 27 January 2012. He went to work on his autobiography, *Spacewalker: My Journey in Space and Faith as NASA's Record-Setting Frequent Flyer*. He did not take to a rocking chair. Much to his chagrin, however, his son Scott, daughter Amy, and a friend were all faster during a session at the Dale Jarrett Racing Adventure at Talladega Superspeedway.

Step by slow step, one foot in front of the other, Scott Parazynski edged up Mount Everest. At 29,035 feet above sea level, it is the biggest of them all, the tallest mountain in the world. In the pitch darkness of the predawn hours on 20 May 2009, he could see only by the glow of small lamp lights on his and other

34. A year after his first attempt, Scott Parazynski finally made it to the summit of Mount Everest on 20 May 2009. Courtesy Scott Parazynski.

climbers' helmets. This work was something far beyond arduous in lethal subzero temperatures, but Parazynski could not afford to quit now. Not this time, not after he had made it so far just the year before, only to be forced into turning back due to a ruptured disc in his back. Now, he had to concentrate. Just hold onto the fixed rope and keep going no matter what. Just. Keep. Going.

Parazynski had already been on the mountain for more than eight weeks, first at base camp to allow his body the opportunity to get acclimated to a new environment that featured decreased air pressure and oxygen. He had made it through the Khumbu Icefall, up the Western Cwm, across the Lhotse Face, past the South Col, and into the death zone at 26,000 feet, the point at which there was not enough oxygen in the air to sustain human life for any significant length of time. In a final push came the Triangular Face, the Balcony, the Southeast Ridge, the dicey pitches beneath the South Summit, the Cornice Traverse, and the last major hurdle, a sixty-foot vertical rock wall known as the Hillary Step. Just a few more measured paces . . . one, two, three . . . finally, there it was, pointed out by a Sherpa companion, the summit. He had made it, and exhausted, he sat down to savor the moment. "Top of the world!" he told a camera crew on hand to document the feat for the Discovery Channel's *Everest: Beyond the Limits* television show. "Oh, man, one of the happiest days of my life. This is the roof of the world, baby! That was a lot harder than I ever thought it would be."

With him were mementoes of his previous life as an astronaut. On the front of Parazynski's jacket was the American flag patch he had worn while performing one of the most hazardous EVAs of the shuttle program during his final spaceflight, STS-120. It was a nod to his own journeys, but he also had carried with him tiny rock samples from the Sea of Tranquility. Still in the moment, Parazynski continued, "Two years in the making, actually forty-seven years in the dreaming." Scott Parazynski was a man who had dreamed many dreams in those forty-seven years, and he lived most of them.

Karl G. Henize was not as fortunate. A member of the crew of STS-51F, he was a veteran of not only a launch-pad abort but the first and only abort-to-orbit call of the shuttle program when *Challenger*'s center main engine shut down prematurely. A few months short of his fifty-ninth birthday, Henize was the oldest rookie ever to fly in space. He retired from the astronaut corps the following year to become a senior scientist at NASA, and while he would never return to space, the Cincinnati, Ohio, native did not sit still. He instead eventually became part of a team dedicated to researching the possibility of a parachute jump from the lower edges of space, an endeavor that took him to Mount Everest in late August–early September 1993. Any such jumper would fall at supersonic speeds and would need the same basic thermal protection as a shuttle during reentry, as well as a defense against high g-force loads.

"Karl was super excited to be part of all this," said Pat Falvey, an Irishman who was at the time a rookie member of an international expedition to Everest working in conjunction with Henize's team. "His objective was never to get to the summit, but basically to do the science. There was a certain amount of—I won't say secrecy—but there was a certain amount of vagueness in the whole thing." Falvey continued, "Karl was a very down-to-earth person. His excitement was from the fact of being with adventurers. There were no prima donnas. He was a quiet man, but when you got to talking to him about the subject of what he was there for—we were all there doing something, and then all of a sudden, this Pandora's Box comes out about what Karl was doing. We were completely and utterly fascinated about what it was that Karl was about."

At age sixty-six, and despite the fact he did not plan to summit, Henize managed to make it to Everest's advanced base camp at 21,000 feet above sea level. "Karl wouldn't have been the fittest bloke on the block," Falvey said. "He felt the altitude, very much so. I'll never forget it. Karl had arrived at advanced base camp and was probably already suffering from a bit of altitude sickness."

As Falvey left for a summit attempt, Henize gave him a hug and admitted that he was not feeling well. On 4 October 1993, Henize collapsed at Everest's advanced base camp, a victim of high-altitude pulmonary edema—a highly dangerous accumulation of fluid in the lungs caused by the constriction of oxygen-starved blood vessels. He died the next day.

Falvey abandoned his summit attempt and headed back down the mountain to assist in Henize's evacuation; he did not try again that year to make it to the top. After his death, the former astronaut was buried in a crevasse near the Chanste Glacier at a little more than 20,000 feet above sea level. In all, Falvey took part in approximately seventy expeditions that saw him twice climb the highest mountains on each of the seven continents and venture to the South Pole. The memories of Henize and other friends who have not returned from such adventures have been constant companions:

[Henize] felt a part of my spirit. He was with us, and we were a team. That creates a bond that's special, even if it's for a week or two, when you're there and you know that each other's lives are in your hands. We all go to these places and we all realize that we may die. I never got a chance to talk to Karl's family, but you know the way you're leaving at the airport and people are hugging and loving you? Karl had left home and kissed his wife and his kids, or whatever he had at the time. That realization hits you when you have to bury a person on the mountain and leave them there, and know that his wife and children will never, ever see him back again. It's that thought that I carry with me on every major expedition.

Parazynski was almost completely unaccustomed to failure. Though he missed the cut for the 1988 Olympic luge contingent, he made it to the Calgary games anyway as coach for Ray Ocampo, the lone competitor from the Philippines. Years later, he served as honorary captain of the American luge team during the 2010 Vancouver Olympics. He had lived in strife-ridden Lebanon and Iran. He flew five shuttle missions. He climbed Mount

Everest, twice. Surely, he is an adrenaline junkie with some sort of bizarre death wish, right? Wrong, he maintained. "I'm not a risk taker, and that might come as a surprise," Parazynski began. "I'm not a daredevil. I manage risk, and I treat the things that I do very, very seriously." In an interview they did about Parazynski's return to Everest, the astronaut told Miles O'Brien:

People look at some of the things I have done in my life and they think, "Wow! You are a risk taker, aren't you?" And that's absolutely not the case. In fact, I guess the proper term would be risk manager. I know the environments that I go into, I understand them, I am trained for them, and I don't do things without a lot of preparation and good headwork. These are exciting places to be and that's what appeals to me is the adventure of it and the exploration. I think we can send people on to these types of environments very successfully and safely with the right preparation.

The goal of conquering Everest was not the result of any kind of inner turmoil caused by a midlife crisis and the end of Parazynski's career at NASA. He had been aware of the mountain since early childhood and began serious rock climbing at the age of fifteen. He made it to the top of Argentina's Cerro Aconcagua, at 22,841 feet tall, the highest mountain outside of Asia. In Colorado Parazynski eventually scaled no fewer than fifty-nine peaks that topped 14,000 feet. The fact of the matter was that it took his selection to the astronaut office to delay—but not completely derail—his drive to climb Everest. He had laid the groundwork to join a British expedition to the mountain in the early 1990s, backing out only after getting the call from NASA. "I hoped to be able to go to Everest in 1993 with my team, but my bosses had different ideas for me. They assigned me to my first shuttle flight at that point," Parazynski said. "Everest had to wait. It went on the backburner for a number of years."

Although it was clear that he likely would not fly again, Parazynski had not yet officially retired from NASA when he went after Everest the first time during its 2008 climbing season. He arrived on the south side of Everest in late March during the pre-monsoon season, and fifty-nine days later, he had made it to an altitude of 24,500 feet above sea level. He could see the summit, just a day's journey away. He did not make it. "I was within twenty-four hours of the summit when I ruptured a disc in my low back," Parazynski said, the impact of the disappointment clear from the tone of his voice. "I can't really attribute the injury to any one action. I've had chronic back injuries being a tall person and athletic for many years. I did injure my back, whether it was lifting my backpack funny or sleeping on it the wrong way. I had to turn around." He later blogged:

With extra determination, I gave my pack a hoist, wincing in sharp pain in the process. Just the day before I'd awoken with low back spasms (something I've dealt with intermittently in the past), but I had still managed to climb the very steep Lhotse face between Camps II and III in a very respectable four and a half hours, cinching my climbing harness like a weight lifter's belt. The night at Camp II had been hard, unable to find a comfortable position for my low back for more than a minute or two. I told myself to persevere, the summit was tantalizingly close—by morning all would be well, else I'd just "ignore" the stabbing pain and press on to the top.

Fellow climbers and Sherpas helped Parazynski ready for the trip to Camp IV, the final stop before a summit attempt. They hooked him up with a fresh oxygen supply, and they even went so far as to rig his boots with the spiked crampons he would need to gain traction on the icy mountain slopes. He tried to make a test run but was stopped in his tracks by the severe pain radiating from his lower back. "Within ten paces I did an about-face and told my friends, 'I'm done,' averting my wet eyes from probably some of theirs," Parazynski said. "I knew that if I continued up with them I'd slow them dramatically, possibly compromising their summit success and conceivably place them in a rescue situation (mine). After fifty-nine days on this expedition, and a lifetime of dreaming about it, it was a painful but easy decision to turn away from the summit."

Other members of his team made it to the mountain crest, and Parazynski could hear their exhausted but elated radio calls. As soon as he returned home to the United States, Parazynski underwent surgery. Within two hours he was upright and walking, and eight weeks after that, he was back in the gym. He was headed back to Everest, and this time, he would not be forced into unsuccessfully coming back down the mountain. According to Parazynski, no one ever tried to talk him out of another summit attempt. "I had no restrictions physically," he insisted. "I wanted to get the job done. I didn't want to waste any time." That, in a nutshell, summed up perfectly what it means to be Scott Parazynski. Get the job done, and do not waste time doing it, because there is something else to accomplish right around the corner.

He began his journey back to the mountain in late March 2009, leaving Houston to head to Katmandu. Parazynski had once been a rookie astronaut, and the year before, he had been inexperienced in the ways of Everest. That was no longer the case:

I had unfinished business there. I had thought about the mountain every single day since my failed attempt. It was really exciting to be back. I knew all of the Sherpas. I knew the major Everest players. I knew the route—except for that very last day. I felt very much at home coming back to Everest. I knew what to expect. I was a veteran. It was like going back to space. Your first mission, you're so overwhelmed by that very first taste of weightlessness and living and working in space. When you go back on subsequent missions, you have much more confidence. You plan for performance in much better ways. That's how it was for me going back to Everest. I really felt I had a leg up on the expedition.

Dawes Eddy was sixty-six years old, the same age as Henize at the time of his passing, when he headed to Everest to join Parazynski's climbing expedition. He had never heard of his fellow climber before, so Eddy was not particularly starstruck. Parazynski, who served as the team's acting physician, was instead just one of the guys. He was a comrade, a teammate trying to make it to the summit.

As the doctor in the house, Parazynski was a mother hen watching over his flock as much as possible. As shown on *Everest: Beyond the Limit*, his role on the team placed Parazynski in the difficult position of helping to determine that a team member, English businessman Paul Samuel, would be better off physically not continuing with a summit attempt. Parazynski had been there himself, so he knew the kind of numbing letdown such a decision would be.

As part of the acclimatization process, any Everest summit bid is an exercise in climbing up and down various portions of the mountain. From base camp to Camp I and back down again. From base camp to the Khumbu Icefall and back. Another trek to Camp I and back. From Camp II to the Lhotse face and back, then yet another return to base camp. Camp III to the Yellow Band and back, then to Camp II, and on to base camp. Finally, Parazynski and the rest of his team were ready for a push to the summit. At 9 p.m. on the mountain, with temperatures plummeting to minus-thirty degrees Fahrenheit, they left Camp IV just inside the death zone, the uppermost rest stop at 26,000 feet. It was an hour earlier than originally planned, in order to beat the rush-hour traffic with other expedition teams trying to crowd their way onto the summit route.

The going was not easy. At 28,700 feet, Parazynski's party had to scale the South Summit. From there, they descended some 700 feet to walk the tightrope of the Cornice Traverse, with a drop of more than a mile on either side, before getting to the steep rock face of the Hillary Step, named after Sir Edmund Hillary, who along with Sherpa Tenzing Norgay became one of the first two men to summit the mountain in 1953. After that, just a few hundred more feet separated Parazynski from the goal that had eluded him the year before. As the sun began to rise, the view defied an adequate description. Parazynski called it "spectacular," but it was something much more than that:

I saw the entire sunrise from the top of the world. I reached the summit at 4 a.m., and then the orange glow began at about 4:05 in the morning. So my entire descent down the summit ridge was watching the sunrise. It was just gorgeous. You could see very clearly from that vantage point the curvature of the earth. I really felt like I was up on orbit, except that I was moving in super-slow speeds. Up on orbit, when you see a sunrise, they happen just within a few seconds. But on my experience there on Everest, it took many minutes for the sun to rise. It was very, very dramatic.

Eddy was just a few minutes behind Parazynski, and in reaching the summit, he became the oldest American to achieve the Herculean feat. Soon after leaving base camp the next day to head back to the real world, however, he developed severe stomach issues. "On the way out, I had some problems and Scott was very helpful to me, personally," continued Eddy, who was turned back in another Everest summit via the North Col route in 2013. "What happened, my stomach shut down and wasn't emptying out. The only way to solve my problem was to have my stomach pumped, so Scott showed up and kind of saved the day for me. You might say he was a real life saver for me. I had a pretty miserable night there." This was not a procedure conducted in the fully stocked, sterile environment of a hospital.

This was out on the trail, maybe five hours from the nearest actual medical facility.

Everest was a defining moment in Parazynski's life, very much on par with his spaceflight career. Those few precious minutes on the summit stayed with him. Every day, some small memory, the feel of the wind on his face, the indescribable cold, would come back to him. "The strength that it gives me, having overcome adversity to return and summit, that really does give you a lot of confidence that you can be able to handle the tough stuff and get the job done," Parazynski concluded.

This was a man who was a hard act for anyone to follow. His dizzying array of accomplishments made for the ultimate bucket list, but after five shuttle flights and two trips to Everest, there was nevertheless much left for the uber-motivated Parazynski to do both personally and professionally after retiring from NASA in March 2009.

The Challenger Center for Space Science Education was formally established just three months after the loss of the STS-51L crew. Not content with the thought of an ordinary memorial plaque that would tarnish and quite possibly vanish over time, the *Challenger* families were determined to honor the memories of their loved ones in a real and vibrant way. The first center opened at Houston's Museum of Natural Science in 1988, and more than fifty followed in thirty-one states, Canada, and the United Kingdom. Its extensive reach was the truest measure of Challenger Center's success—some four hundred thousand students a year took part in its various programs, with training of one kind or another for twenty-five thousand teachers.

Long involved with its efforts, Parazynski took over from former astronaut Bill Readdy as chairman of the organization's board of directors in November 2010. His own son, Luke, took part in a simulated shuttle mission at the Houston center in which he helped put together a satellite probe. "It's just such a tangible, unique experience," Parazynski said. "It's one of those memories that last for a long, long time. It's not just the cut-and-dried reading a book or listening to a lecture. It's actual application of math, science, engineering, physics, chemistry, problem solving, and working as a team that really resonates with kids."

Another major career move was to the Methodist Hospital Research Institute in Houston, where he began serving as chief medical and technology officer in February 2011. In that role Parazynski helped spearhead research into the development of medical and surgical devices, diagnostic tools, and various other aids that he hoped would reshape the future of medicine. Parazynski was working in conjunction with some of the best and brightest physicians, scientists, and engineers in the world. His goal in "translational science" was to get the research and development they were doing to patients' bedsides as safely and quickly as possible. It was if his entire career had vectored him toward this unique environment, where he could leverage his medical and NASA backgrounds into helping solve real-world problems. He elaborated:

I think one of the most exciting things in my life, outside of raising my kids and seeing them happy, healthy, and prosperous, is the exploration of the human body. The Fantastic Voyage *was a science fiction movie with Raquel Welch back when she was in her prime, and I would love to see that type of miniaturization of technology to advance healthcare with my hand on it. I feel like I'm in a great environment now to deliver what skills I do have to the betterment of medical care and helping large numbers of people. That's what I see the next few years of my life related to, trying to improve healthcare for everyone.*

That was Parazynski's professional side. Personally, he was thinking in the early fall of 2011 that he might one day finish climbing the tallest summits on each of the world's seven continents. Then again, he had already been to space and the top of Mount Everest, so why not head the other way and take a submersible to the depths of the Marianas Trench in the western Pacific Ocean? At its deepest point, the 1,580-mile-long fissure is nearly seven miles down.

After returning from his long-duration mission on the ISS, Doug Wheelock served as the head of NASA's Army Space and Missile Defense Command detachment. Headquartered at Redstone Arsenal in Huntsville, Alabama, the command was a direct descendant of the group that launched *Explorer 1*, the first American satellite.

The job kept his irons in the army fire while he got back in line for another command on the ISS. With at least a two-year wait to start another training flow, Wheelock had to juggle possibilities. With a board meeting for promotion to brigadier general coming up, could he possibly get back out with the troops and fly? The army wanted him full time for a couple of years, but that was not going to work with his NASA plans. He went back to his commanding general, who said he would come up with something else. That something else turned out to be a handful of temporary deployments to fly a research and reconnaissance plane housed and maintained in Houston. "It's being used to support our troops in Afghanistan, and so I made the decision to go back to active service," Wheelock said. "The army has bent over backward to help me. Because of them, I got a chance to fly in space. So it's the best way I know to just give back a little bit."

On 1 October 2007, NASA issued a press release saying that Joan Higginbotham had been named as a mission specialist on the flight of STS-126. Fifty-one days later, on 21 November, came another announcement saying that she had left the agency to take a job in the private sector and that Don Pettit was taking her place.

According to Higginbotham, she had not been looking for an exit door. She was not necessarily interested in a long-duration mission on the station, and moving up the management ladder might or might not have been an option. She was trying to figure out what to do with her life, she said, when she sat down with someone who had made the transition from a government job to private industry. One thing led to another, and she was offered a job with Marathon Oil as a senior consultant. "About the same time, I got assigned to the second mission, which to me was just the most horrible thing in the world," she continued. "I didn't want those two paths to collide. I asked Marathon to give me a week to decide, and I woke up every morning at 3:30 and couldn't go back to sleep because I was trying to figure out what to do."

She took the Marathon job, and along the way she worked with proprietary technology, in commercial negotiations, in the budget office, and, right in her wheelhouse, as an electrical engineer. That landed her in the African nation of Equatorial Guinea, and it was there that she became involved in the company's social responsibility efforts. She came back to the States knowing, at last, where she was supposed to be—in that office. She spent two years there, and ended up managing its malaria program in the country.

It took true love to get her to move. She met James Mitchell, a member of the Charlotte city council, in 2009 and in October 2010 moved to the

North Carolina city to be closer to her boyfriend and soon-to-be fiancé. "My mom was like, 'You're moving to Charlotte and leaving your job. You have a very successful career. Have you lost your mind?'" Higginbotham said. "I'm her baby, and she was a little concerned." Mitchell and Higginbotham were married in mid-June 2012.

Other than the astronauts themselves, United Space Alliance shuttle technician Les Hanks had for nearly twenty years what many would consider one of the greatest jobs on the planet. At one time or another, he worked throughout the innards of the spacecraft. As a safety engineer, he was part of a group that was the last to leave the pad before launch and the first to return to both the launch site and the orbiter itself after landing. He used instruments to sniff for toxic fumes, opened the hatch so the astronauts' egress could begin, and hooked the shuttle up to a tow vehicle for the trip back to the OPF so the whole process could start all over again. On those occasions when the shuttle's return was diverted to Edwards, Hanks hopped a plane to help get it ready for a return to KSC. He was intimately familiar with virtually every nook, cranny, piece, and part within his areas of responsibility, and a good many outside them.

To say that Hanks was proud of what he and hundreds of coworkers at KSC were able to accomplish barely scratched the surface of his true emotion. Call it fulfillment to the nth degree—fulfillment cubed, if you will. "I'm going to remember it as the most amazing vehicle," Hanks started, looking for just the right words. "I don't think that there will be a vehicle in my life that will do what the shuttle does in the capacity that the shuttle does it. Its legacy will have to go down as one of the greatest space vehicles that ever existed."

As for its safety, the loss of *Columbia* had touched him very deeply. Working in the OPF was not just a job for Les Hanks and so many others just like him—it was something far more meaningful than just a paycheck and personal satisfaction. "It's the safest vehicle, I can tell you that," Hanks maintained, his voice thick with emphasis. "I mean, neither one of those accidents were related to that orbiter itself. The orbiter itself, which carried the crew, was an awesome machine. It's a flying dream. I hope they come in my life with something that can outdo it, but I don't think they will."

When the Constellation program and the end of the shuttle were announced, Hanks had held out hope that he would be able to continue working in capacities similar to what he had already been doing. Yet as more time passed, the outlook was getting murkier and murkier. Some thought the shuttle would continue to fly for an indefinite amount of time, while others took heed of the storm clouds that were most definitely brewing. "A lot of people's thoughts were that they would keep flying and add five more years once we got closer to the end," Hanks recalled. "There are some people who are still there today who thought that this one on the pad right now wasn't going to be the last one. Now, they're realizing, 'Wow. It really is going to be the last one.' For me, it probably took a good three years or so before I really started thinking the end was right around the corner and that I needed to prepare myself."

Hanks landed a job with Siemens Energy Incorporated as a quality engineer in its wind power division. As jobs go, it was the perfect opportunity for Hanks to continue providing for himself and his family just as he always had. But was it the right thing to do? He was walking away from both an incentive bonus and a severance package. How could he possibly face his supervisor at USA to turn in a two-week notice? "It was tough after nearly twenty years on the shuttle program and then having to walk away from it," Hanks acknowledged. "It was a tough decision. I almost didn't do it."

His last day on the shuttle program was 30 September 2010, a little more than nine months before the last flight. The very next day, approximately 20 percent of USA's shuttle workforce at KSC—maybe eight hundred to one thousand people—were laid off. The timing was no accident. As badly as it hurt Hanks to leave, he had something solid lined up and waiting for him. It would have been unimaginable to get caught up among so many friends whose futures were far from certain. As Hanks was leaving the center that day, he caught a glimpse of the famous VAB. It would be the last time he ever saw the building as he saw it then, as his workplace, as the place where he was able to live so many incredible moments only to experience more the next day.

Hanks quickly snapped a picture of the VAB in his rearview mirror and began to cry.

Sources

Books

Cernan, Eugene, and Donald A. Davis. *The Last Man on the Moon: Astronaut Eugene Cernan and America's Race in Space*. New York: St. Martin's Griffin, 2000.

Hale, Wayne, et al., eds. *Wings in Orbit: Scientific and Engineering Legacies of the Space Shuttle, 1971–2010*. NASA SP-2010-3409. http://er.jsc.nasa.gov/seh/536821main_Wings-ch1-2.pdf.

Husband, Evelyn, with Donna Vanliere. *High Calling: The Courageous Life and Faith of Space Shuttle* Columbia *Commander Rick Husband*. Nashville: Thomas Nelson, 2003.

Lenehan, Anne E. *Story: The Way of Water*. Hornsby Westfield, Australia: Communications Agency, 2004.

Linenger, Jerry M. *Off the Planet: Surviving Five Perilous Months aboard the Space Station* Mir. New York: McGraw Hill, 2000.

Logsdon, John A. "Return to Flight: Richard H. Truly and the Recovery from the *Challenger* Accident." In *From Engineering Science to Big Science: The NACA and NASA Collier Trophy Research Project Winners*, edited by Pamela E. Mack. NASA SP-4219. http://history.nasa.gov/SP-4219/Chapter15.html.

Montgomery, Scott, and Timothy R. Gaffney. *Back in Orbit: John Glenn's Return to Space*. Atlanta: Longstreet Press, 1998.

Morgan, Clay. *Shuttle-Mir: The U.S. and Russia Share History's Highest Stage*. NASA SP-2001-4225. http://history.nasa.gov/SP-4225/toc/info.htm.

Mullane, Mike. *Riding Rockets: The Outrageous Tales of a Space Shuttle Astronaut*. New York: Scribner, 2006.

Portree, David S. F., and Robert C. Trevino. *Walking to Olympus: An EVA Chronology*. Monographs in Aerospace History 7. Washington DC: NASA History Office, 1997. http://history.nasa.gov/monograph7.pdf.

Reichardt, Tony, ed. *Space Shuttle: The First 20 Years*. New York: DK Publishing, 2002.

Tatarewicz, Joseph N. "The Hubble Space Telescope Servicing Mission." In *From Engineering Science to Big Science: The NACA and NASA Collier Trophy Research Project Winners*, edited by Pamela E. Mack. NASA SP-4219. http://history.nasa.gov/SP-4219/Chapter16.html.

Periodicals and Online Articles

Armstrong, Neil, Jim Lovell, and Gene Cernan. "Is Obama Grounding JFK's Space Legacy?" *USA Today*, 24 May 2011. http://www.usatoday.com/news/opinion/forum/2011-05-24-Obama-grounding-JFK-space-legacy_n.htm.

Ashby, Jeff. "Diana Merriweather Ashby; March 22, 1963–May 2, 1997; Beloved Wife, Daughter and Friend, Founder of the Melanoma Research Foundation; Dedicated to Finding a Cure . . . Because Tomorrow Is Too Late." Melanoma Research Foundation. http://www.melanoma.org/community/diana-merriweather-ashby.

Associated Press. "STS-118: The *Endeavour* Crew." *St. Petersburg Times*, 8 August 2007. http://www.sptimes.com/2007/08/08/news_pf/Worldandnation/STS_118__The_Endeavou.shtml.

"Astronaut Story Musgrave Injured during Training." NASA Release 93-038, Houston, 1 June 1993. http://www.nasa.gov/centers/johnson/news/releases/1993_1995/93-038.html.

"Astronaut Story Musgrave Retires from NASA." NASA Release 97-188, 2 September 1997. http://www.nasa.gov/centers/johnson/news/releases/1996_1998/97-188.html.

Atkinson, Nancy. "Spacewalking: Through an Astronaut's Eyes." Universe Today, 16 March 2010. http://www.universetoday.com/59840/spacewalking-through-an-astronauts-eyes/.

"At Long Last, an Inspiring Future for Space Exploration: Statement from Elon Musk." SpaceX, 15 April 2010. http://www.spacex.com/press.php?page=20100415.

"Backup Crew Member for STS-61, HST Maintenance Mission." NASA Release 93-017, Houston, 9 March 1993. http://www.nasa.gov/centers/johnson/news/releases/1993_1995/93-017.html.

Baker, Deirdre Cox. "Former Astronaut from DeWitt Pursues Medicine." *Quad-City Times*, 23 September 2010. http://qctimes.com/news/local/article_fa56a9da-c78b-11df-8226-001cc4c002e0.html.

Beckwith, Steven. "Servicing Mission 4 Cancelled." Space Telescope Science Institute, 16 January 2004. http://www.stsci.edu/resources/sm4meeting.html.

Ben-Gedalyahu, Tzvi. "Shell-Shocked Nation to Bury Assaf Ramon Next to His Father." Arutz Sheva Israel National News, 14 September 2009. http://www.israelnationalnews.com/News/News.aspx/133411.

Bergin, Chris. "Managers Preparing for July 8 SLS Announcement after SD HLV Victory." NASASpaceFlight.com, 16 June 2011. http://www.nasaspaceflight.com/2011/06/managers-sls-announcement-after-sd-hlv-victory/.

"Boeing Opens Office in Greece." Boeing Company, 12 November 1998. http://www.boeing.com/news/releases/1998/news_release_981112n.htm.

Borenstein, Seth. "Future NASA Rocket to Be Most Powerful Ever Built." Associated Press, 14 September 2011. http://news.yahoo.com/future-nasa-rocket-most-powerful-ever-built-171120699.html.

Broad, William J. "2 Shuttle Veterans Grounded by NASA." *New York Times*, 10 July 1990. www.nytimes.com/1990/07/10/science/2-shuttle-veterans-grounded-by-nasa.html.

Capers, Robert S., and Eric Lipton. "The Looking Glass: How a Flaw Reflects Cracks in Space Science." *Hartford (CT) Courant*, 31 March–3 April 1991.

Cassutt, Michael. "Secret Space Shuttles: When You're 200 Miles Up, It's Easy to Hide What You're Up To." *Air & Space Magazine*, 1 August 2009. http://www.airspacemag.com/space-exploration/Secret-Space-Shuttles.html?c=y&page=1.

"The CGRO Mission (1991–2000)." NASA, CGRO Science Support Center, Goddard Space Flight Center. http://heasarc.gsfc.nasa.gov/docs/cgro/index.html.

Clifford, Rich. "Personal Stories: An Astronaut's Journey with Parkinson Disease." National Parkinson Foundation. http://www.parkinson.org/Personal-Stories/An-Astronaut-s-Journey-with-Parkinson-Disease.

Cowing, Keith. "Everest Itineraries 2009 (Revised)." On Orbit, SpaceRef Interactive, 19 March 2009. http://onorbit.com/node/821.

"David Brown: STS-107 Crew Memorial." NASA, 3 December 2004. http://spaceflight.nasa.gov/shuttle/archives/sts-107/memorial/brown.html.

"EURECA-1." NASA, National Space Science Data Center. http://nssdc.gsfc.nasa.gov/nmc/masterCatalog.do?sc=1992-049B.

"Exploration—the Fire of the Human Spirit: A Tribute—to Fallen Astronauts and Cosmonauts." NASA, 4 August 2005. http://www.nasa.gov/returntoflight/crew/sts114_exp11_tribute.html.

"Flashback: Car Talk in Space." *Car Talk*. http://www.cartalk.com/content/flashback-car-talk-space.

"Free Everest 2008." Scott Parazynski: Former Astronaut, Physician, Inventor, Mountaineer, Pilot, Public Speaker. http://www.parazynski.com/free-everest.html.

Friedman, Steve. "16 Minutes from Home." *Runners World*, 1 December 2005. http://www.runnersworld.com/cda/microsite/article/0,8029,s6-243-297—12436-4-1X2X3X4X5-6,00.html.

Greenfieldboyce, Nell. "Wayne Hale's Insider's Guide to NASA." NPR, 30 June 2006. http://www.npr.org/templates/story/story.php?storyId=5522536.

Hale, Wayne. "After the Barn Burned Down." Wayne Hale's Blog: Space, Exploration, Leadership, 18 August 2011. http://waynehale.wordpress.com/2011/08/18/after-the-barn-burned-down/.

———. "Wayne Hale's Blog: Shutting Down the Shuttle." NASA, 28 August 2008. http://blogs.nasa.gov/cm/blog/waynehalesblog/posts/post_1219932905350.html.

Harwood, William. "Astronauts Now Focusing on Hectic Cargo Transfer Work." CBS News Space Place, used with permission by Spaceflight Now, 13 July 2011. http://www.spaceflightnow.com/shuttle/sts135/110713fd6/index2.html.

———. "Foam Loss Grounds Shuttle Fleet Again." CBS News Space Place, used with permission by Spaceflight Now, 27 July 2005. http://spaceflightnow.com/shuttle/sts114/050727foam/.

———. "Legendary Commander Tells Story of Shuttle's Close Call." CBS News Space Place, used with permission by Spaceflight Now, 27 March 2009. http://spaceflightnow.com/shuttle/sts119/090327sts27/.

———. "Solar Wing Tears during Deployment." Spaceflight Now, 30 October 2007. http://spaceflightnow.com/shuttle/sts120/071030fd8/index7.html.

———. "Station Teams Scramble to Resolve Computer Glitch." CBS News Space Place, used with permission by Spaceflight Now, 14 June 2007. http://spaceflightnow.com/shuttle/sts117/070614computers/.

———. "Video Shows Crew Unaware of Impending Disaster." Spaceflight Now, 28 February 2003. http://spaceflightnow.com/shuttle/sts107/030228onboard/.

"How Do You Know If You Have PD?" National Parkinson Foundation. http://www.parkinson.org/Parkinson-s-Disease/PD-101/How-do-you-know-if-you-have-PD-.

"Hubble Essentials." Space Telescope Science Institute. http://hubblesite.org/the_telescope/hubble_essentials.

"Ilan Ramon." *Hebrew Online*, February 2010. http://news-en.hebrewonline.com/content/issue-83-february-2010-ilan-ramon.

"Illustrated Inventory: STS-135/*Atlantis* Official Flight Kit (OFK)." collectSPACE.com, 6 July 2011. http://www.collectspace.com/news/news-070611b.html.

"Inertial Measurement Units." NASA. http://spaceflight.nasa.gov/shuttle/reference/shutref/orbiter/avionics/gnc/imu.html.

"International Space Station." Encyclopedia Astronautica. http://www.astronautix.com/craft/intation.htm.

"Jan 26—Scott Parazynski and Luge Team: Wyle's Scott Parazynski Serves as Honorary Captain of U.S. Luge Team in Vancouver Games." Wyle.com, 26 January 2010. http://www.wyle.com/News/2010NewsReleases/Pages/01-26-2010.aspx.

Jensen, Marlene. "Astronaut Doug Hurley and the *Columbia* Astronauts." Southern Tier Snapshots, Friends-Partners.org, 7 August 2003. http://www.friends-partners.org/pipermail/fpspace/2003-August/009167.html.

"Kalpana Chawla Told Author Philip Chien about What Happened with Spartan on STS-87." Columbia—Final Voyage. http://sts107.com/crew/Chawla/Chawla.htm.

"KH 11-10." NASA, National Space Science Data Center. http://nssdc.gsfc.nasa.gov/nmc/spacecraftDisplay.do?id=1990-019B.

"*Lacrosse-1*." NASA, National Space Science Data Center. http://nssdc.gsfc.nasa.gov/nmc/spacecraftDisplay.do?id=1988-106B.

Landau, Elizabeth. "Astronaut-Climber: 'I'm Not a Risk Taker.'" CNN.com, 8 July 2011. http://www.cnn.com/2011/HEALTH/07/08/parazynski.astronaut.everest/index.html.

Levin, David. "The Insider Who Knew." *NOVA* podcast transcript, 14 October 2008. http://www.pbs.org/wgbh/nova/space/rocha-space-shuttle.html.

Malik, Tariq. "Documentary Provides Intimate Look at *Columbia*'s Last Crew." Space.com, 13 May 2005. http://www.space.com/1072-documentary-intimate-columbia-crew.html.

Mann, Gil. "Ilan Ramon: An Interview of a Hero." Jewish Federations of North America. http://www.jewishfederations.org/page.aspx?id=38918.

"Merritt Island National Wildlife Refuge." Kennedy Space Center Visitor Complex. http://www.kennedyspacecenter.com/wildlife-refuge.aspx.

"Michael Anderson: STS-107 Crew Memorial." NASA, 28 January 2004. http://spaceflight.nasa.gov/shuttle/archives/sts-107/memorial/anderson.html.

"The Mission Continues: The History of Challenger Center for Space Science Education." Challenger Center for Space Science Education. http://www.challenger.org/about/history/challenger_center.cfm.

"NASA Administrator Honors Katrina Heroes." NASA, 5 January 2006. http://www.nasa.gov/audience/formedia/features/maf_rideout.html.

"NASA Announces Revised Plan for *Mir* Staffing." NASA, 30 July 1997. http://www.nasa.gov/centers/johnson/news/releases/1996_1998/97-163.html.

"NASA Approves Mission and Names Crew for Return to Hubble." NASA Release H06-343, Houston, 31 October 2006. http://www.nasa.gov/centers/johnson/news/releases/2006/H06-343.html.

"NASA Facts: Return to Flight External Tank, ET-119." FS-2006-03-31-MSFC Pub. 8-40557. NASA, George C. Marshall Space Flight Center, Huntsville, Alabama. http://www.nasa.gov/centers/marshall/pdf/150034main_Shuttle_ET-119_FS.pdf.

"NASA Facts: Space Shuttle External Tank ET-128, STS-124." FS-2008-05-87-MSFC 8-368946. NASA, George C. Marshall Space Flight Center, Huntsville, Alabama. http://www.nasa.gov/centers/marshall/pdf/228641main_8-368946_%282%29.pdf.

Obama, Barack. "Remarks by the President on Space Exploration in the Twenty-First Century." White House, Office of the Press Secretary, 15 April 2010. http://www.nasa.gov/news/media/trans/obama_ksc_trans.html.

O'Brien, Miles. "Miles O'Brien Interviews Scott Parazynski about His Return to Everest." On Orbit, SpaceRef Interactive, 8 April 2009, submitted by Keith Cowing. http://www.onorbit.com/node/910.

Parazynski, Scott. "Astronaut Scott Parazynski Update 22 May 2008: Summit So Close, Yet So Far . . ." On Orbit, SpaceRef Interactive, 22 May 2008. http://www.onorbit.com/node/250.

"Parazynski Discontinues *Mir* Training." NASA Release 95-062, 14 October 1995. http://www.nasa.gov/centers/johnson/news/releases/1993_1995/95-062.html.

Pearlman, Robert. "STS-120: Official Flight Kit (OFK)." collectSPACE Messages, 4 November 2007. http://collectspace.com/ubb/Forum14/HTML/000620.html.

"Peres: Today We're All Ramon Family." *Jerusalem Post*, 14 September 2009. http://fr.jpost.com/servlet/Satellite?cid=1251804568953&pagename=JPost%2FJPArticle%2FShowFull.

Pool, Smiley N. "As Launch Draws Nearer, Crew Reports to Quarantine." Chron.com blog, *Houston Chronicle*, 2 July 2011. http://blog.chron.com/finalmission/2011/07/as-launch-draws-nearer-crew-reports-to-quarantine/.

"Preflight Interview: Andy Thomas." NASA, 23 February 2005. http://www.nasa.gov/vision/space/preparingtravel/rtf_interview_thomas.html.

"Preflight Interview: Dave Brown." NASA, 11 December 2002. http://spaceflight.nasa.gov/shuttle/archives/sts-107/crew/intbrown.html.

"Preflight Interview: Doug Hurley, Pilot." NASA, 28 May 2009. http://www.nasa.gov/mission_pages/shuttle/shuttlemissions/sts127/interview_hurley.html.

"Preflight Interview: Ilan Ramon." NASA, 11 December 2002. http://spaceflight.nasa.gov/shuttle/archives/sts-107/crew/intramon.html.

"Preflight Interview: Kalpana Chawla." NASA, 11 December 2002. http://spaceflight.nasa.gov/shuttle/archives/sts-107/crew/intchawla.html.

"Preflight Interview: Laurel Clark." NASA, 11 December 2002. http://spaceflight.nasa.gov/shuttle/archives/sts-107/crew/intclark.html.

"Preflight Interview: Michael Anderson." NASA, 11 December 2002. http://spaceflight.nasa.gov/shuttle/archives/sts-107/crew/intanderson.html.

"Preflight Interview: Rick Husband." NASA, 11 December 2002. http://spaceflight.nasa.gov/shuttle/archives/sts-107/crew/inthusband.html.

"Preflight Interview: William McCool." NASA, 11 December 2002. http://spaceflight.nasa.gov/shuttle/archives/sts-107/crew/intmccool.html.

Ray, Justin. "STS-107 Mission Status Center." Spaceflight Now, 16 January–8 February 2003. http://spaceflightnow.com/shuttle/sts107/status.html.

———. "STS-114 Shuttle Report: Mission Status Center." Spaceflight Now, 22 July–21 August 2005. http://spaceflightnow.com/shuttle/sts114/status.html.

———. "STS-120 Shuttle Report: Mission Status Center." Spaceflight Now, 19 October–7 November 2007. http://spaceflightnow.com/shuttle/sts120/status.html.

"Remarks for Administrator Bolden at the National Press Club." NASA, 1 July 2011. http://www.nasa.gov/pdf/566100main_566100main_11%200701%20Final%20Bolden%20NPC%20.pdf.

Robert, Olivier-Louis. "One-on-One with Chris Hadfield." Sympatico.ca. http://space.sympatico.ca/interview.html.

"Rocket Man." *PBS NewsHour* Online, *NewsHour with Jim Lehrer* transcript, 28 October 1998. http://www.pbs.org/newshour/bb/science/july-dec98/glenn_10-28.html.

Sample, Ian. "Houston, We Have a Problem: NASA Will Struggle When Shuttle Retires, Says Boss: Speaking on Agency's 50th Birthday, Head Warns of Tough Times to Come." *Guardian*, 25 July 2008. http://www.guardian.co.uk/science/2008/jul/26/spaceexploration.spacetechnology.

Shaw, Gwyneth K. "O'Keefe Reacts to Hubble Hubbub: The NASA Chief Agreed to Look at Ways to Extend the Space Telescope's Life." *Orlando Sentinel*, 15 March 2004. http://articles.orlandosentinel.com/2004-03-15/news/0403150077_1_hubble-working-hubble-space-hubble-telescope.

"Shuttle Crew Commanders Reassigned." NASA Release 90-96, 9 July 1990. http://www.nasa.gov/home/hqnews/1990/90-096.txt.

"Shuttle Program Reviewing Re-flight of STS-83 Mission." NASA Release J97-12, 11 April 1997. http://www.nasa.gov/centers/johnson/news/releases/1996_1998/j97-12.html.

"Statement from Buzz Aldrin: A New Direction in Space." White House, Office of Science and Technology Policy, 1 February 2010. http://www.whitehouse.gov/files/documents/ostp/press_release_files/Buzz%20Aldrin%20Statement.pdf.

Stenovec, Timothy. "New Challenger Video: Rare Footage of 1986 Disaster Uncovered (EXCLUSIVE VIDEO)." *Huffington Post*, 1 May 2012. http://www.huffingtonpost.com/2012/05/01/new-challenger-video-space-shuttle-footage_n_1463495.html?ncid=edlinkusaolp00000009.

Tinker, Jonathan. "Climbing the North Ridge of Everest (Plates 4, 6–9)." *Alpine Journal* 100, no. 344 (1995). http://www.alpinejournal.org.uk/Contents/Contents_1995_files/AJ%201995%2025-29%20Tinker%20Everest.pdf.

Wade, Mark. "STS-26." Encyclopedia Astronautica. http://www.astronautix.com/flights/sts26.htm.

Walton, Andy. "Shuttle-*Mir*: An Experiment in Science and Politics: Trouble in Orbit Sparks Tempest in Washington." CNN.com, 1998. http://www.cnn.com/TECH/space/9804/02/mir.retrospective/part3.html.

Welch, Brian. "Limits to Inhibit." *Space News Roundup*, 9 August 1985.

Wilford, John Noble. "Telescope Is Set to Peer at Space and Time." *New York Times*, 9 April 1990. http://query.nytimes.com/gst/fullpage.html?res=9C0CE3D6153AF93AA35757C0A966958260&sec=&spon=&pagewanted=all.

"Wilmore Family Ready for Son's Next Soaring Achievement." Tennessee Tech University. http://www.tntech.edu/pressreleases/wilmore-family-ready-for-sons-next-soaring-achievement/.

Wolf, David A. "Tissue Engineering and the International Space Station." NASA Blogs, 2 February 2011. http://wiki.nasa.gov/cm/blog/ISS%20Science%20Blog/posts/post_1296681712918.html.

Zak, Anatoly. "*Zvezda* Service Module (SM)." Russian Space Web. http://www.russianspaceweb.com/iss_sm.html.

Interviews and Personal Communications

Akers, Thomas A. Telephone interview with Rick Houston, 17 August 2011.

Behnken, Robert L. Telephone interview with Rick Houston, 27 June 2011.

Blaha, John E. JSC Oral History interview conducted by Rebecca Wright, Paul Rollins, and Andrea Hollman, Houston, 24 August 1998.

Camarda, Charles J. Telephone interview with Rick Houston, 25 February 2011.

Clark, Laurel B. University of Wisconsin–Madison interview conducted by Cheryl Porior-Mayhew and Phillip Certain, 28 June 2002. http://www.news.wisc.edu/releases/8232.

Clervoy, Jean-François. E-mail correspondence with Rick Houston, 16 February 2011.

———. E-mail correspondence with Rick Houston, 30 July 2011.

———. E-mail correspondence with Rick Houston, 28 December 2011.

———. Skype interview with Rick Houston, 17 September 2010.

Collins, Eileen M. E-mail correspondence with Rick Houston, 22 March 2010.

———. E-mail correspondence with Rick Houston, 28 December 2011.

———. Telephone interview with Rick Houston, 20 July 2010.

———. Telephone interview with Rick Houston, 23 February 2011.

Covey, Richard O. JSC Oral History interview conducted by Jennifer Ross-Nazzal, Houston, 15 November 2006.
———. Telephone interview with Rick Houston, 15 September 2009.
Culbertson, Frank L. JSC Oral History interview conducted by Mark Davison, Rebecca Wright, and Paul Rollins, 24 March 1998.
Duffy, Brian. E-mail correspondence with Rick Houston, 17 February 2012.
———. E-mail correspondence with Rick Houston, 21 March 2012.
———. Telephone interview with Rick Houston, 13 January 2012.
Eddy, Dawes. Telephone interview with Rick Houston, 23 September 2011.
Falvey, Pat. Skype interview with Rick Houston, 27 October 2011.
Ferguson, Christopher J. Telephone interview with Rick Houston, 12 August 2011.
Floyd, Barry. Facebook correspondence with Rick Houston, 6 August 2009.
Fossum, Michael E. Telephone interview with Rick Houston, 23 August 2006.
———. Telephone interview with Rick Houston, 21 December 2011.
Garan, Ronald J., Jr. Telephone interview with Rick Houston, 16 December 2011.
Glenn, John H., Jr. JSC Oral History interview conducted by Sheree Scarborough, Houston, 25 August 1997.
Godges, Mary Jo. Telephone interview with Rick Houston, 17 October 2011.
Griffin, Michael D. E-mail correspondence with Rick Houston, 16 December 2011.
———. Telephone interview with Rick Houston, 19 July 2011.
Grunsfeld, John M. Telephone interview with Rick Houston, 12 November 2010.
Gundy, Cheryl S. E-mail correspondence with Rick Houston, 14 February 2011.
Hale, N. Wayne. E-mail correspondence with Rick Houston, 2 August 2011.
———. Telephone interview with Rick Houston, 17 March 2011.
Hanks, Roy Lester, Jr. Telephone interview with Rick Houston, 1 June 2011.
Hauck, Frederick H. E-mail correspondence with Rick Houston, 17 August 2009.
———. E-mail correspondence with Rick Houston, 7 June 2010.
———. JSC Oral History interview conducted by Jennifer Ross-Nazzal, Bethesda, Maryland, 17 March 2004.
———. Telephone interview with Rick Houston, 11 August 2009.
Hawley, Steven A. JSC Oral History interview conducted by Sandra Johnson, Houston, 17 December 2002.
———. JSC Oral History interview conducted by Sandra Johnson, Houston, 14 January 2003.
Heflin, J. Milton. E-mail correspondence with Rick Houston, 9 February 2012.
———. Telephone interview with Rick Houston, 6 January 2012.
———. Telephone interview with Rick Houston, 28 February 2012.
Higginbotham, Joan E. Telephone interview with Rick Houston, 12 March 2012.

Hilmers, David C. E-mail correspondence with Rick Houston, 13 July 2011.
———. E-mail correspondence with Rick Houston, 11 March 2012.
———. Interview with Rick Houston, Houston, Texas, 12 June 2012.
———. Telephone interview with Rick Houston, 2 April 2010.
Hobaugh, Charles O. Interview with Rick Houston, Kennedy Space Center, 8 July 2011.
Hoffman, Jeffrey A. JSC Oral History interview conducted by Jennifer Ross-Nazzal, Cambridge, Massachusetts, 3 November 2009.
———. Telephone interview with Rick Houston, 18 August 2010.
Hurley, Douglas G. E-mail correspondence with Rick Houston, 22 September 2011.
———. E-mail correspondence with Rick Houston, 26 January 2012.
———. Telephone interview with Rick Houston, 13 August 2009.
———. Telephone interview with Rick Houston, 12 January 2012.
Kelso, Robert M. E-mail correspondence with Rick Houston, 19 March 2012.
———. Telephone interview with Rick Houston, 25 January 2012.
———. Telephone interview with Rick Houston, 27 January 2012.
Lawrence, Wendy. JSC Oral History interview conducted by Rebecca Wright, Paul Rollins, and Frank Tarazona, 21 July 1998.
Leestma, David C. Interview with Rick Houston, Kennedy Space Center, 8 July 2011.
Lichtenberg, Byron K. Telephone interview with Rick Houston, 25 January 2012.
Lounge, John M. Telephone interview with Rick Houston, 6 October 2009.
Magnus, Sandra H. Telephone interview with Rick Houston, 9 January 2012.
Massimino, Michael J. Telephone interview with Rick Houston, 5 April 2011.
McArthur, William S. Telephone interview with Rick Houston, 16 February 2012.
McCandless, Bruce, II. Telephone interview with Rick Houston, 13 August 2010.
Melroy, Pamela A. Telephone interview with Rick Houston, 3 November 2011.
Mullane, Richard M. E-mail correspondence with Rick Houston, 3 June 2010.
———. E-mail correspondence with Rick Houston, 11 July 2011.
Musgrave, F. Story. Telephone interview with Rick Houston, 10 November 2010.
Nelson, George D. E-mail correspondence with Rick Houston, 11 June 2010.
———. JSC Oral History interview conducted by Jennifer Ross-Nazzal, Bellingham, Washington, 6 May 2004.
———. Telephone interview with Rick Houston, 31 August 2009.
Overton, Thomas L. Telephone interview with Rick Houston, 4 May 2010.
Parazynski, Scott E. E-mail correspondence with Rick Houston, 17 November 2011.
———. Telephone interview with Rick Houston, 22 July 2011.
———. Telephone interview with Rick Houston, 5 August 2011.
———. Telephone interview with Rick Houston, 23 October 2011.
———. Telephone interview with Rick Houston, 1 January 2012.

Precourt, Charles J. JSC Oral History interview conducted by Rebecca Wright, Paul Rollins, and Frank Tarazona, 12 July 1998.
Reeves, William D. JSC Oral History interview conducted by Rebecca Wright, Houston, 17 April 2009.
Reilly, James F. Telephone interview with Rick Houston, 26 January 2012.
Robinson, Stephen K. E-mail correspondence with Rick Houston, 10 August 2011.
———. Interview with Rick Houston, Johnson Space Center, 21 June 2010.
Ross, Jerry L. Telephone interview with Rick Houston, 31 January 2012.
Shriver, Loren J. JSC Oral History interview conducted by Rebecca Wright, Houston, 16 and 18 December 2002.
Sotile, Renee. Telephone interview with Rick Houston, 17 October 2011.
Thomas, Andrew S. W. Interview with Rick Houston, Johnson Space Center, 22 June 2010.
Walheim, Rex J. Telephone interview with Rick Houston, 3 November 2011.
Weiler, Edward J. Telephone interview with Rick Houston, 17 December 2010.
Wetherbee, James D. Telephone interview with Rick Houston, 3 February 2012.
Wheelock, Douglas H. E-mail correspondence with Rick Houston, 10 January 2012.
———. Telephone interview with Rick Houston, 12 December 2011.
Wilmore, Barry E. Telephone interview with Rick Houston, 14 December 2011.
Wolf, David A. E-mail correspondence with Rick Houston, 31 July 2011.
———. E-mail correspondence with Rick Houston, 1 August 2011.
———. Telephone interview with Rick Houston, 27 June 2011.
———. Telephone interview with Rick Houston, 3 January 2012.

Other Sources

"ABC News Coverage of the STS-33 Launch." ABC News, 22 November 1989. http://www.youtube.com/watch?v=dJXXBVJbXAs.
Armstrong, Neil. "Written Testimony of Neil A. Armstrong before the Committee on Science, Space, and Technology, United States House of Representatives." 22 September 2011.
Camarda, Dr. Charles J. "Case Study: Recent Shuttle Reinforced Carbon-Carbon (RCC) Wing Leading Edge Anomalies." National Space and Missile Materials Symposium, Henderson, Nevada, 23 June 2008.
Challenger. DVD set. Produced by Mark Gray for Spacecraft Films, 2006.
Columbia Accident Investigation Board Report, vol. 1. NASA. August 2003.
Columbia Accident Investigation Board Report, vols. 1–6. CD-ROM. NASA. August 2003.
"Deadly Countdown." *Everest: Beyond the Limit, Season Three*. Tigress Productions. Discovery Channel, 27 December 2009.

"Episode Five: The Shuttle." *When We Left Earth: The NASA Missions.* Dangerous Films. Discovery Channel, 2008.

"Episode Six: Home in Space." *When We Left Earth: The NASA Missions.* Dangerous Films. Discovery Channel, 2008.

The First Israeli in Space: The Story of Colonel Ilan Ramon, Israel's First Astronaut. DVD. Israel Broadcasting Authority—Channel One, 2003.

The Hubble Space Telescope Optical Systems Failure Report (Allen Commission). NASA Release NASA-TM-103443, Houston, November 1990.

"Investigation Report of the STS-87 SPARTAN Close Call." Johnson Space Center, 2 March 1998.

Legler, R. D., and F. V. Bennett. *Space Shuttle Missions Summary—Book 1: First 100 Flights (STS-1 through STS-92).* Revision S, PCN-1, February 2007.

Massimino, Michael J., guest. *Late Show with David Letterman*, no. 17.164. CBS, 30 June 2010.

Nield, George C., and Pavel Mikhailovich Vorobiev, eds. *Phase 1 Program Joint Report.* NASA SP-1999-6108 (in English), January 1999.

"NOVA: Hubble's Amazing Rescue." DVD and iTunes. Public Broadcasting Service.

"NOVA: Space Shuttle Disaster." DVD and website. Public Broadcasting Service Online. WGBH. http://www.pbs.org/wgbh/nova/space/space-shuttle-disaster.html.

O'Keefe, Sean. Letter to the Honorable Christopher S. Bond, Chairman, Subcommittee on VA-HUD-Independent Agencies, Committee on Appropriations, United States Senate, 10 March 2004. www.nasa.gov/pdf/56694main_Op_Plan_12032004.pdf.

Panotin, Dr. Tina L. "Lesson from the Shuttle Independent Assessment." NASA, 19 September 2002.

Parazynski, Scott. "Shuttle Memories: Scott Parazynski." Spaceflight Now, 15 July 2011. http://www.youtube.com/watch?v=tt3iBpXuUow&feature=player_embedded.

Report of the HST Strategy Panel: A Strategy for Recovery, the Results of a Space Study August–October 1990. NASA Release NASA-CR-187826, Space Telescope Science Institute, Baltimore, Maryland, 1991.

Space Shuttle Mission Chronology: 1981–1999. NASA.

Space Shuttle Mission Chronology, vol. 2: 2000–2003. NASA Release NP-2005-03-02-KSC, Kennedy Space Center, Florida.

Space Shuttle Mission Chronology, vol. 3: 2005–2009. NASA Release IS-2009-05-037-KSC, Kennedy Space Center, Florida.

STS-26 Press Kit. NASA Release 88-121, Houston, September 1988.

STS-26: Return to Flight. DVD set. Produced by Mark Gray for Spacecraft Films, 2007.

STS-31 Press Kit. NASA Release 90-44, Houston, April 1990.

STS-46 Press Kit. NASA Release 92-95, Houston, July 1992.

STS-46 Space Shuttle Mission Report. NASA Release NSTS-08278, Houston, October 1992.

STS-49 Press Kit. NASA Release 92-48, Houston, May 1992.

STS-49 Space Shuttle Mission Report. NASA Release NSTS-08276, Houston, July 1992.

STS-57 Space Shuttle Mission Report. NASA Release JSC-08285, Houston, August 1993.

STS-58 Press Kit. NASA Release 93-135, Houston, September 1993.

STS-66 Press Kit. NASA Release 94-175, Houston, November 1994.

STS-67 Press Kit. NASA Release 95-18, Houston, March 1995.

STS-75 Press Kit. NASA Release 96-27, Houston, February 1996.

STS-75 Space Shuttle Mission Report. NASA Release NSTS-37406, Houston, April 1996.

STS-77 Press Kit. NASA Release 96-83, Houston, May 1996.

STS-77 Space Shuttle Mission Report. NASA Release NSTS-37408, Houston, June 1996.

STS-80 Press Kit. NASA Release 96-206, Houston, November 1996.

STS-80 Press Kit. NASA Release, Houston, 9 April 2001.

STS-93 Space Shuttle Mission Report. NASA Release NSTS-37425, Houston, September 1999.

STS-110 Press Kit: Framework For Expanding Station Research. NASA Release, Houston, updated 20 March 2001.

STS-114 Space Shuttle Mission Report. NASA Release JSC-63290, Johnson Space Center, Houston, June 2006.

Summary Report of the Review of U.S. Human Space Flight Plans Committee. Review of U.S. Human Space Flight Plans Committee, 8 October 2009.

Index

Page numbers in italics refer to illustrations.

In the Outward Odyssey: A People's History of Spaceflight series

Into That Silent Sea: Trailblazers of the Space Era, 1961–1965
Francis French and Colin Burgess
Foreword by Paul Haney

In the Shadow of the Moon: A Challenging Journey to Tranquility, 1965–1969
Francis French and Colin Burgess
Foreword by Walter Cunningham

To a Distant Day: The Rocket Pioneers
Chris Gainor
Foreword by Alfred Worden

Homesteading Space: The Skylab Story
David Hitt, Owen Garriott, and Joe Kerwin
Foreword by Homer Hickam

Ambassadors from Earth: Pioneering Explorations with Unmanned Spacecraft
Jay Gallentine

Footprints in the Dust: The Epic Voyages of Apollo, 1969–1975
Edited by Colin Burgess
Foreword by Richard F. Gordon

Realizing Tomorrow: The Path to Private Spaceflight
Chris Dubbs and Emeline Paat-Dahlstrom
Foreword by Charles D. Walker

The X-15 Rocket Plane: Flying the First Wings into Space
Michelle Evans
Foreword by Joe H. Engle

Wheels Stop: The Tragedies and Triumphs of the Space Shuttle Program, 1986–2011
Rick Houston
Foreword by Jerry Ross

To order or obtain more information on these or other University of Nebraska Press titles, visit www.nebraskapress.unl.edu.